D1212286

Chimpanzee Cultures

Chimpanzee Cultures

edited by

Richard W. Wrangham,

W. C. McGrew, Frans B.M. de Waal,

and ***Paul G. Heltne***

with assistance from ***Linda A. Marquardt***

Published by Harvard University Press
in cooperation with The Chicago Academy of Sciences
Cambridge, Massachusetts, and London, England
1994

Printed in the United States of America

This volume was prepared by the Chicago Academy of Sciences

This book is printed on acid-free paper, and its binding materials have been chosen for strength and durability.

Library of Congress Cataloging-in-Publication Data

Chimpanzee cultures / edited by Richard W. Wrangham . . . [et al.];
foreword by Jane Goodall.
p. cm.
Includes bibliographical references and index.
ISBN 0–674–11662–3
1. Chimpanzees—Behavior. 2. Bonobo—Behavior. 3. Chimpanzees—Ecology.
4. Bonobo— Ecology. 5. Culture. 6. Social behavior in animals. 7. Cognition in animals. I. Wrangham, Richard W., 1948–.

QL737.P96C475 1994
699.88'440451—dc20

94–9080
CIP

The genus *Pan* consists of two species that are endemic to equatorial Africa: *Pan troglodytes* (traditionally known as the common chimpanzee, although "common" is a misnomer now that the implied abundance no longer exists) and *Pan paniscus* (also know as the pygmy chimpanzee, now more often referred to as the bonobo).

Taxonomists generally agree that *Pan troglodytes* can be further divided into three main populations that exhibit mutually exclusive geographical ranges: *P.t. verus* in western Africa, *P.t. troglodytes* in central Africa, and *P.t. schweinfurthii* in eastern Africa.

The historical distribution of chimpanzees was initially perceived as commensurate with the tropical forest belt that once stretched nearly unbroken across most of equatorial Africa, leading many early scholars to conclude that chimpanzees were forest-dwelling, tree-living, plant-eating apes. Today it is evident that chimpanzees also occupy many other habitats and that not all tracts of rain forest contain chimpanzees.

Historically, survival options for chimpanzees in the natural environment were surely enhanced by an exceptional adaptive flexibility based on the successful interplay of reproductive biology and social organization. Now, however, these same reproductive and social characteristics greatly heighten susceptibility to the impact of extrinsic pressures that threaten all wild chimpanzee populations, and survival is not possible unless the requirements for both reproduction and a full social life are available.

Geza Teleki
Understanding Chimpanzees, 1989

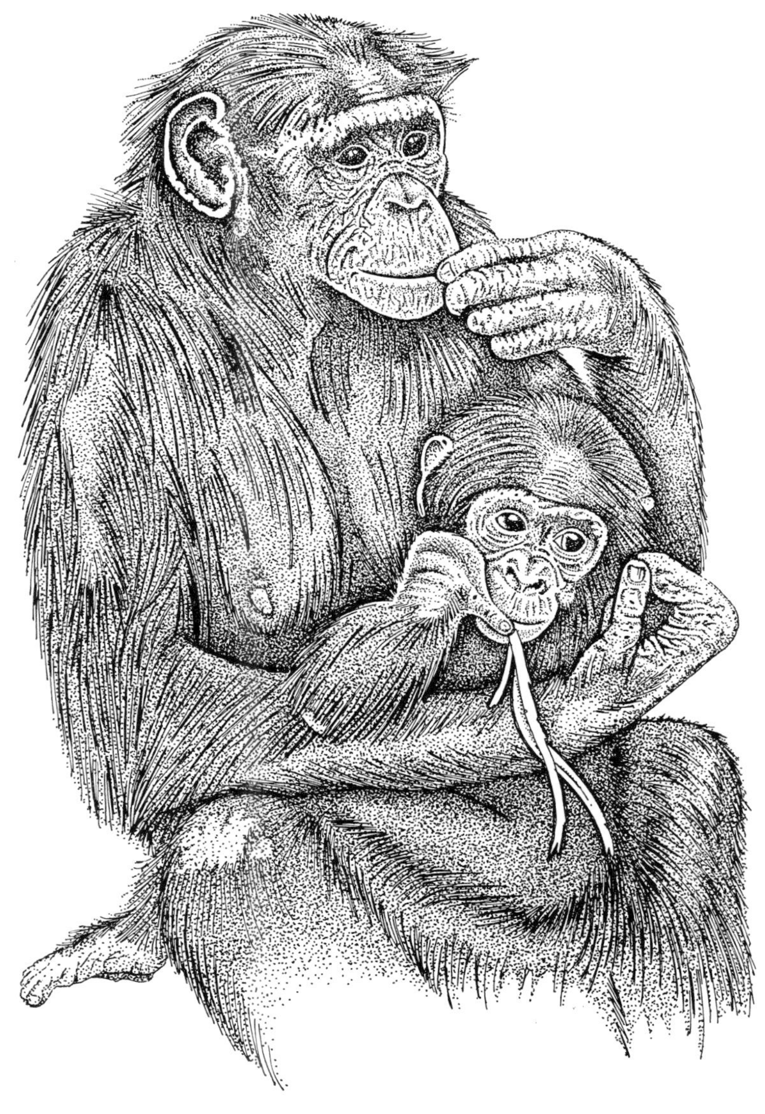

Bonobo Mother and Child

Contents

Section 2 Social Relations 149

Preface

Appreciation of chimpanzees and the other apes began with long-term field studies in the late 1950s. Prior to that time, most humans viewed chimpanzees and the other apes with a mixture of fear, curiosity, amusement, and even derision. By the 1980s, it became clear that the growing number of researchers working to understand chimpanzees and bonobos would benefit tremendously from the opportunity to gather and discuss findings and, perhaps, plan new research around jointly formulated hypotheses. My initial surmise in this regard was warmly confirmed by Jane Goodall. She and other researchers worked diligently to make the first Chicago Academy of Sciences symposium "Understanding Chimpanzees", held in 1986, a milestone for this developing area of research. Researchers from different national and scientific cultures working on the various aspects of chimpanzee culture came together to further the understanding of chimpanzees and bonobos who themselves, as our researchers had shown so clearly and so surprisingly, are capable of considerable understanding in their own right.

The group vowed to come together again in five years to review progress. The second Chicago Academy of Sciences symposium "Understanding Chimpanzees: Behavior and Diversity" was held in 1991 and presented a new realm of comparative studies of chimpanzees, bonobos and, on occasion, gorillas. The comparisons were effective in unraveling questions about the relationships between biology, ecology, and behavior. The results pointed repeatedly to the idea of chimpanzee culture, an idea made clear by the fact that cultural components and emphases varied from site to site and species to species.

This book was born in the conversations that followed the second symposium. Our planning team, which included Bill McGrew, Frans de Waal, Richard Wrangham, Linda Marquardt, and me, determined not to edit the proceedings but instead to approach the researchers whose work did or could embody a comparative approach. We developed tentative topics and suggested them to researchers who could compare and contrast one or more sites, one or more species, or who could draw comparisons between behavior in the wild and in captivity.

These chapters, all of which were contributed specifically for this book, depict an arena of study just now moving into its most productive period. Fertile new hypotheses, a rich commitment to communication among a critical mass of researchers, mutually supportive interactions of long-standing between researchers and host countries, and an urgency created by environmental threats to chimpanzees and bonobos—all these go into making the field depicted by this book stimulating.

The next decade will see great progress in understanding chimpanzees and bonobos, if humans do not continue to drive our sibling species to extinction. A whole book can and should be written about the current status and plight of chimpanzees and bonobos across Africa. All of the contributors to this book feel poignantly: it would be tragic to extinguish chimpanzees and bonobos who, among all other species, are the ones that we are the most likely to understand. All the researchers are working diligently to preserve chimpanzees and bonobos in the wild and to better the lot of chimpanzees in captivity. But while this is a deepfelt concern of our researchers, it is not the focus of the current book.

Rather, we aim to compare and contrast the ecology, social relations, and cognition of chimpanzees, bonobos and, occasionally, gorillas. Only the most pertinent comparisons are drawn to other hominoids, primates, and other large-brained creatures. A collation of the work on other such species, perhaps including birds, would be an apt topic for a symposium and for the construction of a future volume but is not addressed in this book.

Thanks are due to many contributors to this book. Foremost among these is Linda Marquardt who coordinated the two symposia drawing together the leading panologists in the world. She helped to edit *Understanding Chimpanzees* and developed the early stages of this book. She is held in warm regard by all those who have come in contact with her through these collaborative efforts. Also, it is appropriate to thank those who gave financial support to the "Understanding Chimpanzees: Behavior and Diversity" symposium: the Wenner-Gren Foundation and the L.S.B. Leakey Foundation. The cost of preparing this book has been borne by The Chicago Academy of Sciences, the authors of the chapters along with their institutions, and Harvard University Press.

Books like this are produced by a cast of hundreds. Appreciation is paid to many in the acknowledgments of individual chapters. In addition, thanks go to the great work of Bettie Leslie for proofreading, counsel, and a good many other things. Doris Slaven provided continuous clerical support. Lynn Krohn and Susan Messer did the copy editing; Gail Eddington, Cecilia Garibay, and Caron Stein, key-stroked corrections and sometimes whole manuscripts; Santa Rivelli and Donna McClean did proof-reading; Dorothy Hoffman created the index; Marilyn Jacobs helped to compose the author biographies; Jill Koski organized volunteers; David Corona designed the book and produced the print-ready copy; the Jane Goodall Institute and Mark Maglio provided illustrations; Bill McGrew and Richard Wrangham provided valuable knowledge for the study site map and Jon Maples turned it into a visual reality; and Michael Fisher, Ann Downer, and John Walsh of Harvard University Press provided guidance, encouragement, deadlines, and reviewers whom we also thank. Harvard University Press is printing, marketing, and distributing this book. It is obvious that there are many upon whom this book depended.

Thanks are especially due the authors who saw this project as important and the chimpanzees, bonobos and gorillas which provided the important data. The person who at last wove all the threads together into a coherent whole was Elizabeth Altman. Her vast skills in project management created a richer fabric, now in your hands. To Richard, Bill, Frans, and Jane I can only say in deepest gratitude "It's been a hoot".

Chimpanzee Cultures is another in our new attempts to understand chimpanzees and bonobos after millennia during which humans probably worked very hard not to understand and not to recognize how similar we three sibling species are to one another.

We are extremely similar in DNA, in anatomy, and in behavior. Between six and ten million years ago, human ancestors and ape ancestors began to separate. We were very similar. In such evolutionary situations, the emerging species must be physically out of contact or must accentuate behaviors which enforce reproductive isolation. During this phase of the formation of species, the diverging groups may still be potential mates; if members of one species-in-formation encounter members of the other (as our ancestors probably did), the diverging groups must not "understand" one another as potential mates or the two groups may quickly meld into one again. It was imperative that ancestral hominids and ancestral apes evolve behavioral mechanisms not to understand each other and they did.

In these latter days, it is clear that, though we have been separate for millions of years, humans and chimpanzees remain very similar in anatomy, behavior, and DNA. Individual chimpanzees and bonobos can, in some very real sense, understand and communicate with us in our human linguistic systems. We struggle to fathom their cognition, their systems of communication, their modes of interacting with the environment, their methods of resolving conflicts, their ways of sustaining a stable relationship with the world around them, and their cultures.

Now we know that to understand ourselves better, we must learn to understand our sibling species—in their own right, in their own places in nature. In their behavior we have a glimpse at our evolutionary roots and see our own, often inexplicable, actions in new ways. We are no longer alone or without illumination. And if, by our action or inaction, our sibling species soon become extinct, what does that portend for *Homo sapiens?* If we are to write much more of the story of our own species, it seems to me that story must have a new kind of coauthorship with our sibling species and with the rest of the world around us.

Paul G. Heltne
The Chicago Academy of Sciences
March, 1994

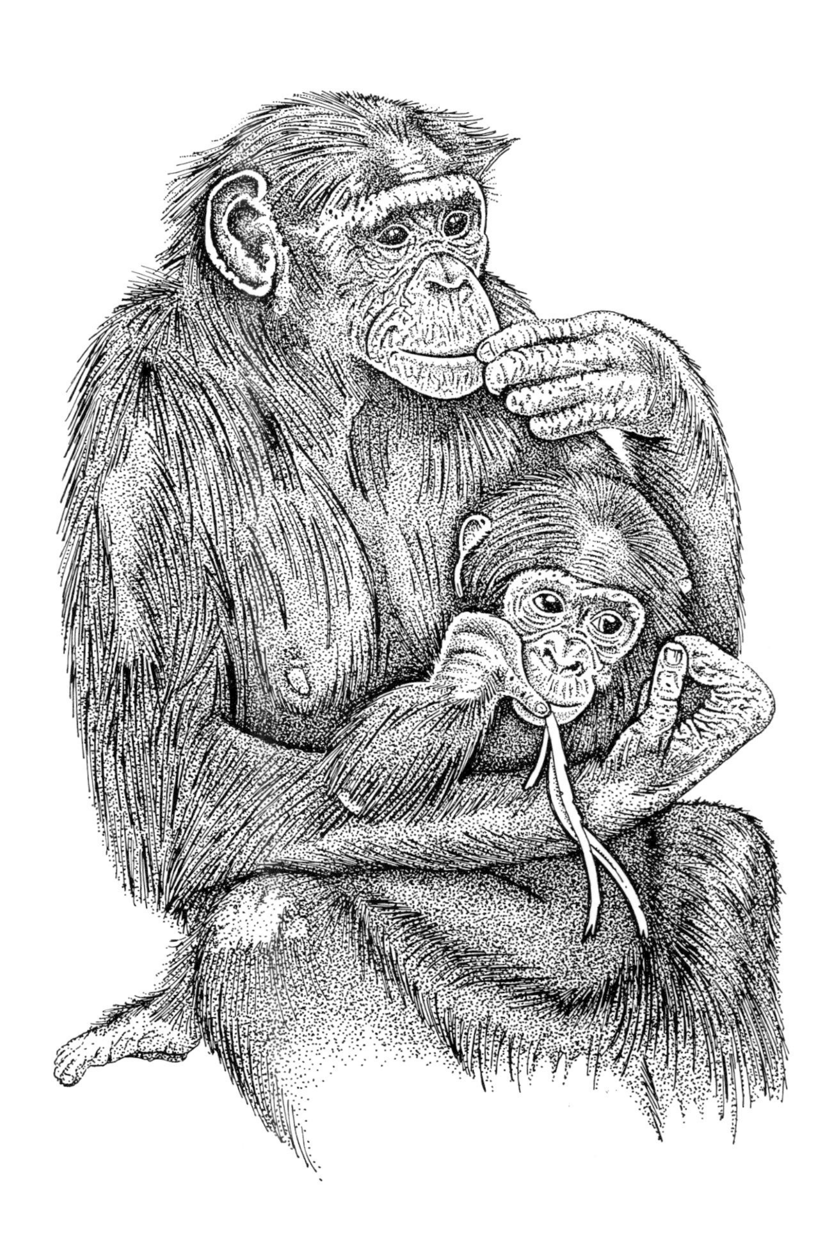

Foreword

Jane Goodall

The first person from the western world to observe chimpanzees in the wild was R. L. Garner in the 1890's. He built a cage in the middle of a West African jungle—not for the chimpanzees, but for himself—so that he could sit in safety and watch any chimpanzees who passed by. He also, incidentally, spent some time trying to teach apes to talk—without success. After that, it was 40 years before the next attempt was made to learn something about chimpanzees in their natural habitat. This was in 1930, when Henry Nissen was sent by Robert Yerkes to what was then French Guinea. But Nissen's study, which lasted only four months, was hampered because he insisted on moving through the forests accompanied by a train of porters carrying his equipment. It was hardly surprising that he made very few direct observations, although he did gather some data, especially on feeding and nesting behavior. This information, along with travelers' tales and a few chance observations—such as those by T. S. Savage and J. Wyman in the Ivory Coast in 1842 and by H. Beaty in 1951—provided almost everything that was known of chimpanzee behavior in the wild (although the African peoples who lived in or near the forests could have told us more) until the flurry of field studies began after the Second World War in the early sixties.

Research in the Field

The Sixties and Seventies

In 1960, Adriaan Kortlandt began the field research and field experiments in Beni, eastern Zaire, that he would later continue in Guinea. In the same year, I began studying chimpanzees in the Gombe National Park, Tanzania (it was a game reserve at the time). These studies were preceded in 1958 by a brief survey by Junichero Itani and Kosei Izawa. Itani came to Gombe to visit me—my very first scientific visitor—at the end of his survey. In 1961, Itani, Kinji Imanishi, Shigeru Azuma, and Akisato Toyashima began the first study of chimpanzees by Japanese scientists. This occurred in Tanzania at Kabogo Point and, then, in 1963, Itani and Izawa surveyed the chimpanzee habitat in Kasakati Basin south of Kigoma. Two years later, Toshisada Nishida started long term observations about 100 miles south of Gombe in what is now the Mahale Mountains National Park.

In 1962, another population of East African chimpanzees was observed by Vernon and Francesca Reynolds during a nine-month study in the Budongo Forest, Uganda. Then, in the following year, Jorge Sabater Pi began investigating the use of tools in various chimpanzee populations in Equatorial Guinea (called Rio Muni in those days). In 1966, Yukimaru Sugiyama spent six months in the Budongo Forest, site of the Reynolds' study; and, in the following year, Akira Suzuki came to continue observations there for another 17 months.

In 1971, Tom Struhsaker and P. Hunkeler published observations they had made in western Cameroon. Then, in 1976, four separate studies were initiated: Michael Ghiglieri began a series of three study periods in the Kibale Forest of Uganda; William McGrew, Caroline Tutin, and Pamela Baldwin began a study at Mount Assirik in the Niokolo-Koba National Park in Senegal that would last four years; Christophe and Hedwige Boesch began the ongoing study of the chimpanzees of the Taï Forest in the Ivory Coast; and Yukimaru Sugiyama began collecting data at regular intervals at Bossou, Guinea—where Kortlandt had made his early observations.

The Eighties and Nineties

In 1980, Caroline Tutin and Michel Fernandez began a two year, nationwide survey of chimpanzee (and gorilla) distribution is Gabon. Then, in 1983, they initiated the ongoing study of chimpanzees and gorillas in the Lopé National Park. In the same year, G. Isabirye-Basuta began a new study in the Kibale Forest of Uganda, observing the same chimpanzee groups as had Ghiglieri. In 1987, Isabirye-Basuta was joined by Richard Wrangham, who has since established a long-term research project at Kibale with students. In 1987, Suehisa Kuroda began observations of the chimpanzees living in the remote Ndoke National Park of northern Congo. In 1989, Dieter Steklis and Jeanne Sept began a study of chimpanzees at Ishasha River in eastern Zaire where the habitat is extremely dry.

In 1990, Vernon Reynolds revived the chimpanzee research project he had begun in 1962 in the Budongo Forest, and Christopher Bakuneeta began working there. Today, they are able to make close-up observations of a number of known, individual chimpanzees. In 1990, Annette Lanjouw began working with chimpanzees in the volcanic Tongo region of eastern Zaire and, in the same year, Rosalind Alp began to try to habituate chimpanzees in Sierra Leone's Outamba-Kilimi National Park. Most recently, Dean Anderson and Nestor Nigaruara started a study on the chimpanzees of Burundi's Kibira National Park—in the area where Peter Trenchard, as a Peace Corps volunteer, had begun to habituate a chimpanzee group six years previously.

Long Term Research

Three of these chimpanzee studies—at Gombe, Mahale Mountains, and Bossou—have already provided many years of continuous observations of known individuals. The sites at Taï in the Ivory Coast and Kibale in Uganda have also provided information about known individuals over a long period. At the time of this writing (March, 1994) the research at Gombe is in its 34th year of continuous observation—thanks to a supportive, stable government, a succession of students, and a team of highly skilled Tanzanian field staff. Three of these Tanzanians, Yahaya Alamasi, Hilali Matama, and Hamisi Mkono, now coordinate the chimpanzee research program along with Janette Wallis. The research at Mahale has been maintained for 28 years by Nishida along with his team of scientists, students, and Tanzanian field staff. Further, Boesch has made observations in the Taï Forest for 17 years with only a few interruptions and work has continued at Bossou since 1976. Finally, the research program at Kibale has collected data since 1976.

Bonobo Field Research

The first bonobo research began in Zaire in the early seventies. In 1972, Nishida conducted a brief survey at Lac Tumba, and Arthur Horn set up a two-year (1972–74) field study. In 1973, Nishida and Takayoshi Kano carried out a survey in the vast area of the Zaire Basin. The following year, Kano began a study at Wamba and, in 1975, data was collected at this site by Suehisa Kuroda. With the exception of a short period in 1975, this study was maintained until September 1991, when political instability brought most research by foreigners in Zaire to an end. After establishing the Wamba site, Kano began work at Yalosidi and continued to work there with Shigeo Uehara and Kohji Kitamura in 1977. Kano then returned to Wamba to continue his research. In 1974, Noel and Alison Badrian went to Lomako with Randall Susman. In the following year the Badrians were joined, first, by Allison Jolly and Nancy Thompson-Handler and, next, by Richard Malenky, Frances White, and Diane Doran. In 1990, Barbara Fruth and Gottfried Hohmann began work in Lomako. The research at Lomako was interrupted by political instability in Zaire in September 1991 but Fruth and Hohmann continued their work in 1992, 1993 and 1994. Meanwhile Sabater Pi and his team spent two years, from 1988 to 1990, in the Lilungu (Ikela) area.

Studies on Captive Chimpanzees and Bonobos

Of course, well before the first field studies of the early sixties, scientists were observing chimpanzees in captivity. Today, research on captive chimpanzees can be divided into three main categories: observations of naturalistic groups, such as those at the better zoos and wildlife parks; psychological testing of chimpanzees in traditional behavior labs; and the more specialized, language acquisition work. The results of this research, taken together, have helped to throw light on the higher cognitive functions of the ape brain as well as social relations and development.

Data on intelligent behavior are frequently collected in the wild, but often in anecdotal form. Carefully controlled work in the lab can sometimes substantiate and often elaborate on field impressions. The precise studies of the social dynamics of captive chimpanzees and bonobos made by Jan van Hooff and by Frans de Waal complement field observations most exquisitely. The ChimpanZoo program of the Jane Goodall Institute, coordinated by Virginia Landau and subscribed to by 14 zoos in the United States, is collecting standardized information on the behavior of over 150 chimpanzees. The sensitive and imaginative work of Sally Boysen and her chimpanzees and of Tetsuro Matsuzawa with his chimpanzee partner Ai, and the skills demonstrated by Kanzi and the other bonobos and chimpanzees in the program of Duane Rumbaugh and Sue Savage-Rumbaugh represent the current endpoints in a struggle to understand the higher reaches of chimpanzee cognitive ability—a struggle that began with Wolfgang Köhler and Robert Yerkes in the early 1900s. These various studies have yielded a wealth of information about the behavior of chimpanzees and bonobos as well as fascinating glimpses into the way their minds work.

The Study of Diversity

Without doubt, we shall learn a great deal more about chimpanzee and bonobo intelligence from laboratory studies, but I feel that even more lessons are to be learned in the natural habitat. It is fortunate that there is no lack of talented students and scientists who are eager to work in the field. Many young primatologists spend up to three years at field sites, collecting data on specialized aspects of behavior for doctoral dissertations. Some of them have demonstrated long-term commitment to fieldwork. New data are continuously being collected and new information published. The picture of our closest relatives living in the wild is continually changing as we assimilate new information into the body of collective knowledge.

Two things stand out. First, the behavior of chimpanzees and bonobos differs in a variety of ways from one field site to another. Second, only when data are collected over time on known individuals can we appreciate the range of behavior in a single social group. This variation—now documented for some sites with respect to ranging patterns, food selection, tool use (objects used and purposes to which they are put), communication postures and gestures, and social structure—is not at all surprising to those of us who know the great apes and are familiar with the inventiveness and resourcefulness of individuals. Given the different environmental challenges faced by chimpanzees in different habitats across Africa, given the innovative performances that have been observed in many individuals, and given the fact that chimpanzees can learn by observing and imitating adult behavior, it would be strange indeed if these close relatives of ours with their complex brains did *not* show cultural diversity.

In 1973, I published an article emphasizing the need for additional field studies and increased collaboration between researchers so that we could learn more about one of the most fascinating aspects of chimpanzee behavior—cultural variations across their range in Africa. At that time, information was available from four field study sites (Gombe, Mahale, Budongo, and Beni) but, for a variety of reasons, it was not always easy to compare the results. For one thing, field-study methods for this sort of research were in their infancy. Moreover, the original field-researchers of the sixties, including myself, had worked hard to get funding and to establish camps in a variety of difficult-to-get-to places in Africa. We were, for the most part, fiercely individualistic, each with our own preferred method of collecting and interpreting information. And, of course, we were interested in different aspects of behavior, and we asked different questions of the different kinds of data we had collected. Thus, from simply reading the early reports, it was difficult to compare the behavior of the chimpanzees of Beni, Budongo, Gombe, and Mahale, except in the most general terms. But then, fortunately for our understanding of chimpanzee behavior, the study sites were opened up to other researchers and students. New sites were established. Gradually, it became commonplace for researchers from the different sites to exchange information—often before publication.

The Chicago Conference

The study of field primatology is growing up, and the two chimpanzee/bonobo conferences organized and hosted by the Chicago Academy of Sciences in 1986 and 1991 have played a major role in that process. The 1986 conference brought a great many panologists under the same roof for the first time. Pioneers such as Itani, Kortlandt, Sugiyama, the Gardners, Menzel, and myself came, and, in many cases, met for the first time. There were a smattering of students and a whole array of scientists who had entered primatology through a wide variety of channels and disciplines.

It was Paul Heltne's idea to have this conference coincide with the publication of the Harvard University Press book, *The Chimpanzees of Gombe.* Initially, he planned to use papers on the third African ape, the gorilla, and, by including papers by paleontologists and evolutionary biologists, to set the apes in evolutionary perspective. But I argued that the chimpanzees and bonobos deserved a whole conference to themselves. It was a breakthrough conference, stimulating and exciting. And, as we listened to one paper after another, it was clear just how much cultural diversity existed, how much had been learned and, above all, how much more we had to discover. We agreed that a second conference should be organized in five years.

When scientists cooperate and when they use similar methods to collect information at different study sites, it becomes easier to document variations in the behavior of the different study groups. But it is only when researchers from one site have the opportunity to observe chimpanzees at other sites that they are able to detect the more subtle variations in behavior. Slight differences in communication gestures and postures, feeding methods, ways of using objects, and so on, are difficult if not impossible to describe adequately with words. And, in this time of economic recession, it is not easy to organize study trips from place to place in Africa. How then can we explore this most fascinating aspect of behavior?

In 1972 Harold Bauer used an 8-mm video camera to record chimpanzee behavior at Gombe. He obtained some wonderful footage, but the camera was large, heavy, and not suitable for carrying regularly into the forest. When he left, the camera and the idea of using it left with him. Some 10 years later in 1982, Christopher Boehm reintroduced the 8-mm video camera during a brief visit to Gombe. By this time, cameras were considerably smaller. Boehm chose a model that was tough, durable, and simple to use. I quickly incorporated this new tool into my mother-infant study and found video footage invaluable for recording developmental stages, differences in maternal techniques, changes in mother-child interactions over time, and so on. A number of the Tanzanian field staff showed interest in using these cameras and as a result, 8-mm video became a regular part of the Gombe recording method. The cameras can be carried from dawn to dusk, so behaviors that had never been filmed before—such as male patrolling, encounters with snakes, reactions to death, and so on—have now been recorded on videotape.

Was this a way of studying cultural variation? If as many behaviors as possible were captured on videotape at each research site, could the tapes substitute, at least to some extent, for travel by individual researchers from place to place? What better way of answering this question than to ask the participants of the 1991 Chicago conference to record behavior at their sites, to edit into five-minute clips on specific topics—tool use, sexual behavior, hunting, and so on—and to bring these clips with them? This they did: researchers brought material from eight different studies in five countries. Without question, the video sessions scheduled throughout the conference were incredibly informative and made the three days extremely stimulating. I think we all gained new insight into fascinating variations in behavior when we saw these illustrated on film.

At the end of the 1991 conference two things were very clear. First, a great deal of new information had been gathered in the five years that had elapsed since the previous Chicago conference, Understanding Chimpanzees. New sites had been opened up and many young, dedicated, and intellectually bright students had eagerly seized the opportunities created for fieldwork. Moreover, the new willingness to share information between field and laboratory and zoo created an intellectually stimulating environment so that ideas proliferated and were then investigated. Not all that information has been published yet, although some appears in this volume. When it has, and when the information has been pooled, even more about chimpanzee cultures and cultural diversity will be revealed. Each new study and each new publication raises new questions highlighting how much more work needs to be done in the field and how many questions are as yet unanswered.

Jane Goodall
The Jane Goodall Institute
March, 1994

(Reference to the studies mentioned in this foreword will be found either in this volume or in *The Chimpanzees of Gombe: Patterns of Behavior* published by Belknap Press.)

Study Sites in Africa

P.t.v. = *Pan troglodytes verus,* West Africa chimpanzee
P.t.t. = *Pan troglodytes troglodytes,* Central Africa chimpanzee
P.t.s. = *Pan troglodytes schweinfurthii,* East Africa chimpanzee
P.p. = *Pan paniscus,* bonobo (also known as pygmy chimpanzee)

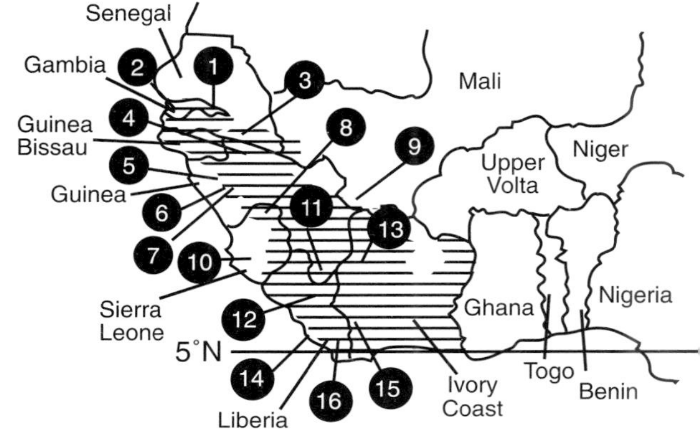

Alphabetical Listing

Map No.	Study Site/Communities	Species or Subspecies
2	Abuko, Gambia	*P.t.v.*
3	Assirik (Niokolo-Koba), Senegal	*P.t.v.*
20	Ayamiken, Equatorial Guinea	*P.t.t.*
1	Baboon, Gambia	Mixed-formerly captive
9	Bafing, Mali	*P.t.v.*
14	Bassa, Liberia	*P.t.v.*
22	Belinga, Gabon	*P.t.t.*
34	Beni, Zaire	*P.t.s.*
11	Bossou, Guinea	*P.t.v.*
35	Budongo, Uganda	*P.t.s.*
17	Campo, Cameroon	*P.t.t.*
18	Dipikar, Equatorial Guinea	*P.t.t.*
43	Filibanga, Tanzania	*P.t.s.*
4	Fouta Djallon, Guinea	*P.t.v.*
39	Gombe, Tanzania	*P.t.s.*
	Kasakela	
	Kahama	
	Mitumba	
24	Ipassa, Gabon	*P.t.t.*
36	Ishasha, Zaire	*P.t.s.*
41	Kabogo, Tanzania	*P.t.s.*
38	Kahuzi-Biega, Zaire	*P.t.s.*
7	Kanka Sili, Guinea	*P.t.v.*
12	Kanton, Liberia	*P.t.v.*
42	Kasakati, Tanzania	*P.t.s.*
33	Kibale, Uganda	*P.t.s.*
	Kanyawara	
	Ngogo	
6	Kindia, Guinea	*P.t.v.*
29	Lomako, Zaire	*P.p.*
23	Lopé, Gabon	*P.t.t.*
31	Lilunga (Ikela), Zaire	*P.p.*
40	Mahale, Tanzania	*P.t.s.*
	Kasoje	
	Bilenge	
19	Mayang, Equatorial Guinea	*P.t.t.*
25	Mbomo, Congo	*P.t.t.*
26	Ndakan, Central African Republic	*P.t.t.*
28	Ndoki, Congo	*P.t.t.*
5	Neribili, Guinea	*P.t.v.*
27	Ngoubunga, Central African Republic	*P.t.t.*
13	Nimba, Ivory Coast	*P.t.v.*
21	Okorobiko, Equatorial Guinea	*P.t.t.*
8	Outamba Kilimi, Sierra Leone	*P.t.v.*
	Kilimi	
	Tenkere	
45	Rubondo, Tanzania	Mixed-formerly captive
16	Sapo, Liberia	*P.t.v.*
15	Taï, Ivory Coast	*P.t.v.*
10	Tiwai, Sierra Leone	*P.t.v.*
37	Tongo, Zaire	*P.t.s.*
44	Ugalla, Tanzania	*P.t.s.*
30	Wamba, Zaire	*P.p.*
32	Yalosidi, Zaire	*P.p.*

Numerical Listing

Map No.	Study Site/Communities
1	Baboon, Gambia
2	Abuko, Gambia
3	Assirik (Niokolo-Koba), Senegal
4	Fouta Djallon, Guinea
5	Neribili, Guinea
6	Kindia, Guinea
7	Kanka Sili, Guinea
8	Outamba Kilimi, Sierra Leone
	Kilimi
	Tenkere
9	Bafing, Mali
10	Tiwai, Sierra Leone
11	Bossou, Guinea
12	Kanton, Liberia
13	Nimba, Ivory Coast
14	Bassa, Liberia
15	Taï, Ivory Coast
16	Sapo, Liberia
17	Campo, Cameroon
18	Dipikar, Equatorial Guinea
19	Mayang, Equatorial Guinea
20	Ayamiken, Equatorial Guinea
21	Okorobiko, Equatorial Guinea
22	Belinga, Gabon
23	Lopé, Gabon
24	Ipassa, Gabon
25	Mbomo, Congo
26	Ndakan, Central African Republic
27	Ngoubunga, Central African Republic
28	Ndoki, Congo
29	Lomako, Zaire
30	Wamba, Zaire

(continued)

Map No.	Study Site/Communities *(continued)*
31	Lilunga (Ikela), Zaire
32	Yalosidi, Zaire
33	Kibale, Uganda
	Kanyawara
	Ngogo
34	Beni, Zaire
35	Budongo, Uganda
36	Ishasha, Zaire
37	Tongo, Zaire
38	Kahuzi-Biega, Zaire
39	Gombe, Tanzania
	Kasakela
	Kahama
	Mitumba
40	Mahale, Tanzania
	Kasoje
	Bilenge
41	Kabogo, Tanzania
42	Kasakati, Tanzania
43	Filibanga, Tanzania
44	Ugalla, Tanzania
45	Rubondo, Tanzania

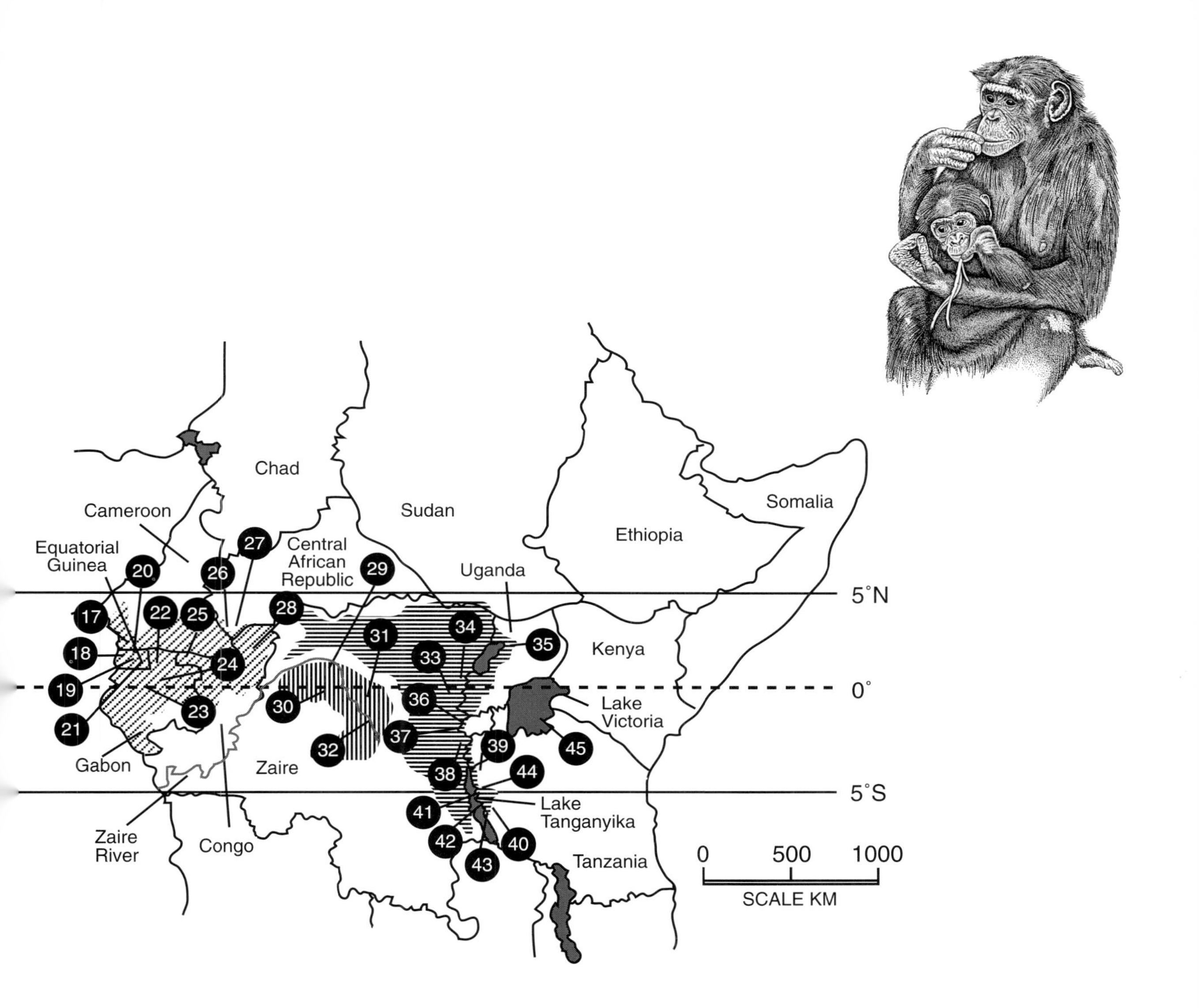
Chad
Sudan
Cameroon
Equatorial Guinea
Central African Republic
Uganda
Ethiopia
Somalia
Kenya
Lake Victoria
Gabon
Zaire
Zaire River
Congo
Lake Tanganyika
Tanzania
5˚N
0˚
5˚S
0
500
1000
SCALE KM
17
18
19
20
21
22
23
24
25
26
27
28
29
30
31
32
33
34
35
36
37
38
39
40
41
42
43
44
45

The Challenge of Behavioral Diversity

Richard W. Wrangham, Frans B.M. de Waal, and W. C. McGrew

A major aspect underlying behavioral diversity is cultural transmission. Cultural transmission of behavior is contrasted with the acquisition of behavior through individual learning or through genetic transmission: cultural transmission implies learning from others and, in its most effective form, teaching by others. Cultural transmission may be particularly relevant to members of the genus *Pan,* because these species are our closest relatives, hence the most likely candidates for human-like behavior. Overviews of early data and concepts in the study of animal culture can be found in Kummer (1971), Menzel (1973), Bonner (1980), Nishida (1987), and McGrew (1992).

Virtually every definition of culture in the social sciences premises human uniqueness. Even a book entitled *The Evolution of Culture* claims that "man and culture originated simultaneously; this by definition" (White 1959:5), thus barring any thought of continuity with other species. The reliance on pre-Darwinian philosophical principles in the study of culture (Count 1973) explains the traditional tension between social and cultural anthropology, on the one hand, and physical or biological anthropology, on the other. Ironically, as this volume testifies, the latter field is now embarking on its own investigation of the evolution of culture, using the term *evolution* in its Darwinian sense, through the study of nonhuman primates. A notable attempt at the modeling of the evolution of culture was made by Boyd and Richardson (1985).

Nonetheless, human culture clearly contains a number of unique elements: anyone looking for the full-blown phenomenon in other animals will come up empty-handed. An endeavor that we may label *cultural primatology* requires a broad definition that includes culture-like, pre-cultural, or proto-cultural manifestations. Such inclusive definitions have been around since the pioneering work by Japanese primatologists such as Imanishi who in 1952 defined *culture* as "socially transmitted adjustable behavior"

(Nishida 1987:462). Of course, one of the hallmarks of human culture is its diversity from group to group, individual acquisition of cultural traits from others and, very frequently, active teaching of the culture to younger individuals.

Behavior varies somewhat in all animals, but in chimpanzees *(Pan troglodytes)* and in bonobos *(P. paniscus)* behavior is so variable from population to population that, even when only one aspect such as tool use is considered, every chimpanzee population studied to date has proved to have its own unique combination of tools and techniques (McGrew 1992). Few if any other species show such diversity not only in tool use but also in communication, in cooperative strategies, in response to other species, and in use of medicinal plants. Finding ways to compare the variation found in chimpanzees and bonobos with variations in other animals is an important challenge for the future.

One way to examine behavioral diversity is to ask how behavioral variants, such as termite fishing or nut smashing or hand-clasp grooming, are maintained across generations. The cognitive abilities of chimpanzees and bonobos suggest that variants could be passed on through learning by imitation and through other forms of social transmission (Goodall 1970). Cultural transmission seems likely because infants watch mothers closely and, occasionally, mothers appear to instruct their young (Boesch 1991). Yet no convincing experimental evidence supports the hypothesis of cultural transmission, and the fact that individuals often vary substantially in technique or skill suggests that cultural transmission among chimpanzees is, at best, inefficient and possibly absent. Furthermore, experimental data from other species show that social traditions are sometimes maintained without imitation. Some chickadees, *Parus atricapillus* for example, have a tradition of opening milk bottles. The tradition is transmitted passively across generations through the two processes of so-called local enhancement: naive birds learn that milk bottles contain something they like, and their attention is drawn to the bottles by experienced openers (Sherry and Galef 1990). They do not learn from others how to open bottles, but merely that bottles are worth opening. Therefore, maintenance of the variant depends on repeated invention, with succeeding generations copying the recognition of the opportunity rather than the technique. A similar process accounts for food washing in brown capuchins, *Cebus apella,* and long-tailed macaques, *Macaca fascicularis,* (Visalberghi and Fragaszy 1990). By extension, this could also explain behavioral traditions in chimpanzees and bonobos (Galef 1992; Tomasello 1990).

Thus, we are offered two contrasting perspectives on the significance of behavioral traditions such as those observed in chimpanzees and bonobos. One suggests that individuals model their behavior on that of others; the alternative portrays individuals as restricted to private conceptual worlds.

To resolve this issue we may need experimental evidence such as Aisner and Terkel (1992) obtained to show that observation was essential for maintenance of the tradition of opening pine cones among black rats, *Rattus rattus*. However, this kind of evidence is difficult to obtain in the field.

Another way to examine behavioral diversity is to ask why a particular behavioral variant occurs in one population but not in another. Differences between species are to be expected as a result of differences in evolutionary adaptation. These differences reflect effects of the evolutionary past—of the environment of adaptation—on a species' genetic makeup. Nongenetic adjustment to current environments is a different, more flexible form of adaptation found in a wide range of animals (Lott 1984). For example, where predators are abundant, chimpanzees build their night nests high in trees (Baldwin et al. 1981). But many cases, including variation in communication patterns, tool use, social relationships, and food choice, are less easily understood because behavioral variants do not show a strong correlation with environmental variables. What does this say about adaptive variation? Does the fact that some populations have edible nuts in their environs but fail to smash them mean that they do not need the extra calories, or does it mean that they have never developed the right skill? If not, why not? More generally, how effectively have the inventive capacities of chimpanzees and bonobos been harnessed for adaptive ends? This book provides a step forward for answering such questions by bringing together a rich compilation of behavioral variation in chimpanzees and bonobos.

The reasons for most differences between chimpanzees and bonobos, such as in their social behavior and ecology, remain a matter of debate. It is certain that the impact of both nature and culture is important for understanding the evolution of each species and as a background for understanding variation within each species. Even less clear are the reasons for variation within and between populations. We will briefly consider some of the more important sources of diversity, including genetics, anatomy, demography, ecology, diffusion boundaries, tool use, feeding, communication, and social organization. To set the scene, we begin by reviewing the evolutionary relationship between chimpanzees, bonobos, and humans.

A Trio of Relatives: Chimpanzees, Bonobos, Humans

An account of evolutionary relationships as they were understood in the 1950s is useful, because the morphological evidence that dominated the debate at that time is still helpful in reconstructing the body plan of our ape ancestors. Genetic analysis had not yet been attempted in the 1950s.

With evidence about evolutionary relationships based solely on morphology, the chimpanzees and bonobos were considered to be more closely related to gorillas *(Gorilla gorilla)* and orangutans *(Pongo pygmaeus)* than to humans (Pilbeam 1989) (see figure 1).

A close evolutionary relationship between *Pan* and *Gorilla* was considered likely because some specialized characteristics such as thinly enameled teeth and knuckle walking occur in both *Pan* and *Gorilla* but not in humans (Groves and Paterson 1991). Furthermore, many morphological differences between *Pan* and *Gorilla* are size related; that is, determined merely by the point at which each species stops growing—resulting in morphological material from big chimpanzees and small gorillas being easily confused with each other (Shea 1984, 1985). Even differences in their diet, social behavior, and sexual behavior may be ultimately attributable to size (Short 1979; Wrangham 1979).

Only in the last decade have the evolutionary conclusions from morphology been seriously challenged. The first genetic evidence placing the genera *Homo* and *Pan* as each other's closest relatives came from nuclear DNA hybridization (Sibley and Ahlquist 1984, 1990) in a study convincingly replicated by Caccone and Powell (1989). Further support for this new relationship comes from analysis of nuclear DNA sequencing (Williams and Goodman 1989; Ueda et al. 1989; Gonzalez et al. 1990) and, most importantly, from the analysis of mitochondrial DNA (Ruvolo et al. 1991; Horai et al. 1992). Our ability to read the genetic code has overturned

Figure 1
Traditional and current evolutionary trees for living apes. Modified from Li and Graur (1991).

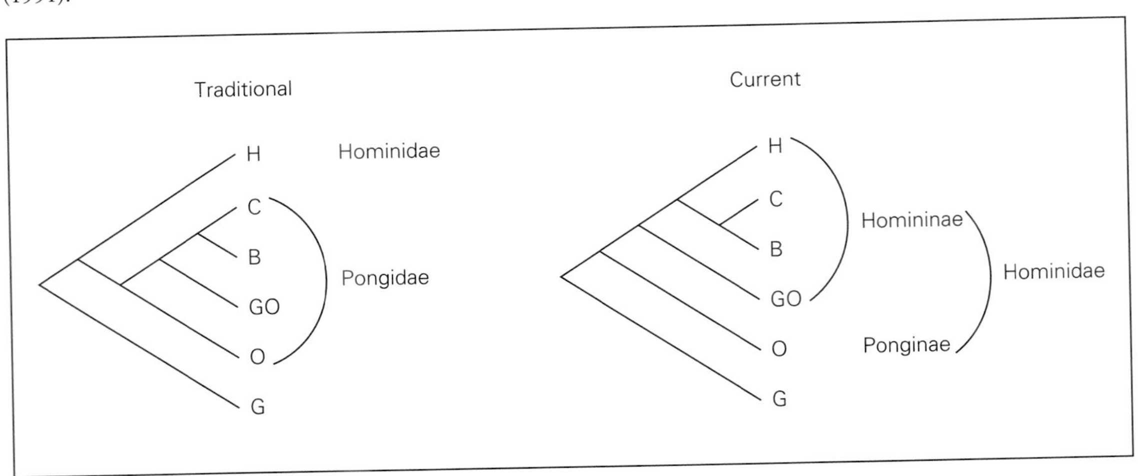

Notes: H = human; C = chimpanzee; B = bonobo; GO = gorilla *(Gorilla gorilla);* O = orangutan *(Pongo pygmaeus);* G = gibbons *(Hylobatidae).*

Photo 1 and 2

The genus *Pan* includes two different species: the chimpanzee *(P. troglodytes)* and bonobo *(P. paniscus)*. The bonobo is also known as the pygmy chimpanzee even though the two species overlap in size. These two adult males show a chimpanzee on the left and a bonobo on the right. Note the bonobo's broader, almost gorilla-like nose; longer and finer hair on top of his head; and light-colored lips compared to the chimpanzee. (Photographs by Frans B.M. de Waal.)

the old picture of evolutionary relationships. It is now generally agreed that chimpanzees and bonobos are our closest living relatives, and we are theirs (see figure 1).

Despite the priority given to molecular evidence, morphological data remain useful for the reconstruction of evolutionary history. These data need to be analyzed further to determine which of the three African apes—chimpanzee, gorilla, or bonobo—has retained most features of the basic body plan of the common ancestor (Hill and Ward 1988). There have been contradictory claims in this regard: Coolidge (1933) and Zihlman (1984), for example, view the bonobo as most similar to *Australopithecus afarensis,* a prime candidate for the earliest hominid ancestor. This claim is supported by the ecology of the three species; bonobos are currently entirely restricted to the rain forest, whereas the other two species have ventured outside the protection of this habitat (Kano 1992). Facing a wider range of environmental pressures, chimpanzees and gorillas may have accumulated more adaptations and, hence, diverged further from the common ancestor. Other scientists, however, have argued that many bonobo features—chromosomes, blood groups, skeletal anatomy, sexual anatomy and sexual physiology—appear derived when compared with chimpanzees (Stanyon et al. 1986).

This argument supports the notion, based on mitochondrial DNA analyses, that the last common ancestor of *Homo* and *Pan* was a chimpanzee-like animal (Ruvolo et al. 1991).

Chimpanzees and bonobos provide two kinds of connections between humans and the rest of the animal world. One is historical: they suggest what our ancestors were like and how we have changed since we evolved from a chimpanzee-like form around 6,000,000 years ago. The other is conceptual: they offer the opportunity to understand the functional significance of traits they share with us. The flowering of behavioral diversity between populations is one such trait.

Sources of Behavioral Diversity

The explanation of behavioral diversity must satisfy various questions, including proximate (questions about how variants occur) and ultimate (questions of why they occur). The following sources of behavioral diversity are not mutually exclusive, but each source provides the kinds of explanations that must be considered when we find individuals or populations varying in behavior. Some explanations, such as the one based on genetic variation, can apply to any animal species. Others, such as diffusion, can apply only where specific cognitive abilities are present. However, in order to accept an explanation based on a complex, cognitive mechanism, it is normally necessary to rule out cognitively lower levels of explanation.

Genetics

Genetic variation could in principle underlie behavioral variation not only between species but also between populations of the same species. The most recent genetic evidence places the time of the chimpanzee-bonobo divergence as 40% to 50% of the age of the *Homo-Pan* divergence (Horai et al. 1992). This means that chimpanzees and bonobos became separate species at about the same time that the genus *Homo* evolved from *Australopithecus*, two-and-a-half to three million years ago. Thus, there has been ample time for behavioral evolution.

Among bonobos, which have no subspecies, relatively little genetic variation has so far been found (Ruvolo, pers. com.). Chimpanzees, by contrast, show substantial genetic variation, at least four times as much as found among humans (Kocher and Wilson 1991). A particularly strong split may occur between the western subspecies *(Pan troglodytes verus)* and the central and western subspecies *(P. t. troglodytes* and *P. t. schweinfurthii)* with a suggested time of separation of 1,000,000 years (Morin et al. 1992, 1993).

This genetic divergence means that if consistent behavioral differences are found between *P. t. verus* and other chimpanzees, genetic variation must be considered a possible explanation for those differences. For example, reports of female chimpanzees showing strong patterns of affiliation with each other are principally from *P. t. verus* at Bossou (Sugiyama 1988) and Taï (Boesch, pers. com.). Admittedly, the theoretical thrust of behavioral studies suggests that ecological and social contexts are more likely than genetic heritage to explain such a finding, as suggested by the high affiliation between female chimpanzees in captive settings (de Waal 1982, this volume). However, if most captive chimpanzees come from far-Western Africa stock, then the variables of genotype and captivity may be confounded. Nevertheless, genetic influences need to be ruled out before explanations based on nongenetic adaptation to local context can be accepted. The same applies to other kinds of behavioral differences, such as tool use. It is in theory possible that *P. t. verus* is genetically predisposed to smash nuts with tools in a way that the other subspecies of chimpanzee are not.

Anatomy

Morphological differences between chimpanzees and bonobos—including aspects of skeletal, cranio-dental, and sexual anatomy—have been well studied. Although bonobos tend to be small, the phrase *small-headed, long-legged chimpanzee* would be a more appropriate morphological description of *P. paniscus* than pygmy chimpanzee. Bonobos' bodies are on average slightly larger than the smallest chimpanzees, such as those living in Gombe (Jungers and Susman 1984; Morbeck and Zihlman 1989). But at comparable body sizes, bonobos have relatively smaller skulls, faces, and teeth, with no sexual dimorphism except in the canines, where the sexual difference is of the same degree as in chimpanzees. The small head of the bonobo has been attributed to *paedomorphism* (which is retention of juvenile characteristics into adulthood). It remains unclear whether the size difference in relation to chimpanzees has adaptive significance and, if so, what this significance might be (Shea 1985; Uchida 1992).

Bonobos differ from chimpanzees in their longer legs, shorter arms, shorter clavicles, narrower pelvis, more horizontal back, larger sexual swelling, more ventrally located clitoris, smaller ears, centrally parted head hair, pink lips, and vocal apparatus that creates remarkably high-pitched calls compared with the chimpanzee (Shea and Groves 1987; de Waal 1988; Kano 1992). All those traits are independent of body size, whereas the trait of scapular proportions depends on the body size (Shea 1986) and tooth morphology depends on the size of the head (Shea 1984). Size-independent variation can be regarded as species specific, whereas size-dependent variation cannot.

Little is known about morphological variation within either species. The largest and most sexually dimorphic subspecies of chimpanzee appears to be *P. t. troglodytes* (Jungers and Susman 1984; Morbeck and Zihlman 1989). There is sufficient variation within subspecies that population differences in size and dimorphism cannot be predicted on the basis of subspecies (Morbeck and Zihlman 1989). An interesting case of size variation is represented by the population of *P. t. schweinfurthii* in Gombe (Tanzania), the smallest chimpanzees known (both body weight and limb length have been measured [Morbeck and Zihlman 1989]). The nearby Mahale population of the same subspecies is noticeably larger (Uehara and Nishida 1987).

In some aspects of chimpanzee morphology, more variation has been described than in bonobos. For example, in bonobos the second molar is normally the longest of the upper molars, whereas this feature is variable in chimpanzees (Uchida 1992). Shea and Coolidge (1988), Groves et al. (1992) and Shea et al. (1993) note cranial differences between the three *P. troglodytes* subspecies and, especially, between *P. troglodytes* and *P. paniscus.* Despite overlap between groups, multivariate analyses of cranial measurements, particularly when adjusted for size differences (using cranial length), show that *P. t. verus* and *P. t. troglodytes* are closer than *P. t. troglodytes* and *P. t. schweinfurthii*, with greater difference between *P. t. verus* and *P. t. schweinfurthii.* Of the three subspecies, *P. t. schweinfurthii* is the closest to *Pan paniscus.* However, the genetic basis for morphological differences needs to be analyzed carefully when populations are compared. For example, differences in tooth structure might be adaptations to differences in diet or reflect random genetic effects such as the peculiar genotypes of the founders of a local population. A genetic basis for morphological differences needs to be examined, rather than assumed, when populations are compared.

The morphological differences between chimpanzee subspecies are notably less pronounced than those separating gorilla subspecies (Shea and Coolidge 1988) and are more similar in degree to those separating demes within gorilla subspecies (Groves 1970). This finding corresponds to the relative amount of genetic variation within the two species.

An example of how apparent subspecific differences in morphology could be misleading comes from variation in chimpanzee relationships with red colobus monkeys *(Colobus badius).* Chimpanzees in Gombe often flee as a result of being mobbed by the red colobus (Boesch, this volume). In Taï, by contrast, chimpanzees are so dominant to red colobus monkeys that the descriptions of Gombe chimpanzees avoiding the red colobus monkeys sounded "incredible" to Boesch and Boesch (1989). Gombe chimpanzees are small-bodied *(P. t. schweinfurthii),* whereas Taï chimpanzees are large-bodied *(P. t. verus).* Does the smaller size of Gombe chimpanzees, explain their timidity in relation to the red colobus monkeys? An answer is provided by data from Kibale, where *P. t. schweinfurthii* are also regularly mobbed and supplanted by the red colobus monkeys (Wrangham, pers. observ.). Contrary to expectations based on the body-size hypothesis, Kibale chimpanzees are large, an estimated 25%–35% larger than Gombe chimpanzees (Kerbis et al. 1993). Body-size differences, therefore, are inadequate for explaining population differences in response to mobbing by the red colobus monkeys. The chimpanzee-colobus relationships are discussed by Boesch (this volume).

Demography

In a number of primates, differences in demographic structure between populations or groups have important effects on social behavior. For example, high rates of juvenile survival can lead to larger kin groups with more cooperative behavior (Dunbar 1988). Chimpanzees are so long-lived, and demographic data so slowly gathered, that demographic differences between populations are still not well described. But rates of infant birth and death clearly vary. Mean interbirth intervals range from 4.4 years at Bossou (Sugiyama 1989; Tutin, this volume) to 6.0 years at Mahale (Nishida et al. 1990), while the probability that infants will survive to four years almost doubles from Mahale (0.42, calculated from Nishida et al. 1990) to Bossou (0.81, Sugiyama 1989). These differences appear large enough that they could have long-term ramifications for social behavior as a result of effects on group structure.

The data from Bossou suggest how group structure could influence behavior. For example, Bossou females form bonds with each other, defined through differential association and grooming, more than do Gombe or Mahale females (Sugiyama 1988). The Bossou population is an isolated group with high infant survival rates and little opportunity for successful emigration. The relationships between females could, thus, be influenced by the high survival and restricted emigration opportunities.

Ecology

Many cases of differences in behavior between populations, including very complex patterns, appear related to local environmental differences. In Tongo, Zaire, for example, chimpanzees use "sponges" of vegetation to gather water from tree holes almost daily (Lanjouw, pers. com.). The rate of sponging is clearly higher than in other populations and is easily explained; the Tongo forest is on a lava flow where rain drains rapidly into the ground, and there are no pools or streams. The Tongo chimpanzees dig for large, water-filled roots *(Clematis),* which they carry and share like a bottle (Lanjouw, pers. com.). These inventive practices are clearly responses to the lack of permanent water. In a similar way, many examples of the behavioral diversity described in this book can be attributed to specific environmental differences.

The relationship, however, is not always as simple as it might appear. For example, foods are sometimes eaten at one site but not at another site, despite being available at both. As McGrew et al. (1988) suggested, this may occur because of differences in overall levels of food availability: where food is less abundant, a population may accept foods of lower quality, such as the stinging ants *(Megaponera foetens)* eaten by Mt. Assirik chimpanzees but not by Gombe chimpanzees. This suggestion is important for thinking about variation in the use of foraging tools. For example, Kibale chimpanzees fail to eat the abundant army ant *(Dorylus nigricans),* which is eaten by tool-using chimpanzees in Bossou, Taï, and Gombe. Does this mean that Kibale chimpanzees have failed to develop the use of tools, or does it mean that Kibale chimpanzees have such abundant food that they do not need to eat ants of low profitability? Present data suggest that Kibale chimpanzees would benefit from using ant-dipping tools, because they appear to have less food available than chimpanzees in Taï, for example. So the fact that they do not dip for ants is a puzzle.

Environmental determinism rarely yields complete, fine-grained explanations. It does not explain, for example, why the Tongo sponges are normally made of moss (Lanjouw, pers. com.), unlike the leaf sponges of Gombe (Goodall 1986) or the stem sponges of Kibale (Wrangham pers. observ.) Similarly, it does not explain why Gombe chimpanzees prefer to fish for termites with grass stems (see figure 2), while Mt. Assirik (Senegal) chimpanzees prefer twig tools (McGrew 1992).

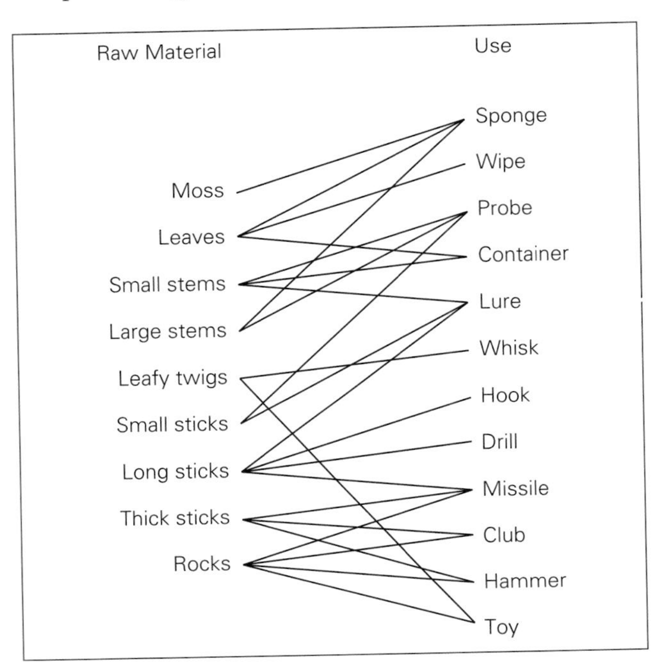

Figure 2
Raw materials used as tools by chimpanzees in the wild. Data are principally from Goodall (1986).

Ecological influences on social organization are not understood well enough to use for comparisons either within or between species. Current data suggest that aspects of social organization—such as community size, sex ratio, and nature of social bonds—correlate poorly with gross habitat type (such as forest, woodland, or savannah). Dynamic processes on a smaller scale, such as those that regulate the intensity of feeding competition, may be more helpful in providing explanations for population differences in social organization.

Diffusion Boundaries

Behavioral variants that are transmitted through observational learning, even if the transmission does not involve imitation, may sometimes be limited to certain populations because these variants are invented rarely and, as a result, have few chances to disseminate within groups and to diffuse across groups. This restriction may explain the distribution of nut cracking: nut cracking occurs in several, but not all, populations of *P. t. verus,* but has not been found in any population of *P. t. troglodytes* or *P. t. schweinfurthii* despite decades of field study. Although adequate supplies of nuts and of hammer stones and anvils are found for *P. t. troglodytes* at Lopé and *P. t. schweinfurthii* at Gombe, the chimpanzees at these sites have not invented nut-cracking technology (McGrew et al. unpubl. data). In addition, diffusion of nut cracking from far-Western Africa to the east has probably been stymied by the major geographical barriers of the Dahomey Gap and the Niger River (McGrew 1992).

Diffusion boundary variants probably represent a continuum. At one extreme, some variants appear to have trivial effects on fitness, such as the choice of raw material for sponging, or whether or not leaf grooming, hand-clasp grooming, or leaf clipping occur. Other variants, such as tool use in subsistence, may have important effects on fitness, in that differences may have energetic or nutritional consequences for natural selection.

An interesting problem is why there are spotty distributions of variants. Consider hand-clasp grooming. Has it been invented independently in populations of *P. t. schweinfurthii* north and south of Gombe (such as Kibale and Mahale)? Or was it continuous, and then went extinct? The appearance of this behavior in a captive group (de Waal, this volume) suggests multiple independent inventions.

Finally, the history of a population needs to be taken into account, because recently isolated populations may prove more likely to share traditions with their neighbors than those separated for longer periods. The period over which populations have become isolated varies from a few hundred years (caused by human habitat destruction, in some western Ugandan forests) to tens of thousands of years (caused by climatic changes that lead to forest loss [Colyn et al. 1991]). This suggests there has been ample time for traditions to be lost between some neighbors, but not between others. Additionally, older populations may prove to have more diverse cultures. East African chimpanzees have been studied principally in small, isolated forests that are unlikely to have survived the last glaciation (Mahale may be the exception). It will be interesting to compare the cultures in these recently colonized forests with those in refuge areas such as the Ituri Forest of Zaire.

Tool Use

The nature of chimpanzee culture was most dramatically shown when Goodall (1970) found that her subjects not only used but also made tools. The presence of both behavior and artifacts encouraged early comparison across sites of study throughout Africa, first for probes of vegetation to get social insects, and later for hammers of wood or of stone to crack nuts. Enough ethnographic data on elementary technology are now available to allow multiple, point-by-point comparisons across many populations of chimpanzees (McGrew 1992). In addition, a real technological difference appears to exist between the more terrestrial tool-using chimpanzee and the arboreal bonobo, which has not been shown to use any tools.

Feeding

As soon as the dietary repertoires of different populations were described, more diversity than would be attributable to differences in the resources of the biotic environment began to emerge. Nishida et al. (1983) listed a host of dietary differences between Mahale and Gombe. Instances of variety in plant and animal food species and in acquisition and processing techniques now abound. Less obvious forms of ingestion—apparently for medicinal and not just nutritional purposes—have been recognized (Wrangham and Nishida 1983). Such examples now appear to be widespread and common among chimpanzee populations (Huffman and Wrangham, this volume).

Communication

With humans as a notable exception, imitation of vocal expression appears rather limited in primates. For example, cross-fostered macaques of one species, who are completely integrated into the society of another species, continue to use calls typical of their own species without modification (Owren et al. 1992). Based on the failure of efforts to teach human speech to apes (Hayes 1951), the same limitation is often assumed for chimpanzees. Recently, however, the notion that ape vocalizations are hard-wired into their nervous systems has been challenged by findings in both field and laboratory. A bonobo reared with humans developed a number of calls substantially different from its species-typical vocal repertoire (Hopkins and Savage-Rumbaugh 1991). In another instance, long distance calls were found to vary across chimpanzee populations, implicating vocal learning as one possible mechanism (Mitani, this volume).

Variation in communication patterns is likely to be more prominent in the domain of visual signals. For example, Köhler (1925) described how captive chimpanzees followed fads, such as a particular style of locomotion. Similarly, juveniles in the Arnhem Zoo chimpanzee colony walked in single file behind a crippled adult female, all adopting the same pathetic body posture (de Waal 1982). Of course, fads and traditions need not be limited to modes of locomotion: careful observation might reveal a wide range of *community-typical behavior patterns,* which are patterns widespread within one community and absent in other communities. Examples of community-typical behavior patterns are leaf clipping as a sexual invitation (Nishida 1980), hand-clasp grooming (McGrew and Tutin 1978), and hand- and foot- clapping behaviors during grooming (de Waal 1989).

Social organization

Social organization is perhaps the most difficult to study in the context of cultural variation. The difficulties arise because many aspects of social structure vary without any necessary involvement of tradition or of social learning. If, for example, group A exhibits more aggression than group B, this difference may simply be the result of scarcity of resources in group A's environment. Moreover, this situation may be temporary: under conditions similar to group B's, group A may show a sharp decline in aggression.

When groups live under particular conditions for a long time, rules of conduct may be adopted that suit those conditions. As a result, not only are rules of conduct affected, but group members develop expectations about each other's behavior that are different from expectations in other groups. Social rules or norms, an essential component of human culture and moral systems, have been little studied in nonhuman primates. Hall (1964) writing about the punishment of transgressions among baboons, was the first to speak of group norms. De Waal reviews the differences between descriptive and prescriptive rules in chimpanzees. He defines this species' sense of *social regularity* as "a set of expectations about the way in which oneself should be treated and how resources should be divided, a deviation of which expectations to one's disadvantage evokes a negative reaction, most commonly protest in subordinate individuals and punishment in dominant individuals" (1991:336). It is in this domain of expectation of reciprocity that chimpanzee societies may vary substantially. Yet, such variations can only be confirmed with very different data collection techniques than those employed thus far. Instead of recording how chimpanzees behave, observers will need to focus more on how behavior is *reacted to* (accepted or not accepted) by the rest of the group.

Some of the most remarkable differences in social customs can be observed between captive and wild chimpanzees. With more free time on their hands than their wild counterparts and because of their continuous association, captive chimpanzees develop a more intense social life than chimpanzees in the wild (Kummer and Goodall 1985). This intense social life may include different rules of conduct, such as strong limits on violence by males against females and their offspring. The curbing of male power and the resulting change in what is acceptable male behavior is enforced by a group-wide female coalition unknown to wild chimpanzees (de Waal, this volume; Boehm, this volume).

We began this introduction with a reference to social and cultural anthropology and to human primates, and we end by returning to the same point. The arguments presented here and in the chapters that follow—on socioecology, social relations, and the mind—compel us to think in terms of *ethnography.* That is, we need to turn to the tools of the cultural sciences to understand fully the complex natural phenomena of our species' nearest relatives. This need makes it all the more imperative that chimpanzees in the wild be allowed to persist and to survive in all their variety and that their counterparts in captivity be accorded the respect and the opportunities they need to express their potential. They still have much to teach us.

Acknowledgments

We are grateful to Paul Heltne and Linda Marquardt for their invitation to edit this volume on behalf of the Chicago Academy of Sciences; to Linda Marchant, David Pilbeam, and Maryellen Ruvolo for essential and insightful comments on the manuscript, and to Carol Kist for patient word processing.

References

Aisner, R., and J. Terkel. 1992. Ontogeny of pine cone opening behaviour in the black rat, *Rattus rattus. Anim. Behav.* 44:327–36.

Baldwin, P.J., J. Sabater Pi, W.C. McGrew, and C.E.G. Tutin. 1981. Comparisons of nests made by different populations of chimpanzees *(Pan troglodytes). Primates* 22:474–86.

Boesch, C. 1991. Teaching among wild chimpanzees. *Anim. Behav.* 41:530–2.

Boesch, C., and H. Boesch. 1989. Hunting behavior of wild chimpanzees in the Taï National Park. *Am. J. Phys. Anthro.* 78:547–73.
Bonner, J.T. 1980. *The Evolution of Culture in Animals.* Princeton, N.J.: Princeton Univ. Press.
Boyd, R., and P.J. Richerson. 1985. *Culture and the Evolutionary Process.* Chicago, Univ. of Chicago Press.
Caccone, A., and J.R. Powell. 1989. DNA divergence among hominoids. *Evolution* 43:925–42.
Colyn, M., A. Gautier-Hion, and W. Verheyen. 1991. A re-appraisal of paleoenvironmental history in Central Africa: Evidence for a major fluvial refuge in the Zaire Basin. *J. Biogeogr.*18:403–7.
Coolidge, H.J. 1933. *Pan paniscus:* Pygmy chimpanzee from south of the Congo River. *Am. J. Phys. Anthro.* 18:1–57.
Count, E.W. 1973. On the idea of protoculture. In E.W. Menzel, ed., *Precultural Primate Behavior,* pp. 1–25. Basel: Karger.
Dunbar, R.I.M. 1988. *Primate Social Systems.* New York: Comstock.
Galef, B.G. 1992. The question of animal culture. *Human Nature* 3:157–78.
Gonzalez, I. L., J. E. Sylvester, T. F. Smith, D. Stambolian, and R. D. Schmickel. 1990. Ribosomal RNA gene sequences and hominoid phylogeny. *Mol. Biol. Evol.* 7:203–19.
Goodall, J. 1970. Tool-using in primates and other vertebrates. *Adv. Study Behav.* 3:195–250.
Goodall, J. 1986. *The Chimpanzees of Gombe: Patterns of Behavior.* Cambridge, Mass.: Belknap Press.
Groves, C.P. 1970. Population systematics of the gorilla. *J. Zool.* 161:287–300
Groves, C.P., and J.D. Paterson. 1991. Testing hominoid phylogeny with the PHYLIP programs. *J. Human Evol.* 20:167–83.
Groves, C.P., C. Westwood, and B.T. Shea. 1992. Mahalanobis and a clockwork orang. *J. Human Evol.* 22:327–40.
Hall, K.R.L. 1964. Aggression in monkey and ape societies. In J. Carthy and F. Ebling, eds., *The Natural History of Aggression,* pp. 51–64. London: Academic Press.
Hayes, C. 1951. *The Ape in Our House.* New York: Harper and Row.
Hill, A., and S. Ward. 1988. Origin of the Hominidae: The record of African large hominoid evolution between 14 My and 4 My. *Yearbook Phys. Anthro.* 31:49–83.
Hopkins, W.D., and E.S. Savage-Rumbaugh. 1991. Vocal communication as a function of differential rearing experiences in *Pan paniscus:* A preliminary report. *Internat. J. Primatol.* 12:559–83.
Horai, S., Y. Satta, K. Hayasaka, R. Kondo, T. Inoue, T. Ishida, S. Hayashi, and N. Takahata. 1992. Man's place in Hominoidea revealed by mitochondrial DNA genealogy. *Mol. Biol. Evol.* 35:32–43.
Jungers, W.L. and R.L. Susman. 1984. Body and skeletal allometry in African apes. In R. L Susman, ed., *The Pygmy Chimpanzee: Evolutionary Biology and Behavior,* pp. 131–77. New York: Plenum Press.
Kano, T. 1992. *The Last Ape: Pygmy Chimpanzee Behavior and Ecology.* Stanford, Calif.: Stanford Univ. Press.
Kerbis Perterhans, J. C., R.W. Wrangham, M.L. Carter, and M.D. Hauser. 1993. A contribution to tropical rain forest taphonomy: retrieval and documentation of chimpanzee remains from Kibale Forest, Uganda. *J. Human Evol.* 25:485–514.

Kocher, T.D., and A.C. Wilson. 1991. Sequence evolution of mitochondrial DNA in humans and chimpanzees: Control region and a protein-coding region. In S. Osawa and T. Honjo, eds., *Evolution of Life: Fossils, Molecules and Culture,* pp. 391–413. Tokyo: Springer.

Köhler, W. 1925. *The Mentality of Apes.* New York: Harcourt, Brace and Co.

Kummer, H. 1971. *Primate Societies.* Chicago: Univ. of Chicago Press.

Kummer, H., and J. Goodall. 1985. Conditions of innovative behaviour in primates. *Phil. Trans. R. Soc. London B* 308:203–14.

Li W. -H., and D. Graur. 1991. *Fundamentals of Molecular Evolution.* Sunderland, Mass.: Sinauer.

Lott, D.F. 1984. Intraspecific variation in the social systems of wild vertebrates. *Behav.* 88:266–25.

McGrew, W.C. 1992. *Chimpanzee Material Culture: Implications for Human Evolution.* Cambridge, England: Cambridge Univ. Press.

McGrew, W.C., and C.E.G. Tutin. 1978. Evidence for a social custom in wild chimpanzees? *Man* 13:234–51.

McGrew, W.C., P.J. Baldwin, and C.E.G. Tutin. 1988. Diet of wild chimpanzees *(Pan troglodytes verus)* at Mt. Assirik, Senegal: *Am. J. Primatol.* 16:213–26.

Menzel, E.W., ed. 1973. *Precultural Primate Behavior.* Basel: Karger.

Morbeck, M.E., and A.L. Zihlman. 1989. Body size and proportions in chimpanzees, with special reference to *Pan troglodytes schweinfurthii* from Gombe National Park, Tanzania. *Primates* 30:369–82.

Morin, P.A., J.J. Moore, and D.S. Woodruff. 1992. Identification of chimpanzee subspecies with DNA from hair and allele-specific probes. *Proc. R. Soc. London B* 249:293–7.

Morin, P.A., J. Wallis, J.J. Moore, R. Chakraborty, and D.S. Woodruff. 1993. Non-invasive sampling and DNA amplification for paternity exclusion, community structure, and phylogeography in wild chimpanzees. *Primates* 34:347-56.

Nishida, T. 1980. The leaf-clipping display: A newly-discovered expressive gesture in wild chimpanzees. *J. Human Evol.* 9:117–28.

———. 1987. Local traditions and cultural transmission. In B. Smuts, D. L. Cheney, R. M. Seyfarth, Rw. W. Wrangham, and T. T. Struhasker, eds., *Primate Societies,* pp. 462–74. Chicago: Univ. of Chicago Press.

Nishida, T., H. Takasaki, and Y. Takahata. 1990. Demography and reproductive profiles. In T. Nishida, ed., *The Chimpanzees of the Mahale Mountains: Sexual and Life History Strategies,* pp. 63–98. Tokyo: Univ. of Tokyo Press.

Nishida, T., R.W. Wrangham, J. Goodall, and S. Uehara. 1983. Local differences in plant-feeding habits of chimpanzees between the Mahale Mountains and Gombe National Park. *J. Human Evol.* 12:467–80.

Owren, M.J., J.A. Dieter, R.M. Seyfarth, and D.L. Cheney. 1992. Evidence of limited modification in the vocalizations of cross-fostered rhesus *(Macaca mulatta)* and Japanese *(M. fuscata)* macaques. In T. Nishida, W.C. McGrew, P. Marler, M. Pickford, and F.B.M. de Waal, eds., *Topics in Primatology Vol. I, Human Origins,* pp. 257–70. Tokyo: Univ. of Tokyo Press.

Pilbeam, D. 1989. Human fossil history and evolutionary paradigms. In M.K. Hecht, ed., *Evolutionary Biology at the Crossroads,* pp. 117–38. Queens College Press.

Ruvolo, M., T.R. Disotell, M.W. Allard, W.M. Brown, and R.L. Honeycutt. 1991. Resolution of the African hominoid trichotomy by use of a mitochondrial gene sequence. *Proc. Nat. Acad. Sci.* 88:1570–4.

Shea, B.T. 1984. Between the gorilla and the chimpanzee: A history of debate concerning the existence of the *kooloo-kamba* or gorilla-like chimpanzee. *J. Ethnobiol.* 4:1–13.

———. 1985. Ontogenetic allometry and scaling: a discussion based on growth and form of the skull in African apes. In W.L. Jungers, ed., *Size and Scaling in Primate Biology,* pp. 175–205. New York: Plenum Press.

———. 1986. Scapula form and locomotion in chimpanzee evolution. *Am. J. Phys. Anthro.* 70:475–88.

Shea, B.T., and C.P. Groves. 1987. Evolutionary implications of size and shape variation in the genus *Pan. Am. J. Phys. Anthro.* 72:253.

Shea, B.T., and H.J. Coolidge. 1988. Cranial differentiation and systematics in *Pan. J. Human Evol.* 17:671–85.

Shea, B.T., S.R. Leigh, and C.P. Groves. 1993. Multivariate craniometric variation in chimpanzees: Implications for species identification in paleoanthropology. In W.H. Kimbel and L. Martin, eds., *Species, Species Concepts, and Primate Evolution.* New York: Plenum Press.

Sherry, D.F., and B.G. Galef. 1990. Social learning without imitation: More about milk bottle opening by birds. *Anim. Behav.* 40:987–89.

Short, R.V. 1979. Sexual selection and its component parts, somatic and genital selection, as illustrated by man and the great apes. *Adv.Study Behav.* 9:131–58.

Sibley, C.G., J.A. Comstock, and J.E. Ahlquist. 1990. DNA hybridization evidence of hominoid phylogeny: a re-analysis of the data. *J. Mol. Evol.* 30:202–36.

Stanyon, R., B. Chiarelli, K. Gottlieb, and W.H. Patton. 1986. The phylogenetic and taxonomic status of *Pan paniscus*: a chromosomal perspective. *Am. J. Phys. Anthro.* 69:489–498.

Sugiyama, Y. 1988. Grooming interactions among adult chimpanzees at Bossou, Guinea, with special reference to social structure. *Internat. J. Primatol.* 9:393–407.

———. 1989. Population dynamics of chimpanzees at Bossou, Guinea. In P.G. Heltne and L.A. Marquardt, eds., *Understanding Chimpanzees,* pp. 134–45. Cambridge, Mass.: Harvard Univ. Press.

Tomasello, M. 1990. Cultural transmission in the tool use and communication signaling of chimpanzees. In S. T. Parker and K. K. Gibson, eds., *"Language" and Intelligence in Monkeys and Apes: Comparative Developmental Perspectives,* pp. 274–311. Cambridge, England: Cambridge Univ Press.

Uchida, A. 1992. Intra-species variation among the great apes: implications for taxonomy of fossil hominoids. Ph.D. diss., Harvard University.

Ueda, S., Y. Watanabe, N. Saitou, K. Omoto, H. Hayashida, T. Miyata, H. Hisajima, and T. Honjo. 1989. Nucleotide sequences of immunoglobulin-epsilon pseudogenes in man and apes and their phylogenetic relationships. *J. Mol. Biol.* 205:85–90.

Uehara, S., and T. Nishida. 1987. Body weights of wild chimpanzees *(Pan troglodytes schweinfurthii)* of the Mahale Mountains National Park, Tanzania. *Am. J. Phys. Anthro.* 72:315–21.

Visalberghi, E., and D.M. Fragaszy. 1990. Food-washing behaviour in tufted capuchin monkeys, *Cebus apella,* and crabeating macaques, *Macaca fascicularis. Anim. Behav.* 40:829–36.

de Waal, F.B.M. 1982. *Chimpanzee Politics: Power and Sex among Apes.* New York: Harper and Row.

———. 1988. The communicative repertoire of captive bonobos *(Pan paniscus)* compared to that of chimpanzees. *Behav.* 106:183–251.

———. 1989. Behavioral contrasts between bonobo and chimpanzee. In P. Heltne and L. Marquardt, eds., *Understanding Chimpanzees,* pp. 154–75. Cambridge, Mass.: Harvard Univ. Press.

———. 1991. The chimpanzee's sense of social regularity and its relation to the human sense of justice. *Am. Behav. Scientist* 34:335–49.

White, L.A. 1959. *The Evolution of Culture: The Development of Civilization to the Fall of Rome.* New York: McGraw-Hill.

Williams, S.A., and M. Goodman. 1989. A statistical test that supports a human/chimpanzee clade based on noncoding DNA sequence data. *Mol. Biol. Evol.* 6:325–30.

Wrangham, R.W. 1979. On the evolution of ape social systems. *Social Sci. Information* 18:334–68.

Wrangham, R.W., and T. Nishida. 1983. *Aspilia* spp. leaves: A puzzle in the feeding behavior of wild chimpanzees. *Primates* 24:276–82

Zihlman, A. 1984. Body build and tissue composition in *Pan paniscus* and *Pan troglodytes,* with comparisons to other Hominoids. In R. Susman, ed., *The Pygmy Chimpanzee: Evolutionary Biology and Behavior,* pp. 179–200. New York: Plenum Press.

Section 1

Ecology

Frodo

Overview—Ecology, Diversity, and Culture

Richard W. Wrangham

The classic problems of survival in the wild are finding food, escaping predators and parasites, and avoiding bad weather. For chimpanzees the set of solutions adopted by even a single population is surprisingly diverse, since it can involve many types of tools, social strategies, or food-handling techniques, for example. As the following chapters make clear, many differences also occur between populations. Some differences are relatively well known, such as the contrasting tool kits of Gombe and Mahale, where chimpanzees have been studied for about 30 years in each site. Other differences are still in the early stages of documentation, either because they involve more recently established sites or because little investigation of the relevant behaviors has yet occurred in any site. The first aim of this section, therefore, is simply descriptive, detailing some of the variety that chimpanzees and bonobos use to exploit their environment. The second aim builds on the first, attempting to find out whether behavioral differences result from ecological pressures and, if so, how. Answering these questions is a key step toward understanding the origins and functions of variants in both individual survival activities and complex social relationships. The chapters represent current research trends rather than a synthetic overview.

As is true for many primates, most of the attention in studies of chimpanzees has been given to feeding behavior, the principal subject of the first four chapters. The fact that chimpanzees have a variety of inventive techniques for exploiting their environment is illustrated by McGrew's overview of tool-use patterns. Chimpanzee tools are used both for feeding and for other purposes, such as body care and signaling, but the subsistence tools used for obtaining food show with special clarity the problem of explaining behavioral variants as ecological adaptations. In particular it might be expected that patterns of subsistence tool use would be predictable rather easily from the availability of the relevant food, but this is not so.

The occurrence or type of nut-smashing, ant-dipping, and termite-fishing technology bears little obvious relationship to the availability of nuts, ants, or termites. As discussed by McGrew, this lack of obvious relationship does not necessarily mean that ecological factors are unimportant. It may mean merely that the relationship between, say, ant density and wand use is mediated by some other variable, such as total food abundance in the habitat. What it clearly indicates, however, is that the explanation of chimpanzee behavioral diversity will not be easy. This message is emphasized by a second point emerging from McGrew's chapter: bonobos use few tools in the wild. It is too early to say whether this dissimilarity in tool use results from differences in the cognitive makeup of chimpanzees and bonobos, divergent environmental pressures, diverse history, or chance. However, when we turn to subjects that are known less well than tool use, we must be open to a variety of explanations for diversity.

The next two chapters ask in complementary ways whether differences in the grouping patterns of chimpanzees and bonobos are explicable as adaptations to the distribution and density of food. The diets of both species are dominated by sweet fruits whenever such fruits are abundant. It has been suggested previously that the reason Gombe chimpanzees have smaller parties than Lomako bonobos is that the fruit trees are smaller in Gombe than in Lomako. By adding a third population, Kibale chimpanzees, to this comparison, Chapman et al. show that the relationship of party size to tree size does not apply to chimpanzees and bonobos in general. The key observation is that only trivial differences exist between Kibale and Lomako in the density and size of fruiting trees, whereas the grouping pattern in Kibale resembles the grouping pattern in Gombe more than in Lomako. The implication is that if ecological factors are to account for differences in the grouping patterns of the two species, they must be found in adaptations other than frugivory.

Malenky et al. take up the challenge of uncovering determinant ecological factors by examining the role of herbs eaten by both chimpanzees and bonobos when tree fruits are scarce. These herbs, which provide mainly pith and leaf foods, sometimes form in large patches that allow many individuals to forage together. It has been suggested that these herbs tend to occur at higher density in bonobo habitats than in chimpanzee habitats and, therefore, allow bonobos to forage together more often than chimpanzees can. New data reported from Ndoki by Malenky et al. contribute to the hypothesis that gorillas may compete with chimpanzees for these nonfruit foods, because bonobos and gorillas appear more similar to each other in their exploitation of herbs than bonobos and chimpanzees. If this hypothesis is confirmed, it implies that chimpanzees living sympatrically with gorillas may suffer reduced herb availability. An important next step will be to explore more directly how the distribution of herbs influences grouping in these species. Another is to find out whether the interspecific pattern is replicated within species. For example, do chimpanzees with relatively greater access to herbs have less variance in party sizes? Data from the West African sites of Bossou and Taï may be helpful, because chimpanzees in these areas appear to travel alone less often than those in East Africa, where most comparisons have come from to date.

Boesch completes the feeding topics by suggesting that the distribution of animal food can influence how often, and in what manner, it is exploited. His previous work showed that Taï chimpanzees routinely take complementary roles while hunting monkeys; that is, they collaborate extensively. By virtue of his own comparative work, Boesch now reports that although such collaboration is rare in Gombe, it does occur. Related to the rarity of collaboration among Gombe chimpanzees is the finding that their principal prey, the red colobus monkey, is more aggressive in Gombe than in Taï. Boesch's direct observations sharpen the clarity of the Taï-Gombe comparison considerably and also offer a simple environmental explanation for the difference. He suggests that because Gombe trees are shorter than those in Taï, the red colobus is more vulnerable, hence more aggressive, and therefore gives chimpanzees less reinforcement for collaboration. As Boesch notes, this hypothesis provides a clear opportunity for tests in other habitats.

The relative height of trees is also an important theme in the chapters by Doran and Hunt and by Fruth and Hohmann, which look at nonfeeding aspects of using the environment. Chimpanzees and bonobos are sufficiently similar in the shape and size of their bodies that it is tempting to assume that the two are equally efficient in the ways they move through the forest. This assumption is implicit, for example, in hypotheses that relate absolute measures of food distribution to grouping patterns. But Doran and Hunt show that we have to be careful, because bonobos are more adept in trees than chimpanzees are, while body size differences between chimpanzee populations could influence their efficiency in using various forms of arboreal locomotion. The relative roles of habitat structure, tree size, and tree morphology as influences on positional behavior have yet to be clearly established, but Doran and Hunt show that tree structure is likely to be an important factor differentiating populations. Their work implies that comparative data on forest structure would be helpful in sorting out the sources of variation in locomotor behavior.

Data on forest structure would also contribute to understanding the population and species differences in nesting (or "bedding") behavior described by Fruth and Hohmann. Considering that the daily construction of nests is a behavior unique to the great apes, it is remarkable that Fruth and Hohmann's chapter is the first comparative review of this behavior. This chapter draws attention to many gaps in the data and indicates that forest structure alone is inadequate to explain population differences. This chapter, and Doran and Hunt's, are primarily descriptive and should stimulate further work. Perhaps the most interesting cultural form proposed by Fruth and Hohmann is that bonobos occasionally use nests as social "refuges," or socially protected areas, a convention not yet described among chimpanzees.

Finally, Huffman and Wrangham provide an update on the use of medicinal plants, which great apes ingest by chewing bitter pith or swallowing whole leaves. Until now the only great apes reported to ingest plants without obtaining nutritional value were chimpanzees in Mahale, Gombe, and Kibale. Huffman and Wrangham's review extends observations of leaf swallowing to other chimpanzee populations and to bonobos and gorillas. Direct observations of the physiological effect of either behavior remain few. Although each population uses its own small repertoire of medicinal plants, the hypothesis that ingestion of these plants represents self-doctoring remains unproven. However, it is now clear that the behavior is more widespread than once appreciated. Experimental work will be difficult but may be necessary to resolve the physiological effects.

The possibility that great apes can treat their own ailments reminds us of the variety and complexity of the ways in which they may have adapted to their environments. It thereby epitomizes the special challenge of studying the ecology of cognitively sophisticated species and raises further questions. For example, why does tradition impose major constraints on the diets of chimpanzees, as Nishida et al. (1983) suggest? How do local patterns of tool use or collaboration influence chimpanzees' ability to repel predators and, therefore, how does the presence of predators affect grouping patterns in different areas (Boesch 1991)? How much are demographic differences between populations the results of differences in nutrition, and how much are they due to varying cultures of body care or medicinal plant use? Why do bonobos show only occasional attempts to protect themselves from the rain by constructing roofs, and chimpanzees even more rarely (McGrew 1992)?

The range of ecological adaptations shown by chimpanzees and bonobos complicates the use of environmental factors as driving variables, but gives special interest to the puzzle of behavioral ecology in these species.

References

Boesch, C. 1991. The effects of leopard predation on grouping patterns in forest chimpanzees. *Behav.* 117:220-42.

McGrew, W.C. 1992. *Chimpanzee Material Culture: Implications for Human Evolution.* Cambridge, England: Cambridge Univ. Press.

Nishida, T., R.W. Wrangham, J. Goodall, S. Uehara. 1983. Local differences in plant-feeding habits of chimpanzees between the Mahale Mountains and Gombe National Park, Tanzania. *J. Human Evol.* 12:467–80.

Tools Compared

The Material of Culture

W. C. McGrew

Introduction

The time has long past since any informed primatologist talked of the chimpanzee as if the behavior of this species were fixed and uniform. Instead, primatologists recognize the rich behavioral diversity of chimpanzees as crucial to understanding this species *(Pan troglodytes)*. Accordingly, this chapter has three aims: one, to survey the diversity of tool use in free-ranging chimpanzees, stressing the use of tools in subsistence; two, to compare the tool kits of different populations to seek species-typical patterns; and three, to question the origins of diversity in tool use across chimpanzee populations.

Tools are a useful aspect of chimpanzee life upon which to base the ethnographic exercise attempted here: tools are tangible and, thus, readily accessible; and tools are common and, thus, basic to chimpanzee behavior. This is not the case for bonobos, for whom not a single case of the habitual use of tools has been recorded in the wild. Thus, the bonobo *(Pan paniscus)* is not considered in this study.

The Extent of Tool Use

How many populations of free-ranging (either wild or released from captivity) chimpanzees across Africa use tools? A survey of published catalogs shows that as we study more populations we identify more populations (see table 1). Goodall (1973) produced the first systematic catalog of populations listing 10. Teleki (1974) listed 12 wild (but not released)

populations a year later. By 1980, Beck in his massive compendium listed 20 populations of both wild and released chimpanzees. Goodall (1986) listed only 16 populations, but these were all wild populations, not released ones. Most recently, McGrew (1992) listed 32 populations, the reports on many of which are not as yet published.

To consider this diversity, it is useful to list these populations by geographical region or by taxonomic subspecies. In far-West Africa (beyond the Dahomey Gap), populations of *Pan troglodytes verus* in seven countries use tools (see table 2). These populations include four of the five released, free-ranging populations for which data are available.

Table 1
Catalogs of tool use by free-ranging chimpanzees.

Source	Wild	Released	Number of populations
Goodall 1973	+	+	10
Teleki 1974	+		12
Beck 1980	+	+	20
Goodall 1986	+		16
McGrew 1992	+	+	32

Table 2
Sites of tool use by free-ranging chimpanzees: far-West Africa—*Pan troglodytes verus.*

Site	Country	Released	Source
Abuko	Gambia	+	Brewer 1978
Assirik	Senegal		McGrew et al. 1979
Assirik	Senegal	+	Brewer 1978
Baboon Island	Gambia	+	Brewer and McGrew 1990
Banco	Ivory Coast		Hladik and Viroben 1974
Bassa Island	Liberia	+	Hannah and McGrew 1987
Bossou	Guinea		Sugiyama 1989
Cape Palmas	Liberia (?)		Savage and Wyman 1844
Kanka Sili	Guinea		Albrecht and Dunnett 1971
Kanton	Liberia		Kortlandt and Holzhaus 1987
Nimba	Ivory Coast		Yamakoshi and Matsuzawa 1993
no site identified	Liberia		Beatty 1951
Sapo	Liberia		Anderson et al. 1983
Taï	Ivory Coast		Boesch and Boesch 1990
Tenkere	Sierra Leone		Alp 1993
Tiwai	Sierra Leone		Whitesides 1985

In Central West Africa (east of the Dahomey Gap but west of the Oubangi River), populations of *Pan troglodytes troglodytes* in five countries use tools (see table 3). Only one of these was a population of released chimpanzees (Hladik 1973).

In East Africa (east of the Oubangi), seven populations, all wild, of *Pan troglodytes schweinfurthii* in only these countries show tool use (see table 4). However, five of the six populations are subjects of long-term studies, and two of these populations, at Gombe (Goodall 1986) and at Mahale (Nishida 1990), provide the bulk of what is known.

Table 3
Sites of tool use by free-ranging chimpanzees: Central West-Africa—*Pan troglodytes troglodytes.*

Site	Country	Released	Source
Ayamiken	Equatorial Guinea		Jones and Sabater Pi 1969
Belinga	Gabon		McGrew and Rogers 1983
no site identified	Cameroon		Merfield and Miller 1956
Campo	Cameroon		Sugiyama 1985
Dipikar	Equatorial Guinea		Jones and Sabater Pi 1969
Ipassa	Gabon	+	Hladik 1973
Lopé	Gabon		Tutin and Wrogemann 1994
Mbomo	Congo		Fay and Carroll, pers. com.
Mayang	Equatorial Guinea		Gonzalez-Kirchner and Sainz de la Maza 1992
Ndakan	Central African Republic		Fay and Carroll, pers. com.
Ngoubunga	Central African Republic		Fay and Carroll, pers. com.
Okorobiko	Equatorial Guinea		Sabater Pi 1974
West Cameroon	Cameroon		Struhsaker and Hunkeler 1971

Table 4
Sites of tool use by free-ranging chimpanzees: East-Africa—*Pan troglodytes schweinfurthii.*

Site	Country	Source
Budongo	Uganda	Sugiyama 1969
Filabanga	Tanzania	Itani and Suzuki 1967
Gombe	Tanzania	Goodall 1968
Kasakati	Tanzania	Suzuki 1966
Kibale	Uganda	Wrangham, pers. com.
Mahale	Tanzania	Nishida 1990
Kahuzi-Biega	Zaire	Yamagiwa et al. 1988

Note: All populations listed are wild.

In summary, the most extensive set of tool-using populations, whether wild or released, are to be found in Far West Africa (see table 5). However, apart from nut-cracking populations, the most extensive data come from East Africa. Studies from the equatorial forests of Central West Africa remain mostly preliminary. Until populations there are habituated, knowledge of their tool use will be limited largely to indirect evidence from artifacts.

Not all free-ranging chimpanzees use tools. Eight populations still show no evidence of tool use (see table 6). However, none of these populations is habituated, and most studies of these populations were surveys or short-term, seasonal observations. Only one year-long study (Azuma and Toyoshima 1962) has failed to find the use of tools in chimpanzee populations.

After more than 30 years of field study of chimpanzees in Africa, the accumulated data suggest that tool use is likely to be typical of the species.

Table 5
Tool use by free-ranging chimpanzee populations across Africa.

Region and subspecies	Number of populations	Released	Long-term studies
far West-Africa—*P.t.v.*	15	4	3
Central West-Africa—*P.t.t.*	13	1	2
East Africa—*P.t.s.*	7	0	5
Totals	36	5	10

Table 6
Populations of free-ranging chimpanzees in which no tool use has been observed to date.

Site	Country	Comments	Main source
Bafing	Mali	short survey	Moore 1985
Beni	Zaire	short study	Kortlandt 1962
Fouta Djallon	Guinea	long survey	de Bournonville 1967
Ishasha	Zaire	short study	Sept 1992
Kabogo	Tanzania	long study	Azuma and Toyoshima 1962
Neribili	Guinea	short study	Nissen 1931
Rubondo*	Tanzania	unstudied	Borner 1985
Ugalla	Tanzania	short surveys	Itani 1979

Notes: * Released population.

Species Typicality

The question of typicality can be approached on a general level by comparing the tool kits of different populations to seek commonalities. Patterns of tool use that appear in all populations, universal technologies, are presumed to be basic to chimpanzee nature. That such universal technologies exist is borne out by *nest building,* the daily making of overnight sleeping platforms. However, a problem in assessing the commonalty of tool kits arises because of incomplete data. In some cases, only a single tool-use event performed by a single individual is known (Brewer and McGrew 1989). Such tool use may be idiosyncratic or accidental.

Valid comparisons of tool kits rely on *habitual patterns,* actions seen to be performed repeatedly by several individuals in a population. When exact information about frequencies and subjects is not known, I have arbitrarily considered a minimum of 10 instances to be sufficient to establish the existence of a habitual pattern. This criterion is admittedly crude, but the data are varied, and some threshold for scoring is needed. Sugiyama (1989) made a similar distinction between established tool use and other types of tool use.

Figure 1
Two infant chimpanzees at Gombe use twigs to start social play. (Photo by Caroline Tutin.)

Table 7
Habitual patterns of tool use by wild chimpanzees.

Pattern	Gombe	Bossou	Mahale	Taï	Kibale	Kanka Sili	Assirik	Others
Termite-fish	X		X				X	
Ant-dip	X	X		X			X	Tenkere
Honey-dip	X			X				
Leaf-sponge	X	X			X			
Leaf-napkin	X				X			
Stick-flail	X	X	X			X		
Stick-club	X		X	X		X		
Missile-throw	X	X	X	X		X		
Self-tickle	X							
Play-start	X		X					
Leaf-groom	X		X		X			
Ant-fish			X					
Leaf-clip		X	X					
Gum-gouge		X						
Nut-hammer		X		X				Kanton, Sapo, Tiwai
Marrow-pick				X				
Bee-probe				X				
Branch-haul		X						
Termite-dig								Campo, Okorobiko
Total	11	8	8	7	3	3	2	6

Note: X = present. Source: McGrew (1992).

Table 7 lists in matrix form the habitual patterns of tool use shown by wild (but not released) chimpanzees from at least one population. These patterns total 19.

Looking down columns, it is obvious that no population of chimpanzees has a comprehensive, habitual tool kit. Gombe chimpanzees, with 11 habitual patterns, show the highest number. This tool kit is likely to be complete, as no new habitual pattern has been added in many years of observation. In contrast, six of the 12 populations show only a single habitual pattern.

These results are most likely gross underestimates, that is, the table should indicate more tool use. If one ranks the populations or study sites by size of tool kit, there appears to be a positive correlation both with the degree of habituation and with the cumulative length of study. These two factors probably have a positive correlation with each other, too. Thus, the totals for Kibale and Taï Forest will likely increase with more close-range observation, but that of Assirik will not increase, unless habituation is somehow achieved.

Some results are likely to be biased. For example, Kanka Sili chimpanzees were studied only in a cleared feeding area so that tool use was possibly underestimated. On the other hand, these chimpanzees were

Table 8

Free-ranging chimpanzees eating termites, with tools or by hand only, across Africa.

Region	*Macrotermes*	Other termites	Source
East Africa—*P.t.s.*			
Budongo*	—	—	Reynolds and Reynolds 1965
Gombe*	Fish	Hand	Goodall 1968
Kabogo*	—	?	Azuma and Toyoshima 1961–62
Kasakati*	Fish?	—	Suzuki 1966
Kibale*	—	Hand?	Ghiglieri 1984
Mahale* (B group)	Fish	—	McGrew and Collins 1985
Mahale* (K and M groups)	—	Fish, Hand	Uehara 1982
Central West Africa—*P.t.t.*			
Belinga	Fish/Brush/Probe	—	McGrew and Rogers 1983
Campo	Brush	—	Sugiyama 1985
Ipassa	—	Hand	Hladik 1973
Lopé*	—	—	Tutin and Wrogemann 1994
Mayang	Probe?	—	Gonzalez-Kirchner and Sainz de la Maza 1992
Ndakan	Brush/Fish/Probe	—	Fay, pers. com.
Okorobiko*	Probe	—	Jones and Sabater Pi 1969
far West-Africa—*P.t.v.*			
Assirik*	Fish	—	McGrew et al. 1979
Bossou*	—	Squash	Sugiyama and Koman 1987
Taï*	—	Hand	Boesch and Boesch 1990

Notes: * = long-term study. — = no termite eating reported.

supplied with tools artificially and so the use of tools could also be overestimated (Albrecht and Dunnett 1971). Finally, sites such as Kibale and Mahale have more than one community, or unit group, and these show some differences between communities in tool kits.

On the specific level, focusing on probing for underground termites, the most famous type of tool use, yields limited results in support of species typicality (see figure 3). Only three populations, at Assirik, Gombe, and Mahale, show habitual termite fishing, and two other populations at Campo and Okorobiko show habitual termite digging.

Table 8 seeks to provide a comprehensive list of those free-ranging populations, both wild and released, for which there is mention of eating termites, either with or without tools. The prey can be divided into *Macrotermes* (a genus of large-bodied, mound-building termites that is pan-African in distribution is widely sympatric in the range of chimpanzees) and other termites. All known records of termite probing, habitual or otherwise, are included in table 8.

At Budongo and Lopé, chimpanzees ignore termites; at Ipassa, Kibale, and Taï Forest termites are taken only by hand, albeit rarely; at other sites only one tool-use technique is used, whether for fishing as at Assirik, digging or probing as at Okorobiko, squashing as at Bossou, or brushing as at

Figure 2
Adult female chimpanzee at Gombe uses a stick to dip for driver ants; her infant watches closely. (Photo by Jim Moore.)

Campo. Only Belinga and Ndakan in Central West Africa show multiple techniques. Most technological solutions to the challenge of getting termites are focused on *Macrotermes*. The sole exception is the K group at Kasoje, Mahale, who fished for *Pseudacanthotermes* in an area lacking *Macrotermes* (Uehara 1982), while their neighbors, the B group, fished for *Macrotermes* (McGrew and Collins 1985).

In summary, the tool-use repertoire of *Pan troglodytes* is far from species typical, as there is great diversity both across and within geographical regions. However, most of what we need to know before we can make definitive statements awaits more ethnographic data.

Origins of Variation

Ever since diversity of tool use came to be recognized, several authors have invoked culture in opposition to ecological determinism as an explanation for that diversity (Goodall 1973; Sabater Pi 1974; Teleki 1974; McGrew et al. 1979; Nishida 1987). The terminology of this discussion has varied. The phenomenon of tool use is sometimes called either *local tradition* or *social custom,* but the key factor is considered to be some sort of social learning, usually unspecified. Various mechanisms, from social facilitation and stimulus enhancement to imitation, have been mentioned, usually on a speculative basis.

The first prima facie evidence for cultural differences in tool use emerged from regional contrasts. These contrasts suggested dissemination, or diffusion, of tool use patterns within but not across biogeographical boundaries (see table 9). Thus, Struhsaker and Hunkeler (1971) divided wild chimpanzees into nut smashers and termite fishers. Nishida (1973) modified the latter to incorporate ant fishers. Sabater Pi (1974) took a different tack, with a three-way classification scheme based on raw materials. Teleki (1974) proposed another variant, and McGrew et al. hypothesized a more focused dichotomy of termite probing (1979). All these early hypotheses of regional differences in tool use have been falsified by later field data.

The only gross dichotomy still standing is between the nut crackers of far-West Africa and the non-nut crackers of Central West and East Africa. (see figure 4.) However, even this dichotomy could be ecologically determined if only the far-Western forests contained the appropriate species of

Table 9
Hypothesized regional differences in tool use across wild chimpanzee populations.

East Africa—*P.t.s.*	Central West Africa—*P.t.t.*	far West Africa—*P.t.v.*	Source
Termite fishers	Termite fishers	Nut smashers	Struhsaker and Hunkeler 1971
Ant/termite fishers	Ant/termite fishers	Nut smashers	Nishida 1973
Foliage industry	Stick industry	Stone industry	Sabater Pi 1974
Ant/termite probers	Termite probers	Pounders	Teleki 1974
Termite fishers	Termite diggers	Termite fishers	McGrew et al. 1979
Non-nut crackers	Non-nut crackers	Nut crackers	McGrew 1992

◀ **Figure 3**
Adult male chimpanzee at Gombe uses a vine to fish for termites. (Photo by Linda Marchant.)

Table 10
Species of nuts found in Lopé Reserve, Gabon, but *not* eaten there by chimpanzees.

Species of nut in Lopé	Species cracked and eaten elsewhere	Other study sites
Coula edulis	same	Taï
Detarium macrocarpum	*D. senegalense*	Taï, Tiwai
Elaeis guineensis	same	Bossou, Kanton
Klainedoxa gabonensis	single?	single?
Panda oleosa	same	Taï
Parinari excelsa	same	Taï
Saccoglottis gabonensis	same	Taï

Source: Project NUTI: Noisette Untapped Technology Investigation

nut-bearing trees or the raw materials for hammers and anvils. If such differences in resource availability are a factor, then there is no need to propose sociocultural factors in trying to explain regional differences.

However, recent evidence from Lopé in Gabon suggests that arguments of ecological determinism will *not* explain the absence of nut cracking there (McGrew et al. in prep.) (see table 10). The nut-bearing species are abundant at Lopé, as are raw materials for wooden and stone hammers and for stone and root anvils. By exclusion, a more likely explanation for why Lopé chimpanzees do not crack nuts is ignorance, that is, they do not consider these nuts to be edible. Such findings, along with findings on differences in the tool kits of neighboring communities within a region, suggest that culture is involved (McGrew 1992).

◀ **Figure 4**
Adult female chimpanzee at Bassa Island uses stone hammer to crack nuts of the oil palm. (Photo by Alison Hannah.)

Several analysts (Galef 1990; Tomasello 1990, this volume; Whiten 1989) have sought to deny culture or tradition to chimpanzees on the grounds that their ability to imitate is unproven. Similar arguments are applied to teaching as a possible mechanism of transmission. If simple phenomena, such as enhancement plus trial-and-error learning, can account for the performance involved in the use of tools, then culture is deemed unnecessary.

The denial of culture is suspect on three counts. First, ethnographers of human culture define culture in many ways, but few definitions incorporate either imitation or teaching as necessary conditions. Culture is not a monolithic, consensual entity, but rather an amorphous array of constructs. Second, much human behavior, even skilled activities, may be acquired by so-called mindless or rote learning (Wynn 1993). Third, failure of chimpanzees to show imitation or teaching under the contrived and impoverished conditions of the laboratory is hardly generalizable to natural conditions. The challenge for field primatologists is to devise methods of inferring intentions and mental processes in the uncontrolled conditions of the field. Experimental intervention during rehabilitation into free-ranging settings seems an obvious candidate for such research (Hannah and McGrew 1987).

Finally, to seek to *explain* variation in a behavioral pattern by saying that it is cultural is no more useful than saying that lack of variation is natural. Neither statement is helpful nor likely even true. Tackling the thorny question of the relative importance of the various influential factors, or determinants, of variation is much more interesting and challenging.

Summary

At least 36 populations of chimpanzees across East, Central West, and far-West Africa use tools. This rich array of technology suggests that use of tools is typical of *Pan troglodytes.* No particular behavioral pattern is universal, but 11 of 19 patterns, which were observed as habitual in at least one population, concern subsistence. For example, chimpanzees as predators on termites use various techniques to exploit the prey; some techniques involve tools while other techniques do not. Knowledge of chimpanzee tool use traditions continues to mount, while proposed regional distinctions continue to disappear (with the exception of nut-cracking technique in far-West Africa). The mechanisms by which tool use is acquired are likely affected externally by environmental and social factors and affected internally by cognitive processes.

Acknowledgments

Caroline Tutin has shared my study of chimpanzee tool use for more than 20 years, since it began in January 1972, at the Delta Primate Center under the patient guidance of Pal Midgett. Carol Kist and Fay Somerville did the manuscript preparation, cheerfully and efficiently. Alison Hannah, Linda Marchant, Jim Moore, and Caroline Tutin generously donated photographs. Linda Marchant and Richard Wrangham critiqued the manuscript, but the final result is the sole responsibility of the author.

References

Albrecht, H., and S.C. Dunnett. 1971. *Chimpanzees in Western Africa.* München: Piper Verlag.

Alp, R. 1993. Meat eating and ant dipping by wild chimpanzees in Sierra Leone. *Primates* 34:463–68.

Anderson, J.R., E.A. Williamson, and J. Carter. 1983. Chimpanzees of Sapo Forest, Liberia: Density, nests, tools and meat-eating. *Primates* 24:594–601.

Azuma, S., and A. Toyoshima. 1962. Progress report of the survey of chimpanzees in their natural habitat, Kabogo Point area, Tanganyika. *Primates* 3:61–70.

Beatty, H. 1951. A note on the behavior of the chimpanzee. *J. Mammal.* 32:118.

Beck, B.B. 1980. *Animal Tool Behavior: The Use and Manufacture of Tools by Animals.* New York: Garland STPM Press.

Boesch, C., and H. Boesch. 1990. Tool use and tool making in wild chimpanzees. *Folia Primatol.* 54:86–99.

Borner, M. 1985. The rehabilitated chimpanzees of Rubondo Island. *Oryx* 19:151–54.

de Bournonville, D. 1967. Contribution à l'étude du Chimpanzé en République de Guinée. *Bull. Instit. Fond. Afri. Noire* 29A:1188–1269.

Brewer, S. 1978. *The Chimps of Mt. Asserik.* New York: Alfred A. Knopf.

Brewer, S.M., and W.C. McGrew. 1989. Chimpanzee use of a tool-set to get honey. *Folia Primatol.* 54:100–4.

Galef, B.G. 1990. Tradition in animals: Field observations and laboratory analyses. In M. Bekoff and D. Jamieson, eds., *Interpretation and Explanation in the Study of Animal Behavior,* pp. 74–95. Boulder, Colo.: Westview Press.

Ghiglieri, M.P. 1984. *The Chimpanzees of Kibale Forest.* New York: Columbia Univ. Press.

Gonzales-Kirchner, J.P., and M. Sainz de la Maza. 1992. Sticks used by wild chimpanzees: A new locality in Rio Muni. *Folia Primatol.* 58:99–102.

Goodall, J. 1968. The behavior of free-ranging chimpanzees in the Gombe Stream Reserve. *Anim. Behav. Monogr.* 1:161–311.

——. 1973. Cultural elements in a chimpanzee community. In E.W. Menzel, ed., *Precultural Primate Behavior,* pp. 144–84. Basel: S. Karger.

——. 1986. *The Chimpanzees of Gombe.* Cambridge, Mass.: Belknap Press.

Hannah, A.C., and W.C. McGrew. 1987. Chimpanzees using stones to crack open oil palm nuts in Liberia. *Primates* 28:31–46.

Hladik, C.M. 1973. Alimentation et activité d'un group de chimpanzé réintroduits en fôret Gabonaise. *La Terre et la Vie* 27:343–413.

Hladik, C.M., and G. Viroben. 1974 L'alimentation protéique du chimpanzé dans son environement forestier naturel. *Comptes Rendes* D279:1475–8.

Itani, J. 1979. Distribution and adaptation of chimpanzees in an arid area. In D.A. Hamburg and E.R. McCown, eds., *The Great Apes,* pp. 55–71. Menlo Park, Calif.: Benjamin/Cummings.

Itani, J., and A. Suzuki. 1967. The social unit of chimpanzees. *Primates* 8:355–81.

Jones, C., and J. Sabater Pi. 1969. Sticks used by chimpanzees in Rio Muni, West Africa. *Nature* 223:100–1.

Kortlandt, A. 1962. Chimpanzees in the wild. *Sci. Am.* 206(5):128–38.

Kortlandt, A., and E. Holzhaus. 1987. New data on the use of stone tools by chimpanzees in Guinea and Liberia. *Primates* 28:473–96.
McGrew, W.C. 1992. *Chimpanzee Material Culture: Implications for Human Evolution.* Cambridge, England: Cambridge Univ. Press.
McGrew, W.C., and D.A. Collins. 1985. Tool-use by wild chimpanzees *(Pan troglodytes)* to obtain termites *(Macrotermes herus)* in the Mahale Mountains, Tanzania. *Am. J. Primatol.* 9:47–62.
McGrew, W.C., and M.E. Rogers. 1983. Chimpanzees, tools, and termites: New records from Gabon. *Am. J. Primatol.* 5:171–4.
McGrew, W.C., R. Ham, L. White, C.E.G. Tutin, and M. Fernandez. In prep. Why don't chimpanzees in Gabon crack nuts?
McGrew, W.C., C.E.G. Tutin, and P.J. Baldwin. 1979. Chimpanzees, tools and termites: Cross-cultural comparisons of Senegal, Tanzania and Rio Muni. *Man* 14:185–214.
Merfield, .G., and H. Miller. 1957. *Gorillas Were My Neighbours.* London: Companion Book Club.
Moore, J.J. 1985. Chimpanzee survey in Mali, West Africa. *Primate Conservation* 6:59–63.
Nishida, T. 1973. The ant-gathering behavior by the use of tools among wild chimpanzees of the Mahale Mountains. *J. Human Evol.* 2:357–70.
——. 1987. Local traditions and cultural transmission. In B.B. Smuts, D.L. Cheney, R.M. Seyfarth, R.W. Wrangham, and T.T. Struhsaker, eds., *Primate Societies,* pp. 461–74. Chicago, Ill.: Univ. of Chicago Press.
—— ed. 1990. *The Chimpanzees of the Mahale Mountains.* Tokyo: Univ. of Tokyo Press.
Nissen, H.W. 1931. A field study of the chimpanzee. Observations of chimpanzee behavior and environment in Western French Guinea. *Comp. Psychol. Monogr.* 8:1–122.
Reynolds, V., and F. Reynolds. 1965. Chimpanzees of the Budongo Forest. In I. DeVore, ed., *Primate Behavior,* pp. 368–424. New York: Holt, Rinehart and Winston.
Sabater Pi, J. 1974. An elementary industry of the chimpanzees in the Okorobiko Mountains, Rio Muni (Republic of Equatorial Guinea), West Africa. *Primates* 15:351–64.
Savage, T., and J. Wyman. 1844. Observations on the external characters and habits of *Troglodytes niger* and on its organization. *Boston J. Nat. History* 4:362–86.
Sept, J.M. 1992. Was there no place like home? A chimpanzee perspective on early archaeological sites. *Current Anthro.* 33:187–207.
Struhsaker, T.T., and P. Hunkeler. 1971. Evidence of tool-using by chimpanzees in the Ivory Coast. *Folia Primatol.* 15:212–9.
Sugiyama, Y. 1969. Social behavior of chimpanzees in the Budongo Forest, Uganda. *Primates* 10:197–225.
——. 1985. The brush-stick of chimpanzees found in south-west Cameroon and their cultural characteristics. *Primates* 26:361–74.
——. 1989. Description of some characteristic behaviors and discussion on their propagation process among chimpanzees of Bossou, Guinea. In Y. Sugiyama, ed., *Behavioral Studies of Wild Chimpanzees at Bossou, Guinea,* pp. 43–76. Inuyama: Primate Research Institute, Kyoto Univ.
Sugiyama, Y., and J. Koman. 1979. Tool-using and -making behavior in wild chimpanzees at Bossou, Guinea. *Primates* 20:513–24.

—. 1987. A preliminary list of chimpanzees' alimentation at Bossou, Guinea. *Primates* 28:133–47.

Suzuki, A. 1966. On the insect-eating habits among wild chimpanzees living in the savanna woodland of Western Tanzania. *Primates* 7:481–7.

Teleki, G. 1974. Chimpanzee subsistence technology: Materials and skills. *J. Human Evol.* 3:575–94.

Tutin, C.E.G., and D. Wrogemann. 1994. Tool use by chimpanzees *(Pan t. troglodytes)* in the Lopé Reserve, Gabon. In press.

Tomasello, M. 1990. Cultural transmission in the tool use and communicatory signalling of chimpanzees? In S.T. Parker and K.R. Gibson, eds., *"Language" and Intelligence in Monkeys and Apes,* pp. 274–311. Cambridge, England: Cambridge Univ. Press.

Uehara, S. 1982. Seasonal changes in the techniques employed by wild chimpanzees in the Mahale Mountains, Tanzania, to feed on termites *(Pseudocanthotermes spininger). Folia Primatol.* 37:44-76.

Whiten, A. 1989. Transmission mechanisms in primate cultural evolution. *Trends in Ecology and Evol.* 4:61–2.

Whitesides, G.H. 1985. Nut cracking by wild chimpanzees in Sierra Leone, West Africa. *Primates* 26:91–4.

Wynn, T. 1993. Layers of thinking in tool behavior. In K.R. Gibson and T. Ingold, eds., *Tools, Language and Cognition in Human Evolution,* pp. 389–406. Cambridge, England: Cambridge Univ. Press.

Yamagiwa, J., T. Yumoto, M. Ndunda, and T. Maruhashi. 1988. Evidence of tool use by chimpanzees *(Pan troglodytes schweinfurthii)* for digging a bee nest in the Kahuzi-Biega National Park, Zaire. *Primates* 29:405–11.

Yamakoshi, G., and T. Matsuzawa. 1993. Preliminary surveys of the chimpanzee in the Nimba Reserve. Côte d' Ivoire. *Primate Research* 9:13–8.

Use of stone tools.
Photograph courtesy of T. Matsuzawa.

Party Size in Chimpanzees and Bonobos

A Reevaluation of Theory Based on Two Similarly Forested Sites

Colin A. Chapman, Frances J. White, and Richard W. Wrangham

Introduction

When detailed information became available on the social organization and ecology of the bonobos, *Pan paniscus* (Kuroda 1979; Kano 1983; Badrian and Badrian 1984; Badrian and Malenky 1984; White 1986; Kano 1992), it became clear that marked differences existed between this species and the closely related chimpanzees, *Pan troglodytes,* who have been the focus of many earlier long-term studies (Nishida 1968; Goodall 1986). Three ecological hypotheses were proposed to explain observed differences (Badrian and Badrian 1984; White 1986; Wrangham 1986; White and Wrangham 1988; Malenky 1990). All three hypotheses were based on the premise that the larger mean party size reported in *P. paniscus* was a species-specific trait that resulted from less feeding competition between bonobos than between chimpanzees.

According to the first hypothesis, bonobos exhibit less feeding competition than chimpanzees because bonobos have greater access to terrestrial herbaceous vegetation (THV), a food source viewed to be more continuously abundant and widespread than tree fruits (Badrian and Badrian 1984;

Wrangham 1986). Bonobos regularly consume THV at all the major sites (Malenky and Stiles 1991). In contrast, chimpanzees eat THV less frequently and less consistently. For example, THV consumption by chimpanzees at Gombe accounted for 7% of the monthly food intake, while THV consumption by bonobos at Wamba averaged 33% of the monthly food intake (Wrangham 1986). These differences led Wrangham (1986) to suggest that abundant THV available to bonobos reduced feeding competition between bonobos and permitted bonobo females to feed together even when tree fruits were not abundant.

The second hypothesis proposed that feeding competition in bonobos is reduced because bonobos have access to larger patches of tree fruits than chimpanzees (Badrian and Badrian 1984; White 1986). In support of this hypothesis, White and Wrangham (1988) first demonstrated that party size was related to the size of the patch for bonobos at Lomako. Subsequently, the product of the number of animals feeding in a patch multiplied by the length of time in that patch was used as a measure of patch use. This measure was shown to be greater for the bonobos at Lomako than for the chimpanzees at Gombe. White and Wrangham's (1988) analysis used the only comparable data available at that time. This data involved a comparison of two very different habitats: Gombe Stream Reserve in Tanzania and Lomako Forest in Zaire. Gombe is a mosaic of semi-evergreen forest, deciduous forest, and grassland, with 750–1,250 millimeters rainfall (Goodall 1986); Lomako is a tropical evergreen forest, with 2,000 millimeters rainfall (White 1986). Therefore, tree-fruit patch sizes could be expected to differ as a result of habitat differences and may not be one of the determining differences in *Pan* social organization.

A third hypothesis, recently proposed by Malenky (1990), suggests that the tree fruits utilized by the bonobos at Lomako are available consistently throughout the year. Malenky (1990) used rainfall data to suggest that chimpanzee habitats vary more with respect to tree fruits than bonobo habitats. Thus, Malenky suggested that chimpanzees are more likely to experience severe seasonal shortages of tree fruits. Malenky suggested that a uniform production of fruit contributes to the larger size of bonobo parties.

This paper addresses all three hypotheses with special emphasis on the second hypothesis: that bonobos experience reduced feeding competition because bonobos have larger patches of tree fruits than do chimpanzees. We build on White and Wrangham's (1988) examination of this hypothesis. However, the analysis conducted here compares chimpanzee and bonobo populations that inhabit similar forest habitats: chimpanzees in the Kibale Forest Reserve of Uganda and bonobos at the Lomako Forest of Zaire. We describe the size of the fruit trees (patches) and the length of time spent in a patch for both sites. Subsequently, we examine whether Kibale and Lomako differ in the overall density of tree patches and whether this density relates to differences in the distribution of party sizes. We review the literature on the average size of a party from a variety of field studies to assess whether the average party size of each of these species clearly differ.

In addition, we examine both the way in which bonobos and the way in which chimpanzees associate in tree-fruit patches of similar size. Our analysis of this question was influenced by our impression that the two species differ in the degree to which foraging individuals prefer to associate in the same tree. The authors had observed bonobos and chimpanzees during periods when abundant fruiting of a large and relatively common tree species had given individuals the option either of feeding together in the same large tree or of dispersing into neighboring trees to feed. Under such equivalent conditions of food-patch distributions, bonobos appeared more likely to feed together in the same tree such as *Dialium zenkeri* (Kano 1992; White, pers. observ.), while individuals in a party of chimpanzees would disperse over several neighboring trees such as *Mimusops bagshawei* (Chapman, pers. observ.; Wrangham, pers. observ; see figure 1). This observation suggests that differences in the spatial distribution of tree-fruits cannot by themselves account for differences in these grouping patterns, implying that individuals in different species respond to similar environments in different ways. In addition to presenting data concerning size of party; length of time spent in patches; grouping dispersion patterns while feeding; size of fruit tree-patches; and density of fruit trees, we also present data concerning party composition and diet for both study sites.

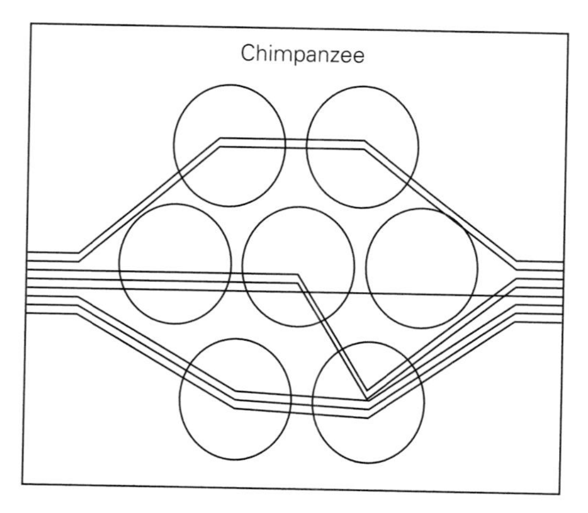

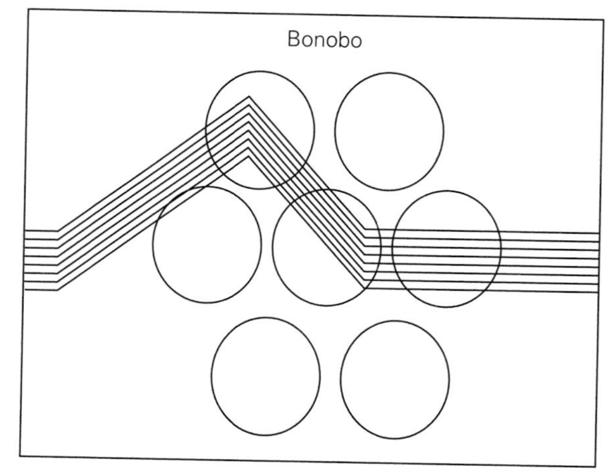

Figure 1
During periods of abundant tree-fruit, we believe that when individual chimpanzees or bonobos have the option to feed together in a single tree or not that bonobos feed together while chimpanzees disperse to feed in neighboring trees. Each circle represents a large, fruiting tree; the lines represent the foraging paths taken by individual bonobos or chimpanzees.

Methods

The Kibale Forest Reserve is located in western Uganda from 0°13' to 0°41' north latitude and from 30°19' to 30°32' east longitude near the base of the Ruwenzori Mountains. About 60% of the Kibale Forest Reserve is a moist evergreen forest, transitional between lowland rain forest and montane forest (Struhsaker 1975; Skorupa 1986, 1988; Wing and Buss 1970). The forest canopy is generally 25–30 meters high (Butynski 1990), although some trees may exceed 55 meters. The remainder of the reserve is a mosaic of swamp, grassland, pine plantation, thicket, and colonizing forest (Wing and Buss 1970; Butynski 1990). The study site, known locally as Kanyawara, is situated at an elevation of 1,500 meters. Its annual mean daily maximum temperature for the years 1977–83 was 23.3 ± 0.6°C. Mean annual rainfall (1987–91) averaged 1,740 millimeters with a range of 1,607–1,864 millimeters. Rainfall tends to be well dispersed, falling on an average of 166 days per year (Kingston 1967); however, there are distinct wet and dry seasons. May through August and December through February tend to be drier than other months, with the May-August dry season of a longer duration than the December-February dry season.

The Lomako Forest is located in the Equator Province of central Zaire from 0°51' north latitude and 21°5' east longitude at an altitude of approximately 300 meters. The study area is predominantly evergreen rain forest, with small areas of swamp and secondary forest (White 1992a). The average annual rainfall is approximately 2,000 millimeters with two peaks, one occurring in October through December and one in March through May. Thus, on a superficial level, Lomako and Kibale appear similar and very suitable for comparisons.

Observations of chimpanzees at Kanyawara in Kibale took place from November 1989 through January 1990, March 1990 through December 1990, May 1991 through October 1991, and May 1992 through October 1992. Additional data from the long-term camp records collected since July 1988 are also presented here. Observations of bonobos at Lomako took place from July 1983 through April 1984, October 1984 through April 1985, and June through August 1991.

Definitions

A number of issues of methodology were considered before testing these hypotheses. First, defining a patch has proven difficult in field situations. A number of theoretical or conceptual definitions have been proposed (Altmann 1974; Hassell and Southwood 1978; Addicott et al. 1987). For the purpose of this study, however, a *patch* was considered an aggregation of food items structured so that animals could use the area without interrupting feeding activity. A patch was considered to be an isolated fruit tree or, on rare occasions, two adjacent trees with crowns touching.

The size of a patch would ideally represent the number of food items that a tree contained; however, the number of patches used by both species precluded an accurate counting of food items. We estimated *patch size* as either the diameter in centimeters of the tree measured at breast height (DBH) or radius of the crown. Previous studies have evaluated the applicability of these measures as indexes of fruit abundance (Chapman et al. 1992; Peters et al. 1988). If the same patch was used more than once, an average party size for that patch was calculated and used in subsequent tests.

Next, there are problems associated with comparisons of party size between various locations. First, it is difficult to count individuals that are dispersed throughout a habitat where visibility is limited. Under such conditions, the farther the animal is from the observer or the denser the vegetation, the higher the probability that it will not be included in a count of party size. Second, our studies differ in duration. As an investigation proceeds, the study animals may become habituated, easier to see and, thus, more likely to be included in a count of party size. Third, the method by which the observer locates animals affects the perception of party size. At Kanyawara, for example, chimpanzee parties located by waiting at fruiting trees are smaller than parties found by searching trails or by locating parties after hearing calls (Wrangham et al. 1992). Thus, variation in party size between studies may reflect differences in the methods used to find the animals or in the proportion of sightings resulting from each method.

Party size varies over time, possibly in relation to periodic changes in food availability throughout the year which may follow supra-annual patterns. Thus, until long-term data are available it is not possible to preclude the possibility that the average party size reported in any one study may reflect a relatively unique ecological or social situation that does not accurately represent the site.

Finally, definitions of what constitutes a party, and procedures to measure it, vary. Past studies of chimpanzees have considered *party size* as the number of independent individuals present when the party was first contacted (White 1986, 1988); or as the focal animal and all others within 100 meters (Wrangham and Smuts 1980); or as the initial party plus new counts made with each change in party composition (White 1988; Boesch 1991); or as a scan taken after a set number of minutes (Wrangham 1977; Wrangham et al. 1992); or as the *acoustic party size,* which includes all those individuals within auditory range (Mitani and Nishida, in prep.; Chapman et al., in prep.).

The definition of party size selected and the criteria used to determine when a party is independent from another party previously seen will influence the average party size reported for a site. For example, if a party is considered to be the initial party plus new counts made following changes in composition and, if large parties are more likely to change composition than small parties, then the average party size determined by this method will tend to be larger than the average party size determined as the composition on first sighting.

For our purposes, unless otherwise stated, we consider party size to be the initial party plus new counts made after each change in composition. We recognize that this does not mean that subsequent data points are statistically independent. The use of the same definition at both sites should permit the assessment of relative differences. We include only individuals traveling independently in our definition of a party, so that infants are not included in the total count.

Samples

Ten-minute, focal animal samples were used to estimate the diet of the chimpanzees at Kanyawara, while two-minute focal animal samples were used to estimate the diet of bonobos at Lomako. Whenever possible, the focal animal to be sampled was selected randomly from the animals present. When the focal animal was feeding, the food item and the species of that item were identified. Patch size at Kanyawara was measured as the tree diameter measured at breast height (DBH); if the animals left the tree rapidly, DBH was estimated visually. The mean error in the visual estimation of DBH is 3.7%; N = 46; mean DBH of the trees = 100.02 centimeters. Patch size at Lomako was indexed as radius of the crown.

As a measure of patch use, the proportion of time that focal animals spent feeding was multiplied by the number of animals in the patch and the total time spent in the patch. This measure is called *chimp-minutes.* Parties at both sites were followed whenever possible. The observer was typically behind the party and, thus, could determine the total duration of stay in the patch for only a small subset of the patches used.

At Kanyawara, the density of chimpanzee food trees was determined by using transects placed randomly along existing trails within strata. Each of the 26 vegetation transects was 200 meters by 10 meters, providing a total sample area of 5.2 hectares. Each tree greater than or equal to 10-centimeters DBH was identified and marked with a numbered metal tag. Phenology data were recorded for each tree during the first 10 days of every month. For the density of bonobo food trees, we rely on the data presented by Malenky (1990) on the floral composition of the mature forest. In Malenky's study, 12 plots totaling 3 hectares were surveyed. As with the

Table 1
The average composition of the parties of bonobos at Lomako Forest, Zaire, and chimpanzees at Kanyawara, Kibale Forest Reserve, Uganda.

Class	Site	Mean	Median	Range
Adult males	Lomako	1.85	2	0–7
	Kibale	2.17	2	0–9
Adult females	Lomako	3.37	3	0–11
	Kibale	1.45	1	0–7
Subadult males	Lomako	0.12	0	0–1
	Kibale	1.47	1	0–5
Subadult females	Lomako	1.27	1	0–3
	Kibale	0.16	0	0–2

Notes: At Lomako, the sample size averaged 141 between classes, while at Kibale, the values presented are the result of 8,198 15-minute scans.

transect vegetation sampling at Kanyawara, all trees greater than or equal to 10-centimeters DBH were identified and permanently tagged. We present data only for those tree species from which chimpanzees or bonobos fed.

Results

At Kanyawara, the average party size from July 1988 to June 1991 was 5.11 (N = 2,414) and ranged between months from 1.22 (N = 9, 49 15-minute scans) to 12.26 (N = 49, 315 15-minute scans). At Lomako, there were on average 5.4 individuals in a party (White 1986, 1992b). Other methods of calculating party size produced different results: at Kanyawara, with scans taken every 15 minutes the mean was 5.24, (N = 8,190); and with acoustic party size the mean was 7.13 (N = 1,208).

Based on independently collected data, Malenky and Stiles (1991) reported an average party size of 5.4 for bonobos at Lomako (N = 135). Wrangham et al. (1992) reported an average party size of 5.6 for chimpanzees at Kanyawara for the period between 1984 and 1985 (based on 1,050 30-minute scans).

The distribution of party sizes differed between the two sites (see figure 2). The mode (or most common party size) at Kanyawara was two individuals, while at Lomako the mode was five individuals. At Kanyawara, the mode varied between months from 1 (N = 9) to 19 (N = 41), while at Lomako it varied from 3 (N = 14) to 6 (N = 7) for months with single

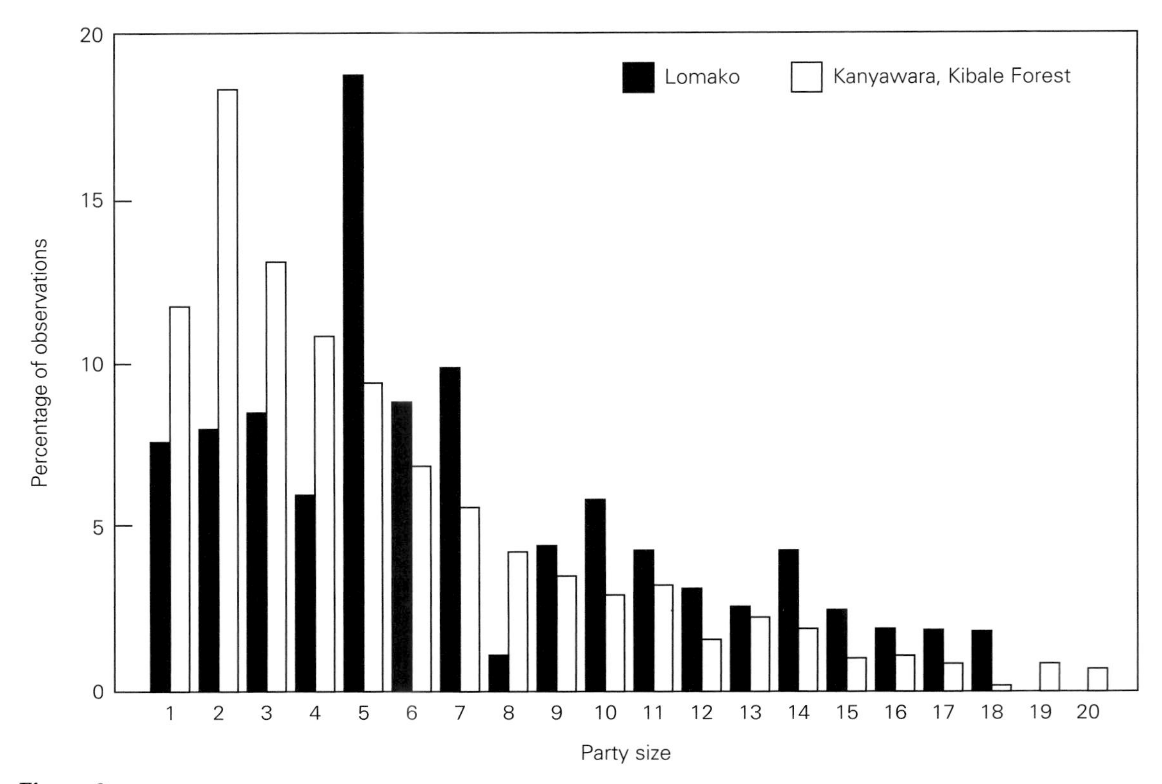

Figure 2

modes. Chimpanzee parties tended to be dominated by males, while females dominated the bonobo parties (see table 1). The number of subadults in each party seems to reflect the composition of the communities, as is evident by the maximum number seen in a party.

Quantitative comparisons of chimpanzee and bonobo diets suffer from a number of problems. For example, until habituation is complete, animals are uncomfortable with observers and are less likely to feed on the ground than they normally would. Moreover, sites may differ in levels of visibility, and, thus differences in ground feeding may simply reflect the fact that in dense vegetation it is difficult to approach close enough to a feeding animal to record data. Both factors would result in an underestimation of ground feeding. Diet may change over time in relation to periodic changes in food availability. Thus, until long-term data are available, we cannot preclude the possibility that the diet recorded from any one study reflects a relatively unique situation that does not adequately reflect the norm for the site.

To illustrate this program, we considered data collected from Kibale and Lomako from different researchers or reporting periods. Isabirye-Basuta (1989) reported that chimpanzees from Kibale ate fruit during 61.3% of their feeding time. During 14 days of observations, Wrangham et al. (1992)

found that members of the same community fed on fruits during 71.7% of their feeding time. At Lomako, Badrian et al. (1981) report that in a seven-month period in 1974–75, fruit accounted for 49% of the foods eaten by the bonobos. In contrast, Badrian and Malenky (1984) found that 54% of the foods eaten were fruit during a subsequent field season. Keeping the limitations and the level of within-site variation in mind, the data collected during these studies confirm that fruit is the most frequently consumed food item for both species (see table 2). Consumption of young leaves occupied about a quarter of the bonobos' feeding time, while young leaves constituted a less important component of the chimpanzees' diet. Consumption of terrestrial herbaceous vegetation (THV) accounted for only 2.1% of the bonobos' feeding time, while it constituted 11.8% of the chimpanzees' feeding time. The fact that chimpanzees rely more heavily on THV than do bonobos contradicts the first hypothesis which suggests that bonobo feeding competition is reduced by their reliance on readily available THV.

The patch size (or food tree size) used for feeding by the chimpanzees at Kanyawara averaged 100.0 centimeters DBH (SD = 49.8, N = 49, range 15–200). White and Wrangham (1988) report that the bonobos of Lomako fed in patches that averaged 94.2 centimeters DBH (SD = 50.2, N = 72, range 15–178), while Malenky (1990) reports the average bonobo patch size as 88.2 centimeters DBH (SD = 49.7, N = 35). Considering all patches visited, chimpanzees at Kanyawara visited a greater proportion of large patches than did the bonobos of Lomako (see table 3).

For chimpanzees at Gombe, the average length of time that a party, or combinations of parties, spent in a food patch was 40 minutes (SD = 39.1, N = 164) (White and Wrangham 1988). In contrast, the bonobos at Lomako spent an average of 1.1 hours in each patch (SD = 1.24, N = 92). Based on this information, White and Wrangham (1988) suggested that bonobos use larger patches than chimpanzees and, therefore, occur in larger parties. However, the average time that the chimpanzees of Kanyawara spent feeding in each patch was 1.5 hours (SD = 1.43, N = 49).

Table 2
Diet of the bonobos of Lomako Forest, Zaire, (White 1992b) and the chimpanzees at Kanyawara, Kibale Forest Reserve, Uganda.

Food Item	Lomako (%)	Kibale (%)
Fruit	72.1	82.1
Leaves	24.9	8.0
Animal	0.1	0.2
THV	2.1	11.7
Other	0.8	0.1

Notes: At Kibale, the values are based on 10–minute samples of focal animals.

Table 3
Comparison of tree-food patch use of the bonobos of Lomako Forest, Zaire, and the chimpanzees of Kanyawara, Kibale Forest, Uganda.

	Chimp-minutes	
Patch-Size Estimate	Lomako	Kibale
1 to 50	38.8	26.5
50 to 100	14.7	8.2
100 to 150	5.4	10.2
150 to 200	7.0	4.0
200 to 250	6.2	2.0
> 250	27.9	49.0
Sample size	72	49

We use the data available from Kanyawara and the data presented by Malenky (1990) from Lomako to test for differences in the overall density of food tree patches between the sites. At Lomako, there were 212 individual bonobo food trees per hectare, while at Kanyawara there were 252 chimpanzee food trees per hectare. Further, at Lomako, the total basal area of tree species that bonobos used as food sources was 142,313 square centimeters per hectare, while at Kanyawara this parameter was 139,618 square centimeters per hectare. This data suggest that overall food availability is similar at the two sites. However, this analysis does not assess the temporal availability of food resources.

We modified a technique used by White and Wrangham (1988) to examine the dispersion patterns of the two species during periods of abundant food from fruit trees. If animals from both populations fed in trees until the available food was depleted and if this occurred in all patches and for all party sizes, then large and small parties should remove equivalent amounts of food from patches of the same size. However, our observation that chimpanzees disperse to feed in neighboring patches while bonobos feed together suggests that chimpanzees may not deplete food in large patches if dispersal to neighboring trees relates to social factors.

To address this question, we performed multiple regressions to predict chimp-minutes (our measure for of the amount of food removed) from patch size (DBH) and party size. Separate regressions were calculated for small and large patches on log-transformed data using different patch size categories (<50 centimeters, >50 centimeters, <90 centimeters, >90 centimeters). Note that this approach is not statistically valid, since party size goes into the calculation of chimp-minutes and, thus, is on both sides of the regression equation. As a result, the magnitude of observed relationships should be viewed with caution. However, we can examine the relative differences between the two species (see table 4).

Table 4
Amount of food removed from a patch (chimp-minutes) in relation to size of patch and size of party for large and small patches for the bonobos of Lomako Forest, Zaire, and the chimpanzees of Kanyawara, Kibale Forest, Uganda.

Patch size		Lomako (t value p)		Kibale (t value p)	
Small (< 50 cm)	Tree Size	4.80	0.001	2.87	0.017
	Party Size	1.87	0.073	2.92	0.184
Large (> 50 cm)	Tree Size	0.41	0.685	0.85	0.725
	Party Size	4.67	0.001	0.36	0.413
Small (< 90 cm)	Tree Size	4.54	0.001	3.76	0.002
	Party Size	1.46	0.153	4.20	0.001
Small (> 90 cm)	Tree Size	−0.31	0.760	−0.15	0.887
	Party Size	4.28	0.001	−0.69	0.511

For bonobos encountering small patches, (defined as either <50 centimeters or <90 centimeters), the results demonstrate that the size of the patch had a stronger influence on the number of minutes spent feeding by all individuals than did party size, implying patch depletion as the most important factor (White and Wrangham 1988). For bonobos encountering large patches (defined as either >50 cm or >90 cm), the reverse was true; variance in the number of minutes spent feeding by all individuals was explained primarily by the number of animals feeding, implying that food was not limited and that feeding time for each individual was limited by some other constraint, such as stomach capacity.

For chimpanzees, this analysis suggests that food availability was limited for small patches. In large patches, the analysis for chimpanzees suggests that food was not limited. However, in contrast to bonobos, the amount of food removed by each chimpanzee in large patches was not related to the number of animals present.

Discussion

Previous hypotheses to explain differences in the social organization of bonobos and chimpanzees were based on the premise that bonobos experience lower levels of feeding competition. We have presented evidence to suggest that the size of the tree-fruit patches used by the chimpanzees of Kanyawara and the bonobos of Lomako are similar. Thus, tree-patch size is not likely to account for differences in lower levels of feeding competition in bonobos as compared to chimpanzees.

A theoretical argument has been developed to explain how the spatial and temporal abundance of food could influence party size (Altmann 1974; Bradbury and Vehrencamp 1976). The logic of this argument is relatively simple. Animals must forage over an area that can meet their energy and nutritional requirements. Assuming that there is an advantage to living in a group and that animals deplete the patches they feed in, it follows that an increase in party size would also increase the area animals need to travel to find adequate food supplies. Thus, over a given period of time, individuals in a group would have traveled farther and spent more energy than they would have if they foraged in smaller parties or alone. With an increasing amount of time spent traveling, a point is reached where the energy spent in travel exceeds that obtained by feeding, and smaller groups become advantageous (Milton 1985; Chapman 1990a, b).

Applying this theoretical argument to the proposed species-specific differences in party size for chimpanzees and bonobos can lead to three interrelated predictions. First, although the size of the patches may be the same, they may differ in density. With respect to the sites we have studied, evidence suggests that this is false. The density of chimpanzee food trees determined for Kanyawara and the density of bonobo food trees presented by Malenky (1990) are very similar.

Second, the spatial distribution of food patches at different sites may differ, with one site having more clumped patches than another site. The spatial distribution of patches may require animals at one site to spend a larger portion of the day traveling than animals at another site. If parties were depleting the patches they used and if individual animals had to travel farther after depleting a patch, then these individuals may elect to reduce travel costs by traveling in smaller parties. Smaller parties would deplete patches more slowly and, thus, require individuals to visit fewer patches. Because we cannot envision any process that would cause radical differences in the spatial distribution of food patches at the two sites, we suggest that this factor is unlikely to operate in isolation to explain differences in chimpanzee and bonobo party size.

Third, although the size, density, and spatial distribution of patches may be similar between two locations, there may be temporal differences in the availability of food resources. For example, one site may have periods of high food abundance alternating with periods of scarcity, while at another site the food availability may remain relatively constant. Based on the seasonality of rainfall and a qualitative measure of number of fruit production (the species fruiting per month), Malenky (1990) suggested that a uniform production of fruit contributed to the large size of bonobo parties at Lomako. The distribution of party sizes (see figure 2) at Kanyawara for chimpanzees and Lomako for bonobos follows a pattern that would support this idea. Although the average party size of these two species is very similar, chimpanzees have many small parties and some very large parties, while bonobos exhibit a less bimodal distribution.

The question of species-specific differences in average party size may be moot. By reviewing studies that report the average party size for chimpanzees and bonobos, it is evident that the variance *within* a species can be as large as the variance *between* species. The average size of chimpanzee parties ranged between study sites from 2.6 to 10.1, while for bonobos it ranged from 5.4 to 16.9 (see table 5).

We propose that discussions of average party size help little in understanding the differences in bonobo and chimpanzee social organization and behavior. Both species show extreme flexibility in the size of their parties. We hypothesize that the social organization and the behavioral strategies of chimpanzees reflect adaptations to extended periods during which individuals are forced to be in small parties. In contrast, we suggest that bonobos are rarely forced to live in small parties.

Table 5
Mean party sizes of chimpanzees and bonobos at various locations in Africa.

Study	Location	Mean party size	N
Chimpanzees:			
Sugiyama 1968	Budongo, Uganda	4.4	514
Reynolds and Reynolds 1965	Budongo, Uganda	3.9	215
Goodall 1968, 1986	Gombe, Tanzania	4.0	—
Nishida 1968[a]	Mahale, Tanzania	6.2	—
Sugiyama 1981	Bossou, Guinea	6.0	133
Sabater Pi 1979	Okorobiko Mountains, Rio Muni	9.9	60
Tutin et al. 1983	Mt. Assirik, Senegal	4.0	267
Boesch 1991[b]	Taï, Ivory Coast	10.1	527
Ghiglieri 1984, 1986[c]	Kibale (Ngogo), Uganda	2.6	667
This study[d]	Kibale (Kanyawara)	5.1	2,414
		5.24	8,190
		7.1	1,208
Bonobos:			
Kuroda 1979	Wamba, Zaire	16.9	147
Kano 1983	Yalosidi, Zaire	8.5	60
Badrian and Badrian 1984	Lomako, Zaire	7.9	268
Malenky and Stiles 1991[e]	Lomako, Zaire	5.4	135
This study[f]	Lomako, Zaire	5.4	164
		6.2	114

Notes: a. Infants excluded. b. Extracted from Boesch (1991) table 1, combining columns A, B, C. c. Traveling parties from the Ngogo study site. d. Calculated using different methods: (1) initial party plus all changes in composition, (2) scans, and (3) acoustic party size. e. The average group size feeding on fruit and THV. f. Calculated using (1) first sighting plus changes in composition, and (2) first sighting only.

Previous studies of bonobos at Lomako have reported a high degree of cohesion and the maintenance of close proximity among females (White 1989). In contrast, Wrangham et al. (1992) described female chimpanzees as being less affiliative with other females than with males. If female chimpanzees were together rarely in comparison to female bonobos, then the bonds between female chimpanzees would be relatively less beneficial. This pattern would have particular impact in cooperative alliances against males. Moreover, this differential in the benefit of female-female bonds may have a number of ramifications throughout the entire social system.

Summary

Previous attempts to explain differences in the social organization of bonobos and chimpanzees suggested that bonobos experience less feeding competition, which permits larger bonobo parties. Here, we tested the hypothesis that any difference in average party size that exists between the chimpanzees of Kanyawara, Kibale Forest, Uganda, and the bonobos of the Lomako Forest, Zaire, results from differences in the size of tree-fruit patches. Our results do not confirm a difference in party size.

In Kanyawara, the average party size was 5.11 (mode = 2), while at Lomako it was 5.4 (mode = 5). The size of the tree-fruit patches used for feeding by the chimpanzees of Kanyawara averaged 100.02 centimeters DBH, while the bonobos of Lomako fed in tree fruit patches that averaged 94.2 centimeters DBH. The average time that the chimpanzees of Kanyawara spent feeding in each patch was 1.5 hours, while the bonobos at Lomako spent an average of 1.1 hours in each patch. At Kanyawara, there are 252 chimpanzee food trees per hectare (139,618 square centimeters per hectare), while at Lomako there are 212 bonobo food trees per hectare (142,313 square centimeters per hectare; Malenky 1990). We suggest that, although the size and overall abundance of patches may be similar between the two locations, there may be temporal differences in the availability of food resources.

We propose that understanding the differences in chimpanzee and bonobo social organization and behavior may have little to do with differences in the average party size. We hypothesize that social organization in chimpanzees reflects adaptations to extended periods in which individuals are forced to travel in small parties, while social organization in bonobos reflects a situation in which individuals are rarely forced to travel in small parties.

Acknowledgments

The Kibale Chimpanzee Project was supported by a National Geographic Grant, USAID Funding, and an NSF grant (BNS-8704458). We thank the Government of Uganda and especially the Forest Department and Makerere University for permission to work in the Kibale Forest Reserve. Francis J. White was supported by NSF (BNS-8311251), the Leakey Foundation, Conservation International, and the Boise Fund. She wishes to thank the Government of Zaire for permission to work in Lomako, and to thank Randall Susman, Richard Malenky, and Nancy Thompson-Handler for all their continuing help and support. We wish to thank Lauren Chapman for assistance in all aspects of this study, from data collection through manuscript preparation. We also thank Leah Gardner, Tony Goldberg, Richard Malenky, and Adrian Treves, who provided helpful comments on earlier drafts.

References

Addicott, J.F., J.M. Aho, M.F. Antolin, D.K. Padilla, J.S. Richardson, and D.A. Soluk. 1987. Ecological neighborhoods: Scaling environmental patterns. *Oikos* 49:340–6.

Altmann, S. 1974. Baboons, space, time, and energy. *Amer. Zool.* 14:221–48.

Badrian, A.J., and N.L. Badrian. 1984. Group composition and social structure of *Pan paniscus* in the Lomako Forest. In R.L. Susman, ed., *The Pygmy Chimpanzee: Evolutionary Biology and Behaviour,* pp. 173–81. New York: Plenum Press.

Badrian, N. L, A. J. Badrian, and R.L. Susman. 1981. Preliminary observations on the feeding behavior of *Pan paniscus* in the Lomako Forest of Central Zaire. *Primates* 22:173–181.

Badrian, A.J., and R.K. Malenky. 1984. Feeding ecology of *Pan paniscus* in the Lomako Forest, Zaire. In R.L. Susman, ed., *The Pygmy Chimpanzee: Evolutionary Biology and Behavior,* pp. 275–99. New York: Plenum Press.

Boesch, C. 1991. The effect of leopard predation on grouping patterns in forest chimpanzees. *Behav.* 117:220–41.

Bradbury, J.W., and S. Vehrencamp. 1976. Social organization and foraging in emballonurid bats: A model for the determination of group size. *Behav. Ecol. Sociobiol.* 1:383–404.

Butynski, T.M. 1990. Comparative ecology of blue monkeys *(Cercopithecus mitis)* in high- and low-density subpopulations. *Ecol. Monogr.* 60:1–26.

Chapman, C.A. 1990a. Ecological constraints on group size in three species of neotropical primates. *Folia Primatol.* 55:1–9.

———. 1990b. Association patterns of spider monkeys: The influence of ecology and sex on social organization. *Behav. Ecol. Sociobiol.* 26:409–14.

Chapman, C.A., L.J. Chapman, R.W. Wrangham, K. Hunt, D. Gebo, and L. Gardner. 1992. Estimators of fruit abundance of tropical trees. *Biotropica* 24:527–31.

Chapman, C.A., R.W. Wrangham, and L.J. Chapman. In prep. Ecological constraints on group size: An analysis of spider monkeys and chimpanzee subgroups.

Chapman, C.A., F.J. White, and R.W. Wrangham. In press. Defining subgroup size in fission-fusion societies. *Folia Primatol.*

Ghiglieri, M.P. 1984. *The Chimpanzees of Kibale Forest.* New York: Columbia University Press.

———. 1986. Feeding ecology and sociality of chimpanzees in Kibale Forest, Uganda. In P.S. Rodman and J.G.H. Cant, eds., *Adaptations for Foraging in Non-Human Primates,* pp. 161–94. New York: Columbia University Press.

Goodall, J. 1968. The behaviour of free-living chimpanzees in the Gombe Stream Reserve. *Anim. Behav. Monogr.* 1:161–331.

———. 1986. *The Chimpanzees of Gombe: Patterns of Behaviour.* Cambridge, Mass.: Belknap Press.

Hassell, M.P., and T.R.E. Southwood. 1978. Foraging strategies of insects. *Ann. Rev. Ecol. Syst.* 9:75–98.

Isabirye-Basuta, G. 1989. Feeding ecology of chimpanzees in the Kibale Forest, Uganda. In P.G. Heltne and L.A. Marquardt, eds., *Understanding Chimpanzees,* pp. 116–127. Cambridge, Mass.: Harvard University Press.

Kano, T. 1983. An ecological study of the pygmy chimpanzees *(Pan paniscus)* of Yalosidi, Republic of Zaire. *Internat. J. Primatol.* 4:1–31.

———. 1992. *The Last Ape: Pygmy Chimpanzee Behaviour and Ecology.* Stanford, Calif.: Stanford University Press.

Kuroda, S. 1979. Grouping of the pygmy chimpanzees. *Primates* 20:161–83.

Malenky, R.K. 1990. Ecological factors affecting food choice and social organization in *Pan paniscus.* Ph.D. diss., State University of New York, Stony Brook.

Malenky, R.K., and E.W. Stiles. 1991. Distribution of terrestrial herbaceous vegetation and its consumption by *Pan paniscus* in the Lomako Forest, Zaire. *Am. J. Primatol.* 23:153–69.

Milton, K. 1985. Mating patterns of wooly spider monkeys, *Brachyteles arachnoides* E. Geoffroyi 1806. *Internat. J. Primatol.* 5:491–514.

Mitani, J.C., and T. Nishida. In prep. Context and social correlates of long distance calling by chimpanzees.

Nishida, T. 1968. The social group of wild chimpanzees in the Mahale Mountains. *Primates* 9:167–224.

Peters, R., S. Cloutier, D. Dube, A. Evans, P. Hastings, D. Kohn, and B. Sawer-Foner. 1988. The ecology of the weight of the fruit on trees and shrubs in Barbados. *Oecologia* 74:612–6.

Reynolds, V., and F. Reynolds. 1965. Chimpanzees of the Budongo Forest. In I. DeVore, ed., *Primate Behaviour,* pp. 368–424. New York: Holt, Rinehard and Winston.

Sabater Pi, J. 1979. Feeding behaviour and diet of chimpanzees *(Pan troglodytes troglodytes)* in the Okorobiko Mountains of Rio Muni (West Africa). *Z. Tierpsychol.* 50:265–81.

Skorupa. J.P. 1986. Responses of rain forest primates to selective logging in Kibale Forest, Uganda: A summary report. In K. Benirschke, ed., *Primates: The Road to Self-Sustaining Populations,* pp. 57–70. New York: Springer-Verlag.

———. 1988. The effect of selective timber harvesting on rain forest primates in Kibale Forest, Uganda. Ph.D. diss., Univ. of Calif., Davis.

Struhsaker, T.T. 1975. *The Red Colobus Monkey.* Chicago: Univ. of Chicago Press.

Sugiyama, Y. 1968. Social organization of chimpanzees in the Budongo Forest, Uganda. *Primates* 9:225–58.

———. 1981. Observations on the population dynamics and behaviour of wild chimpanzees at Bossou, Guinea, in 1979–1980. *Primates* 22:435–44.

Tutin, C.E.G., W.C. McGrew, and P.J. Baldwin. 1983. Social organization of savanna-dwelling chimpanzees, *Pan troglodytes verus,* at Mt. Assirik, Senegal. *Primates.* 24:154–73.

White, F.J. 1986. Behavioural Ecology of the Pygmy Chimpanzee. Ph.D. diss., State Univ. of New York, Stony Brook.

———. 1988. Party composition and dynamics in *Pan paniscus. Internat. J. Primatol.* 9:179–93.

———. 1989. Ecological correlates of pygmy chimpanzee social structure. In V. Standen and R.A. Foley, eds., *Comparative Socioecology: The Behavioural Ecology of Humans and Other Mammals,* pp. 151–64. Oxford: Blackwell.

———. 1992a. Activity budgets, feeding behaviour, and habitat use of pygmy chimpanzees at Lomako, Zaire. *Am. J. Primatol.* 26:215–23.

———. 1992b. Pygmy chimpanzee social organization: Variation with party size and between study sites. *Am. J. Primatol.* 26:203–14.

White, F.J., and A. Lanjouw. 1992. Feeding competition in pygmy chimpanzees: Variation in social cohesion. In T. Nishida, W.C. McGrew, P. Marler, M. Pickford, and F. de Waal, eds., *J. Human Evol.,* pp. 67–79. Tokyo: Tokyo Univ. Press.

White, F.J., and R.W. Wrangham. 1988. Feeding competition and patch size in the chimpanzee species *Pan paniscus* and *Pan troglodytes. Behav.* 105:148–64.

Wing, L.D., and I.O. Buss. 1970. Elephants and forest. *Wildl. Monogr.* No. 19.

Wrangham, R.W. 1977. Feeding behaviour of chimpanzees in Gombe National Park, Tanzania. In T.H. Clutton-Brock, ed., *Primate Ecology,* pp. 503–38. London: Academic Press.

———. 1986. Ecology and social relationships in two species of chimpanzee. In D.I. Rubenstein and R.W. Wrangham, eds., *Ecological Aspects of Social Evolution: Birds and Mammals,* pp. 352–78. Princeton, N.J.: Princeton Univ. Press.

Wrangham, R.W., A.P. Clark, and G. Isabirye-Basuta. 1992. Female social relationships and social organization of Kibale forest chimpanzees. In T. Nishida, W.C. McGrew, P. Marler, M. Pickford, F. de Waal, eds., *Human Origins,* pp. 81–98. Tokyo: Univ. of Tokyo Press.

Wrangham, R.W., and B.B. Smuts. 1980. Sex differences in the behavioural ecology of chimpanzees in the Gombe National Park, Tanzania. *J. Reprod. Fert.* (suppl.). 28:13–31.

Chimpanzee brothers Freud and Frodo.
Photograph courtesy of J. Goodall.

The Significance of Terrestrial Herbaceous Foods For Bonobos, Chimpanzees, and Gorillas

Richard K. Malenky, Suehisa Kuroda, Evelyn Ono Vineberg, and Richard W. Wrangham

Introduction

The spatiotemporal distribution of critical food resources influences the food choices and, by extension, the social organization of foraging animals. This distribution has been proposed as a fundamentally important factor in the evolution of social structures in all African great apes, which include bonobos, chimpanzees and gorillas (Wrangham 1986). For example, Watts (1984) showed that herbaceous food is ubiquitous in time and space in the habitat of the mountain gorilla *(Gorilla gorilla beringei)* and is thought to have allowed the formation of the cohesive social groups in that species.

Early field studies made it clear that the social organization of bonobos *(Pan paniscus)* differs from that of most chimpanzees *(Pan troglodytes)* (Kuroda 1979, 1980; Sugiyama 1988). Although these species have *fission-fusion societies,* the temporary parties of bonobos are sometimes larger than the temporary parties reported for chimpanzees and are often more mixed with animals of both sexes and all ages (Badrian and Badrian 1984; Kano 1980, 1982, 1987). Female bonobos are more cohesive than female chimpanzees, associating with one another regardless of reproductive state and engaging in a variety of affiliative behaviors such as mutual grooming and genito-genital rubbing (Thompson-Handler et al. 1984; Kano 1980, 1987; White 1988). Chimpanzee mothers without sexual swellings are more solitary and less affiliative than are bonobo mothers (Nishida 1979; Wrangham 1979; Wrangham and Smuts 1980; Wrangham et al. 1992). Further, male bonobos groom females more often than they groom other males (Badrian

and Badrian 1984; Kano 1980), while male chimpanzees are more affiliative toward each other than are bonobo males (Nishida 1979; Wrangham 1979; Wrangham and Smuts 1980; Wrangham et al. 1992). Male chimpanzees associate with unrelated females primarily during estrus (Goodall 1986).

Studies of the feeding ecology of bonobos suggest that an important difference between the diets of bonobos and chimpanzees might be a result of either a greater reliance by bonobos on the piths and leaves of herbaceous foods, terrestrial herbaceous vegetation, (Badrian and Malenky 1984) or a greater diversity of fruit species in bonobo habitats (Kano and Mulavwa 1984). The hypothesis of terrestrial herbaceous vegetation was supported by evidence showing that bonobo molars provide greater shearing efficiency than do those of chimpanzees, allowing bonobos to process ingested pith and leaves more efficiently (Kinzey 1984). Bonobos exploited herbaceous food regularly in all three sites—Lomako, Wamba, and Yalasidi—where their feeding ecology was described (Badrian and Malenky 1984; Kano and Mulavwa 1984; Kano 1983). Moreover, these foods were reported to be ubiquitous in distribution in bonobo habitats. Thus, the reportedly more cohesive social structure of bonobos was associated with food resources that were apparently similar to the food resources of the mountain gorilla, and a causal relationship was suggested (Wrangham 1986).

In Central Africa, chimpanzees in lowland rain forests live *sympatrically,* that is, in the same range, with lowland gorillas *(Gorilla gorilla gorilla).* Studies have shown that the lowland gorillas rely heavily on both terrestrial herbaceous vegetation and a variety of fruit and seeds (Jones and Sabater Pi 1971; Kuroda 1992; Mitani 1992; Tutin and Fernandez 1985; Rogers and Williamson 1987; Kuroda 1990; R. Carroll pers. com.), suggesting the possibility of competition between species. In contrast, Wrangham (1986) proposed that the absence of feeding competition from gorillas on the left bank of the Zaire River has allowed the bonobos there to become to be more dependent than chimpanzees on herbaceous food. If this proposal is true, the lower levels of feeding competition between the two species might have allowed bonobos to maintain larger, more cohesive parties, even when tree fruits are scarce. Wrangham (1986) further suggested that larger parties have been an important causal factor in the evolution of the unique sociosexual behaviors, described in the foregoing discussion, that differentiate bonobos from chimpanzees.

All known populations of African great apes consume herbaceous food to some extent. Much less is known about the socioecology of the mountain gorilla, *G. g. beringei,* than about the socioecology of the lowland gorilla, *G. g. gorilla.* Similarly, our knowledge of the socioecology of chimpanzees is dominated by reports on the eastern chimpanzee subspecies, *P.t. schweinfurthii,* and much less is known about chimpanzees that inhabit lowland rain forests, *P.t. troglodytes.* Further, to determine whether bonobos are inherently different from chimpanzees, we need data on chimpanzees in lowland rain forests similar to the lowland rain forests inhabited by bonobos. In general, to evaluate hypotheses concerning the origin of differences in the socioecology of African great apes, quantitative data from lowland rain forests where chimpanzees and gorillas occur sympatrically are essential.

Quantitative data about the distribution of food resources in the habitats of bonobos, chimpanzees, and gorillas have only recently begun to appear in the literature. For example, Malenky and Wrangham (in press) show that the density of herbaceous food in the habitat of bonobos is higher than the density of herbaceous food that in the forests of Kibale, Uganda, where chimpanzees live, and that bonobos consume more herbaceous food than chimpanzees. In this chapter, we add recent data on the herbaceous food of chimpanzees collected from the Ndoki Forest, Congo. We compare and contrast the density and distribution of herbaceous food of chimpanzees both in the Ndoki Forest and in the Kibale Forest to similar data reported for bonobos in the Lomako Forest, Zaire. We also compare the feeding behavior of bonobos, chimpanzees, and gorillas at these three sites based on direct observations, on fecal analysis, and on data on temporary party size. Using this information, we will examine four questions: one, do the densities and distributions of terrestrial herbaceous vegetation differ at the three sites?; two, do party sizes reflect the relative abundance of terrestrial herbaceous vegetation in the three sites?; three, can we infer that lowland gorillas compete for food with chimpanzees in the same range?; and four, do the patterns shown by bonobos suggest that they have been released from competition with gorillas?

Methods

Density and Distribution of Herbaceous Foods

The species of herbaceous plants included in this analysis were known to be exploited as food resources based on direct observation of the animals and based on the presence of these species among feeding remains. The herbaceous plants analyzed here are nonwoody and often canelike.

The layouts of survey plots at each study site were similar in certain important respects. We selected all the sampling areas in accordance with a *stratified-random system* (Grieg-Smith 1964). Moreover, we chose the placement of transects, or plots, randomly within uniform forest types. Within each survey plot, we counted all stems of herbaceous plants as long as the plants were rooted in that plot. However, the size and number of survey plots monitored at each study site were different. The limitations imposed by these differences is addressed in our discussion.

In the Lomako Forest, 12 survey plots, 10 meters by 10 meters, were placed randomly within the primary forest. In that forest, two herbaceous plants, *Haumania liebrechstiana* and *Palisota* species, are significantly more common in both frequency and density than any other herbaceous food plants and are known to be exploited disproportionately to their availability (Malenky and Stiles 1991). These two species were studied in the greatest detail. In all 12 plots, all stems of *H. liebrechstiana* were counted; in seven of the plots, all stems of both species were counted. In the analyses conducted here, the seven plots in which both species were counted were used to maximize comparability with the other two sites.

In the Kibale Forest, each of three main compartments of a gridded research zone had an area of approximately six square kilometers. The three compartments were K30 (unlogged), K14 (lightly logged), and K15 (more heavily logged). The origin and direction of the six 50-meter transects, or sample areas, were chosen at random within each compartment subject to the constraint that each transect was located at least 250 meters from any other transect. Each of the 18 independent plots covered a total area of 50 square meters. The number of all stems was recorded in contiguous plots of one meter by one meter along a 50-meter transect (Wrangham et al. 1993).

In the Ndoki Forest, two major forest types were surveyed: a riverine forest dominated by *Gilbertiodendron dewevrii* and an upland mixed-species forest. In the riverine forest, a 200-meter transect was laid out. Along that transect, 10 sites were chosen at random. At each site, a secondary line was established at a 45-degree angle to the transect; along each secondary line were five survey plots, each two meters by two meters, alternating to the right and to the left of the secondary line. Within each survey plot, all herbaceous plants were identified and counted.

In the mixed-species forest, a similar system was used except that the main transect was one kilometer in length. Each of the 10 secondary lines was 100 meters in length, and 50 survey plots were set along each secondary line. Using a random number generator, 25 survey plots were selected from the riverine forest, and 100 survey plots were selected from the mixed-species forest to compare the data from all three sites directly.

Statistical Comparison

The statistical means for habitatwide density of herbaceous stems at all three study sites were compared using a *Kruskal-Wallis test* because some could not be transformed to a normal distribution. All pairs of study sites were tested using a *Wilcoxon two-sample test.* The distribution with respect to the degree of clumping of herbaceous plants within each study site was estimated using the coefficient of dispersion, that is, the ratio of the variance to the mean (Blackman 1942). Values greater than one indicate clumping and values less than one indicate a regular distribution. Comparisons of variance and dispersion between study sites are complicated by the large difference in size of the plots used (Lomako, 100 square meters; Kibale, 50 square meters; Ndoki, four square meters). To roughly compare variances between study sites, we also calculated the *coefficient of variation* since this measure of variance is unit free and reported as a percentage of the statistical mean (Sokal and Rohlf 1981).

Fecal Analysis

Fecal samples from the Kibale Forest and from the Lomako Forest were collected throughout both the rainy and the dry season and, therefore, incorporate possible seasonal variation. At Ndoki Forest, fecal samples were collected only during the three months of maximum rainfall, August through October. The quantitative analysis of fecal remains was designed to determine which population consumed more terrestrial herbaceous vegetation (Malenky and Wrangham, in press). During consumption, stems are broken open and the pith, the soft, young inner core is extracted and eaten. Young leaves also may be consumed.

At all sites, fecal samples were weighed, dissociated in water, drained, and rinsed through a one millimeter screen mesh. The remaining strands of plant material were considered *fiber*. Fiber was easily distinguished and separated from undigested leaf material, because the reticulate venation pattern of undigested leaves was quite different from the more linear fibrous strands of pith. The fibrous strands were isolated and excess water was squeezed out. At Kibale and at Lomako, the wet fiber was weighed and the percentage of wet weight was calculated for each sample. At Kibale, estimates of fiber content in fecal samples of chimpanzees were based on visual estimates, using a scale of zero to four to rate abundance of fibrous strands (Wrangham et al. 1991).

At Ndoki, fecal samples were washed in a one millimeter screen mesh. The presence or absence of fiber, leaves, and seeds was recorded and the number of seeds counted. However, the weights of the various components were not measured. The contribution of each component to the diet was then calculated as a percentage of all samples (Kuroda 1992).

Party Size

We defined *party size* as the number of independently locomoting animals in a tree or on the ground that were feeding on a given resource. A party was considered to be feeding if at least one individual was consuming fruit or herbaceous food. We then calculated mean party size.

Results

Density and Distribution of Herbaceous Foods

Densities of herbaceous foods for all three study sites differ in a consistent way with climate (see tables 1 and 2). The densities at both lowland forest sites (Ndoki and Lomako) were higher than at Kibale, where annual rainfall is lower. However, the data for densities are not significantly different statistically (Kruskal-Wallis; H = 3.885). Comparing the same data in pairs across sites results in only one significant difference: Lomako and Kibale ($p < .025$; Mann-Whitney U-test).

The difference in coefficient of variation and coefficient of dispersion between Ndoki and the other two sites suggests that individual stems of herbaceous food at Ndoki are clumped in distribution, while those at Kibale and Lomako are evenly dispersed (see table 1). Survey plots used at both

Table 1
Habitatwide densities of herbaceous foods.

Site	D	SD	N	PS	CV	CD	IP
Kibale	0.89	0.43	18	50m^2	48.99	0.20	−0.89
Ndoki	2.25	2.29	125	4m^2	102.80	2.33	0.59
Lomako	2.02	0.65	7	50m^2	32.20	0.21	0.10

D = Density per square meter; SD = Standard deviation; N = Number of plots; PS = Plot size; CV = Coefficient of variation (standard deviation/mean x 100); CD = Coefficient of dispersion (variance/mean); IP = Intensity of pattern.

Table 2
Densities of herbaceous foods by forest type.

Forest Type	D	SD	N	PS	CV	CD	IP
Kibale							
Unlogged	0.94	0.60	6	50m^2	56.4	0.38	−0.66
Lightly logged	1.05	0.30	6	50m^2	28.6	0.09	−0.87
Heavily logged	0.68	0.28	6	50m^2	41.1	0.12	−1.30
Ndoki							
Gilbertiodenon	3.19	3.27	25	4m^2	102.5	3.35	0.73
Mixed species	1.59	1.94	100	4m^2	122.0	2.36	0.86
Lomako							
Mixed species	2.02	0.65	7	100m^2	32.2	0.21	−0.39

D = Density per square meter; SD = Standard deviation; N = Number of plots; PS = Plot size; CV = Coefficient of variation (standard deviation/mean x 100); CD = Coefficient of dispersion (variance/mean); IP = Intensity of pattern; Unlogged = no cutting yet; *Gilbertiodendron* = the dominant species.

the Kibale and Lomako sites were much larger in area (12.5 and 25 times larger respectively) than survey plots used at the Ndoki site. The even distribution of stems suggested by the coefficients of variation and dispersion for Kibale and Lomako may result from the fact that the survey plot size at these sites was larger than the average clump size. As a result, variance that might have been recorded among smaller plots would not be seen when larger plots were surveyed. Thus, these results may reflect the difference of scale at which the data were collected rather than a meaningful difference in the actual distribution of available food.

Fecal Analysis of Fiber Content

At Ndoki, the gorilla fecal samples contained a higher percentage of fiber than the chimpanzee fecal samples (62%, N = 29 and 40%, N = 42, respectively) (see table 3). Comparable data for bonobos in the Lomako Forest are similar to data for gorillas at Ndoki (67.9%, N = 81), but both are much lower than data for the chimpanzees in Kibale (93.8%, N = 839) (Wrangham et al. 1991). Malenky and Wrangham (in press) show that over the course of a year, however, the mean percentage fiber weight for the Lomako population was significantly higher than that of the chimpanzees at Kibale (Mann-Whitney U-test, $p < .05$, two-tailed) (see table 3). Thus, the percentage of samples containing fiber by itself may not indicate higher levels of fiber consumption.

Party Size

Party sizes of the chimpanzees at Ndoki and at Kibale and party sizes of the bonobos at Lomako and at Wamba are compared in table 4. These figures are hard to compare statistically because the Ndoki data include figures only for the rainy season, while data for the other sites were collected over longer periods. Still, the results suggest that party sizes of bonobos, particularly at Wamba, are larger than party sizes of chimpanzees. While some data from Wamba are from studies of provisioned animals, other data are not. Kuroda (1979) reports larger party sizes among free-ranging bonobos

Table 3
Fecal analysis.

	Ndoki				Lomako		Kibale	
	Gorilla		Chimpanzee		Bonobo		Chimpanzee	
	%	N	%	N	%	N	%	N
Feces with fiber	62	29	40	42	67.90	81	93.80	839
Mean fiber weight	NA	NA	NA	NA	10.90	81	2.79	73

Table 4
Chimpanzee party size at selected sites.

Chimpanzees		Party Size		N	
Ndoki					
P.t. troglodytes		7.0 ± 4.9 including independent individual		32	
		5.5 ± 4.0 excluding independent individual		32	
Kibale					
P.t. schweinfurthii		5.07			Wrangham et al. 1992
Bonobos		**Party Size**		**N**	
Lomako					
P. paniscus		6.37	SE = .26	333	Thompson-Handler 1990
		6.21		114	White 1988
		5.81		147	Lanjouw 1987
		7.90		268	Badrian and Badrian 1984
		5.60		128	Malenky and Stiles 1991
	Mean	6.37	SD = 0.9	N = 5	
Wamba					
P. paniscus		19.0		172	Kano 1982
		16.9		147	Kuroda 1979

before provisioning at Wamba began and notes that party size did not increase significantly after provisioning. Therefore, it is not possible to state that larger party sizes at Wamba resulted from provisioning: other factors must be considered.

Discussion

Differences in the density of terrestrial herbaceous vegetation across the three sites are consistent with expectations based on climate and suggest a greater abundance of herbaceous food at the two lowland sites of Ndoki and Lomako than at Kibale. This conclusion, however, must be qualified because of inconsistencies in the methods used to collect the data. Nonetheless, the differences in the mean densities at Ndoki and Lomako probably reflect real differences in the abundance of herbaceous food presented to the foraging animals. If apes in the lowland forests of Ndoki are presented with more abundant herbaceous foods (and probably fruit as well) as compared to other sites, we might expect these apes to be released from the competition that occurs at those other sites. The available data suggest some interesting possibilities in this regard.

Competition Between Chimpanzees and Gorillas at Ndoki

The demonstration of competition in nature is extremely difficult and has always been fraught with controversy (Schoener 1982), especially when controlled manipulations such as removal experiments cannot be performed. Nonetheless, the available data suggest either that gorillas and chimpanzees living sympatrically may experience some degree of feeding competition as in times of food stress, or that competition between chimpanzees and gorillas has spurred divergent evolution.

At Ndoki, no species of herbaceous food is known to be eaten exclusively by chimpanzees, yet at least five species are eaten exclusively by gorillas. This pattern has been observed at Dzanga-Sangha in the Central African Republic (R. Carroll pers. com.). Thus, there appears to be an overlap in the herbaceous food consumed by these sympatric apes.

Gorillas at Ndoki show a clear preference for *Haumania dankelmania* and *Hydrocharis* species based on frequency of consumption (Kuroda 1992; pers. observ.). There have been no directly comparable observations of chimpanzee feeding patterns to compare with Kuroda's study of gorillas, but data on habitat preference, species composition, and fecal analysis of chimpanzees and gorillas suggest some habitat partitioning. For example, Mitani (1992) demonstrates a strong difference in habitat use: chimpanzees forage about 95% of the time in dry, mixed-species forest while gorillas forage 45% of the time in swamp forest and 41% of the time in mixed-species forest. *Hydrocharis* species are found only in swamps where gorillas forage. *H. dankelmania,* common but clumped in its distribution, is found in 57% of plots surveyed in the mixed-species forest where both gorillas and chimpanzees forage (Ono Vineberg, pers. observ.). Furthermore, Kuroda (1992) reports that within the mixed-species forest, chimpanzees foraged more often in areas that did not contain herbaceous plants of the family Marantaceae, while gorilla feeding traces were most often encountered in areas that did contain these herbaceous species. Herbaceous food appears to be far more important to gorillas than to chimpanzees based on percentage of fecal samples with fiber (see table 3) and quantity of fiber in the feces (Kuroda pers. observ.). These data suggest that chimpanzees at Ndoki exploit a subset of the available herbaceous food eaten by gorillas; that both gorillas and chimpanzees exploit different microhabitats in the mixed-species forest; and that gorillas and chimpanzees diverge almost completely in their exploration of swamp forest, where the herbaceous food primarily available is that consumed only by gorillas.

Fruit is a major food resource for both chimpanzees and gorillas in lowland forests. At Ndoki and other lowland forests such as those in Gabon and the Central African Republic, there is significant overlap in the fruit species consumed by both gorillas and chimpanzees (Tutin and Fernandez 1985; R. Carroll, pers. com.). However, herbaceous food may be a critical food resource and, thus, engender competitive interactions, because consumption of herbaceous food increases during times of fruit scarcity. Therefore, herbaceous food is thought to be a fallback resource at Ndoki and elsewhere (Rogers et al. 1990; Wrangham et al. 1991; R. Carroll pers. com.).

Interestingly, gorillas and chimpanzees have been observed feeding together in the same fruiting tree without aggressive interactions at Ndoki (Suzuki and Nishihara 1992), and at sites in the Central African Republic (R. Carroll pers. com.). At Ndoki, these observations occurred during the season of reduced levels of fruit production. By contrast, gorillas and chimpanzees have never been observed feeding on terrestrial herbaceous vegetation simultaneously. This suggests that in the fruiting season, or whenever a large enough tree is in fruit, tree fruits may be sufficiently abundant to preclude competitive interactions. However, at the end of the peak fruiting season and during the dry season, gorillas and chimpanzees concentrate on different types of fruit. Gorillas feed on larger, more fibrous species (Rogers et al. 1990; R. Carroll pers. com.). It is during the dry season, when distinctions between the diets of the two species are greatest and when herbaceous food constitute a more critical dietary component, that competitive interactions may be more important. In the end, herbaceous food may drive competitive interactions and habitat partitioning, because this food resource is the most likely fallback food in times when fruit is scarce and because the two species differ in their preferences at all times.

Are bonobos released from feeding competition?

In some respects, patterns of preference and of consumption of herbaceous foods for the bonobos are more similar to those patterns for the gorillas at Ndoki than to those patterns for chimpanzees. In terms of species preference, bonobos of the Lomako Forest show the strongest preference for *Haumania liebrechstiana* of the family Marantaceae, which constitutes an important component of the diet of bonobos throughout the year (Malenky and Stiles 1991). Among the available herbaceous food at Ndoki, gorillas exhibit the strongest preference, based on frequency of consumption, for a member of the same genus, *Haumania dankelmania* (Kuroda pers. observ.). *H. dankelmania* represents a minuscule part of the diet of chimpanzees at Ndoki based on frequency of consumption and abundance of fibers in fecal samples, (Kuroda pers. observ.).

Bonobos at both Wamba and Lomako unlike chimpanzees at Ndoki, frequently exploit herbaceous food in the swamp forest (Malenky pers. observ.; Kano 1992), but detailed information on which species are consumed is not available. In the swamps at Lilungu (but not at Wamba), *Hydrocharis* provides a major component of the bonobos' diet (Bromejo et al., in prep.) and is the second most important herbaceous plant eaten by gorillas at Ndoki (Kuroda pers. observ.). Thus, in terms of habitat exploitation, food preference, and quantity of herbaceous food consumed, bonobos resemble the gorillas of Ndoki more than they resemble chimpanzees. Assuming that chimpanzees and bonobos are closer physiologically than either is to gorillas, this difference in feeding patterns suggests that the chimpanzees at Ndoki may be constrained by the presence of gorillas.

Clearly, this suggestion is speculative; much more data are required for evaluation. For example, we do not know if the *Hydrocharis* species eaten by bonobos at Lilungu is the same as that eaten at Ndoki. The two species of *Haumania* are neither nutritionally (Calvert 1985) nor structurally (Kuroda pers. observ.) equivalent. Even if they are, we must be sure that these plants are processed in the same way before we can conclude that bonobos feed in a fashion equivalent to that of gorillas. Ultimately, we should take preference into account, since all available food resources may not be eaten with a frequency proportional to their availability (Malenky and Stiles 1991). For example, gorillas at Ndoki show a strong preference for *H. dankelmania* and appear to eat some other herbaceous food at a frequency lower than their availability, although we cannot currently assess this quantitatively. We must also determine how the digestive systems of these apes handle the ingested plants. For example, do gorillas and bonobos extract the same nutrients from these foods?

The nutrient content of available food resources is also critical to an understanding of food choice and foraging behavior (Altmann and Wagner 1978; Belovsky 1978). All species of herbaceous plants probably are not equivalent and, thus, do not represent a single monolithic food resource. *Hydrocharis* is reputed by the local inhabitants of Wamba to contain large amounts of salt; in fact, the local inhabitants occasionally extract salt from this plant. Thus, the gorillas of Ndoki may exploit *Hydrocharis* for its salt content, while the chimpanzees may obtain salt from a different resource, obviating potential competitive interactions with respect to *Hydrocharis.* While gorillas may consume herbaceous food for carbohydrate content, chimpanzees in lowland forests may consume herbaceous food for protein content or for another dietary component. If these speculations are true, then the smaller quantities of herbaceous food consumed by chimpanzees may be adequate for their needs, further lessening the likelihood of competitive interactions.

Finally, we cannot be sure that we are measuring the world from the apes' point of view. Apparent preference varies from one site to another within the same species. While bonobos apparently consume *Hydrocharis* preferentially at Lilungu, they do not show this same preference at Wamba (Kuroda pers. observ.). More extensive surveys as well as direct observations are needed, because we cannot know if this variation occurs as a result of quirks in availability or as a result of cultural differences between sites.

Food Choice and Social Organization: The Implications of Party Size for Bonobos

Discussions of the factors controlling party size among the African great apes have focused on two major food resources: fruit and herbaceous terrestrial vegetation. It is generally thought that the presence of discrete food patches, such as individual trees, limit the group or party size through feeding competition, especially during times of fruit scarcity. Comparison of fruit-patch size in bonobo habitats and chimpanzee habitats have shown that patch size varies from site to site. For instance, White and Wrangham (1988) showed that patch size for the bonobos at Lomako is significantly larger than for the chimpanzees at Gombe. However, Chapman et al. (this volume) found no significant difference when comparing patch size of bonobos' habitat to that of the Kibale chimpanzees. Conversely, herbaceous food in the bonobo's habitat has often been viewed as so ubiquitous as to be "patch free" and as not to engender either feeding competition or limitation of party size (Wrangham 1986). However, based on detailed surveys of terrestrial herbaceous vegetation, examination of the food choices, and nutritional analysis of the diet of bonobos at Lomako, Malenky and Stiles (1991) proposed that herbaceous food might, in fact, limit party size.

Based on the data presented in this chapter, it is not clear that dependence on a reliable food resource such as herbaceous foods is the reason for larger party size and enhanced sociality among bonobos relative to chimpanzees. For example, mean party size at Lomako Forest is only slightly larger than the party size of chimpanzees at Kibale (see table 4) and corresponds roughly to party size in other areas where chimpanzees have been studied (Tutin et al. 1983). At the same time, bonobo party sizes at Wamba are larger than bonobo party size at Lomako or for chimpanzees at Kibale or elsewhere. The Lomako data suggest that, even if bonobos have been released from competition with gorillas and have evolved to rely heavily on abundant herbaceous food, their party size does not seem to have increased dramatically. The data from Wamba suggest that party size in bonobos is so variable as to be difficult to characterize easily. The human population density at and around Wamba is higher than at Lomako. Wamba has more abandoned agricultural fields in which terrestrial herbaceous vegetation, preferred by bonobos, is abundant. Humans exploit the forest at Wamba for a favored species of the herb, *Megaphrynium,* collecting large quantities in short periods of time (Kano and Mulavwa 1984), which is impossible to do at Lomako (Malenky pers. com.). Thus, terrestrial herbaceous vegetation may be significantly more abundant at Wamba than at Lomako, although quantitative data are not currently available to test this hypothesis. Also, the presence of researchers at Wamba and the habituation of the bonobos has resulted in decreased disturbance of the animals (Kano 1992). Thus, party size may increase dramatically when habitat changes affect the distribution of food resources.

Social organization in chimpanzees shows remarkable variability. For example, at Bossou in Guinea, Sugiyama (1988) documents affiliative behaviors and migration patterns of chimpanzees that are more like those of bonobos than is usually the case for chimpanzees (Wrangham and Smuts 1980). Interestingly, the human population density at Bossou is high, and Sugiyama and Koman (1979) report year-round exploitation of 12 species of wild terrestrial herbs, suggesting that terrestrial herbaceous vegetation densities at Bossou are unusually high.

Malenky and Wrangham (in press) show that, for chimpanzees at Kibale, parties feeding on herbaceous food are smaller than parties feeding on fruit trees. Moreover, while party size for bonobos feeding on herbaceous food is also smaller than for fruit trees, recent data from the Lomako suggest that these figures may be an underestimate caused by lack of habituation. Similar problems with unhabituated chimpanzees and gorillas plague researchers at the other study sites in Central Africa.

Future Studies

More reliable data on party size and, perhaps more importantly, on social interactions and party composition are required before the impact of food resources on social organization can be understood. In fact, Chapman et al. (this volume) suggest that the variance in party size, rather than the mean party size, is a better indicator of possible feeding competition. Simply, they suggest that periods of food scarcity force smaller party size and increase variance in party size even when the mean party size remains relatively constant. If terrestrial herbaceous vegetation is more abundant in the bonobo's habitat and if its levels do not vary significantly with season, then we might predict, as the data from Wamba suggest, that while bonobos feed on herbaceous foods, their party size should be equal or larger to the party size of chimpanzees and should show less variance.

Clearly, more data are required to understand the differences in foraging behavior and social organization of African great apes and the significance of these differences. In the absence of controlled experiments, information must come from a wide range of sites, with differing habitats and species composition, to distinguish species-specific patterns from site-specific patterns. The results reported here clearly demonstrate that the methods used to collect the data must be as similar as possible so that the data will be comparable.

The methods used here to assess density varied in plot size, plot layout, and number of survey plots across the three studies. The use of larger survey plots in the Lomako and Kibale studies has two important consequences: it dramatically lowers the sample size (number of survey plots) relative to

Ndoki, where many smaller survey plots were used, and it alters the scale at which the data are collected. This in turn results in large differences for variance and for the range of densities recorded. The smaller sample size along with the restricted range of recorded densities may have resulted in a false lack of statistically significant differences in comparisons between Ndoki and the other two sites. Also, the scale encompassed by larger survey plots hides clumping or variance in density. We do not know the scale at which the animals perceive their environment: individual stems, patches of stems, or, perhaps, groups of patches. With the development of standard methods (Malenky et al. 1993), we will ultimately learn more about not only what the habitat presents to the animals but also how the animals perceive their habitat.

In this chapter, we have, for the first time, presented a quantitative comparison of food availability from three sites encompassing the three great ape species. Despite the preliminary nature of this comparison, we feel that it represents the kind of collaboration that is essential to ultimately answer questions surrounding great ape evolution and social organization.

Summary

We have presented a quantitative comparison of the density and distribution of terrestrial herbaceous foods at three sites inhabited by African great apes: at the Kibale Forest, Uganda *(Pan troglodytes schweinfurthii)*, at the Lomako Forest, Zaire *(Pan paniscus)*; and at the Ndoki Forest, Congo *(Gorilla gorilla gorilla* and *Pan troglodytes troglodytes)*. We use these data in conjunction with observations of habitat preference, food-species preference, and group or party size to examine hypotheses concerning feeding competition and the evolution of differences in social organization of the African great apes.

References

Altmann, S.A., and S.S. Wagner. 1978. A general model of optimal diet. *Rec. Adv. Primatol.* 4:407–14.

Badrian, A., and N. Badrian. 1984. Social organization of *Pan paniscus* in the Lomako Forest, Zaire. In R.L. Susman, ed., *The Pygmy Chimpanzee: Evolutionary Biology and Behavior,* pp. 325–46. New York: Plenum Press.

Badrian, N.L., and R.K. Malenky. 1984. Feeding ecology of *Pan paniscus* in the Lomako Forest, Zaire. In R.L. Susman, ed., *The Pygmy Chimpanzee: Evolutionary Biology and Behavior,* pp. 275–99. New York: Plenum Press.

Belovsky, G.E. 1978. Diet optimization in a generalist herbivore: The moose. *Theoretical Population Biol.* 14:105–34.

Blackman, G.E. 1942. Statistical and ecological studies in the distribution of species in plant communities. Dispersion as a factor in the study of changes in plant populations. *Ann. Bot., London* 6:351–70.

Bromejo, M., G. Illera, and J. Sabater Pi., In prep. Animals and mushrooms consumed by pygmy chimpanzees *(Pan paniscus):* New records from Lilungu (Ikela), Zaire.

Calvert, J. J. 1985. Food selection by western gorillas (G. g. gorilla) in relation to food chemistry. *Oecologia* 65:236–46.

Goodall, J. 1986. *The Chimpanzees of Gombe: Patterns of Behavior.* Cambridge, Mass.: Belknap Press.

Grieg-Smith, P. 1964. *Quantitative Plant Ecology.* London: Butterworths.

Jones, C., and J. Sabater Pi. 1971. Comparative ecology of *Gorilla gorilla* and *Pan troglodytes* in Rio Muni, West Africa. *Bibl. Primatol.* 13:1–96.

Kano, T. 1980. Social behavior of wild pygmy chimpanzees, *Pan paniscus* of Wamba: A preliminary report. *J. Human Evol.* 9:243–60.

———. 1982. The social group of pygmy chimpanzees *(Pan paniscus)* of Wamba. *Primates* 23(2):171–88.

———. 1983. An ecological study of the pygmy chimpanzees *(Pan paniscus)* of Yalosidi, Republic of Zaire. *Internat. J. Primatol.* 4(1):1–31.

———. 1987. Social organization of pygmy chimpanzees and the common chimpanzee: Similarities and differences. In S. Kawano, J.H. Connell, and T. Hidaka, eds., *Evolution and Coadaptation in Biotic Communities,* pp. 53–64. Tokyo: Univ. of Tokyo Press.

———. 1992. *The Last Ape: Pygmy Chimpanzee Behavior and Ecology.* Stanford, Calif.: Stanford Univ. Press.

Kano, T., and M. Mulavwa. 1984. Feeding ecology of the pygmy chimpanzees *(Pan paniscus)* of Wamba. In R.L. Susman, ed., *The Pygmy Chimpanzee: Evolutionary Biology and Behavior,* pp. 233–274. New York: Plenum Press.

Kinzey, W.G. 1984. The dentition of the pygmy chimpanzee, *Pan paniscus.* In R.L. Susman, ed., *The Pygmy Chimpanzee: Evolutionary Biology and Behavior,* pp. 65–88. New York: Plenum Press.

Kuroda, S., 1979. Grouping of the pygmy chimpanzees. *Primates* 20(2):161–183.

———. 1980. Social behavior of the pygmy chimpanzees. *Primates* 21(2):181–97.

———. 1990. Ecological interspecies relations between sympatric gorillas and chimpanzees in Ndoki Reserve, Northern Congo. Abstract of the 13th Congress of International Primatological Society.

———. 1992. Ecological interspecies relationships between gorillas and chimpanzees in the Ndoki-Nouabale Reserve, Northern Congo. In N. Itoigawa, Y. Sugiyama, G.P. Sofkett, and K.R. Thopmson, eds., *Topics in Primatology,* Vol. 2, pp. 385–394. Tokyo: Univ. of Tokyo Press.

Lanjouw, A. 1987. Dynamics of sociality of the pygmy chimpanzee, *Pan paniscus.* Unpublished thesis.

Malenky, R.K., and E.W. Stiles. 1991. Distribution of terrestrial herbaceous vegetation and its consumption by *Pan paniscus* in the Lomako Forest, Zaire. *Am. J. Primatol.* 23:153–69.

Malenky, R.K., and R.W. Wrangham. In Press. A quantitative comparison of terrestrial herbaceous food consumption by *Pan paniscus* in the Lomako Forest, Zaire and *Pan troglodytes* in the Kibale Forest, Uganda. *Am. J. Primatol.*

Malenky, R.K., R.W. Wrangham, C.A. Chapman, and E.O. Vineberg. 1993. Measuring chimpanzee food abundance. *Tropics* 2(4):231–44

Mitani, M. 1992. Feeding behaviors of the Western Lowland Gorillas in the Ndoki Forest, the Ndoki-Nouabale Planning Reserve, in the Congo: Why can they live in high density in this forest? *Rapport Annual.*

Nishida, T. 1979. The social structure of chimpanzees of the Mahale Mountains. In D.A. Hamburg and E.R. McCown, eds., *The Great Apes,* pp. 73–122. Menlo Park, Calif.: Benjamin/Cummings.

Rogers, M.E. and E.A. Williamson. 1987. Density of herbaceous plants eaten by gorillas in Babon: Some preliminary data. *Biotropica* 19(3):278–81.

Rogers, M.E., F. Maisels, E.A. Williamson, M. Fernandez, and C.E.G. Tutin. 1990. Gorilla diet in the Lopé Reserve, Gabon: A nutritional analysis. *Oecologia* 84:326–39.

Schoener, T.W. 1982. The controversy over interspecific competition. *Sci. Am.* 70:586–95.

Sokal, R.R., and F.J. Rohlf. 1981. *Biometry.* 2nd ed. New York: Freeman and Company.

Sugiyama, Y. 1988. Grooming interactions among adult chimpanzees at Bossou, Guinea, with special reference to social structure. *Internat. J. Primatol.* 9:393–407.

Sugiyama, Y., and J. Koman. 1979. Tool using and making behavior in wild chimpanzees at Bossou, Guinea. *Primates* 20:513–24.

Suzuki A., and Nishihara. 1992. Paper presented at the XIV Congress of the International Primatological Society, Strasbourg, France.

Thompson-Handler, N. 1990. The pygmy chimpanzee: sociosexual behavior, reproductive biology and life history patterns. Ph.D. diss. New Haven, Conn.: Yale Univ.

Thompson-Handler, N., R.K. Malenky, and N. Badrian. 1984. Sexual behavior of *Pan paniscus* under natural conditions in the Lomako Forest, Equateur, Zaire. In R.L. Susman, ed., *The Pygmy Chimpanzee: Evolutionary Biology and Behavior,* pp. 347–68. New York: Plenum Press.

Tutin, C.E.G., and M. Fernandez. 1985. Foods consumed by sympatric populations of *Gorilla g. gorilla* and *Pan t. troglodytes* in Gabon: Some preliminary data. *Internat. J. Primatol.* 6:27–43.

Tutin, C.E.G., W.C. McGrew, and P.J. Baldwin. 1983. Social organization of savanna-dwelling chimpanzees *Pan troglodytes verus,* at Mt. Assirik, Senegal. *Primates* 24(2):154–73.

Watts, D.P. 1984. Composition and variability of mountain gorilla diets in the central Virungas. *Am. J. Primatol.* 9:179–93.

White, F. 1988. Party composition and dynamics in *Pan paniscus. Internat. J. Primatol.* 9:179–93.

White, F., and R.W. Wrangham. 1988. Food competition and patch size in the chimpanzee species *Pan paniscus* and *Pan troglodytes. Behav.* 105:148–64.

Williamson, E.A. 1992. Methods used in the evaluation of lowland gorilla habitat in the Lopé Reserve, Gabon. Presented at the XIV Congress of International Primatological Society, Strasbourg, France.

Wrangham, R.W. 1979. Sex differences in chimpanzee dispersion. In D.A. Hamburg and E.R. McCown, eds., *The Great Apes,* pp. 481–9. Menlo Park, Calif.: Benjamin/Cummings.

———. 1986. Ecology and social relationships in two species of chimpanzee. In D.L. Rubenstein and R.W. Wrangham, eds., *Ecological Aspects of Social Evolution,* pp. 352–78. Princeton, N.J.: Princeton Univ. Press.

Wrangham, R.W., and B.B. Smuts. 1980. Sex differences in the behavioural ecology of chimpanzees in the Gombe Stream National Park, Tanzania. *J. Reprod. Fert. (Suppl.)* 28:13–31.

Wrangham, R.W., A.P. Clark, and G. Isabirye-Basuta. 1992. Female social relationships and social organization of Kibale Forest chimpanzees. In T. Nishida, W.C. McGrew, P. Marler, M. Pickford and F.B.M. deWaal, eds., *Topics in Primatology. Vol. 1. Human Origins,* pp. 81–98. Tokyo: Univ. of Tokyo Press.

Wrangham, R.W., N.L. Conklin, C.A. Chapman, and K.D. Hunt. 1991. The significance of fibrous foods for Kibale Forest chimpanzees. *Phil. Trans. of the Royal Soc. of Lond. B,* 334:171–8.

Wrangham, R.W., M.E. Rogers and G. Isabirye-Basuta. 1993. Ape food density in the ground layer in Kibale Forest, Uganda. *African J. Ecol.* 31:49–57.

Wunda, a nine-year-old chimpanzee, with his adopted brother.
Photograph courtesy of J. Goodall.

Hunting Strategies of Gombe and Taï Chimpanzees

Christophe Boesch

Gombe Stream National Park, Tanzania

Once, as three adult male chimpanzees (Figan, Jomeo, and Goblin) started to hunt, three male colobus bounded down toward them, shaking branches and barking. The chimpanzees, screaming, quickly left their tree. The colobus followed them to the ground, and the entire party of chimpanzees (which also included an adolescent male and a mother with her offspring) turned and fled! When the aggressive monkeys climbed into the trees again, the chimpanzees began to move quietly along the ground, looking up at the large and very spread-out monkey troop above. Five minutes after being chased, Figan again climbed to hunt. At once a male rushed toward him, shaking branches and uttering threat calls. Figan whimpered and quickly climbed down. As the chimpanzees sat below, still looking up, three male colobus leaped down to the lowest branches of the tree only a few meters above the chimpanzees' heads and threatened them. Once again, the hunters were routed, this time for good. (Goodall 1986:274)

Taï National Park, Côte d'Ivoire

Five chimpanzees arrived under a mixed group of four monkey species. Ulysse, an adult male chimpanzee climbed up a large vine after scrutinizing the canopy. He slowly pushed some red colobus monkeys toward the south. When he was about to cross into an adjacent tree, a large black-and-white male colobus leaped over and, with huge, roaring calls, jumped on Ulysse's shoulders. Ulysse, screaming aggressively, faced him, and they

fought for a while. The colobus monkey tried to bite Ulysse, who repeatedly tried to grab hold of an arm or a foot of the colobus. Eventually, the colobus monkey ran away, trailed by Ulysse. Ulysse snatched a foot, but the colobus monkey answered by trying to bite. Another noisy quarrel ensued, and the colobus monkey ran away. Ulysse tried, with an ultimate jump, to catch its tail but missed, almost falling to the ground. The colobus monkey escaped. Ulysse followed by moving in the canopy and, two minutes later, the same male colobus fell on him from above. Another struggle ensued, and for the sixth time Ulysse had to release the monkey in order to avoid being bitten. Eventually, Ulysse succeeded in turning the colobus monkey on its back and achieved the capture (pers. observ. 1989).

These two examples are representative of a major difference between Gombe and Taï chimpanzees when they face aggressive colobus monkeys, and this difference has been reported previously (Busse 1978a; Goodall 1986; Boesch and Boesch 1989; Wrangham 1975). Observations similar to those at Gombe are reported from other study sites including Mahale and Kibale, emphasizing that colobus monkeys chase chimpanzees out of trees and even pursue them on the ground (Nishida et al. 1983; Ghiglieri 1984; Wrangham 1991). So far, Taï chimpanzees stand out for their fearlessness when facing aggressive colobus attacks, even the attacks of the large black-and-white colobus monkeys.

Another major difference between Gombe and Taï chimpanzees is the prey they target. Gombe chimpanzees are regularly observed to snatch colobus babies away from their mothers without attempting to catch the mother (Goodall 1986), while Taï chimpanzees kill both mother and baby. This difference between the two populations seems related to the fear of the colobus monkeys shown by Gombe chimpanzees, who specialize in infant prey. Taï chimpanzees prefer to kill adult colobus monkeys (Boesch and Boesch 1989). We shall try to answer the intriguing question of why Taï chimpanzees, unlike other chimpanzee populations, do not fear colobus monkeys.

In a previous study comparing the hunting behavior of chimpanzees in the Gombe, Mahale, and Taï National Parks (Boesch and Boesch 1989), the tendencies of the three populations to hunt in groups for colobus monkeys appeared rather different (see table 1). Taï chimpanzees as a rule hunt in groups and, in most cases, collaborate while hunting. In contrast, Gombe and Mahale chimpanzees generally hunt alone and, when they do hunt in groups, rarely adopt different and complementary roles. How do such differences develop? Do these differences correspond with the observed differences in the fear of the red colobus? These questions lie at the heart of this chapter.

Table 1
Hunting strategies of social predators (Boesch and Boesch 1989).

	Hunts	Group hunts (%)	Collaboration (%)
Primates:			
Chimpanzee			
Gombe	86	36	7
Mahale	34	23	0
Taï	80	92	63
Baboon	147	14	0
Social carnivores:			
Lion	523	52	5
Hyena*	46:164	91:35	0
Wild dog	54	91	0
Wolf	103	86	3

Notes: Group hunts occur frequently in most social predators, including chimpanzees, but the frequencies vary. Group hunts involve more than one individual hunting at the same time. *Collaboration* is observed rarely in social predators. Collaboration occurs when many hunters act together by adopting different but complementary roles to subdue one prey. * Hunts for zebra and wildebeest, respectively.

Sources: Data for Gombe chimpanzees, Teleki (1973) and Busse (1978); for Mahale chimpanzees, Kawanaka (1983), Nishida et al. (1983), and Takahata et al. (1984); for Taï chimpanzees, Boesch and Boesch (1989); for baboons, Strum (1981); for lions, Schaller (1972); for hyenas, Kruuk (1972); for wild dogs, Estes and Goddard (1969); and for wolves, Mech (1970).

Background

Anthropologists often consider hunting and cooperation as the key factors in the evolution of humans. When our human ancestors were forced to live in open, dry environments east of the Rift Valley some five million years ago, they supposedly started to walk bipedally, hunt for meat cooperatively, and share the meat with other group members at the home base to survive (Leakey 1980; Johanson and Edey 1981; Issac 1978). This theory, central to models of human evolution in paleoanthropology, relates the appearance of *cooperation in hunting* in life in an open, dry habitat. Thus, a comparison between chimpanzee populations living in different environments could provide a test of the influence of environment on hominid species (Boesch and Boesch 1989).

On another level, the evolution of cooperation between unrelated individuals has been the center of a lot of theoretical work. Cooperation between unrelated individuals can be explained only if individuals not taking part in the joint action but profiting from that action gain less than the cooperators (Axelrod and Dion 1988; Axelrod and Hamilton 1981; Hamilton 1964; Maynard-Smith 1982; Trivers 1971; Wrangham 1982). Empirical studies must struggle with measuring cost and benefit of different strategies (Mesterton-Gibbons and Dugatkin 1992). In the context of hunting, there are two considerations. One, the benefit of hunting in larger groups should be provision of at least as much meat per hunter as is obtained by hunters in smaller groups. Two, the average amount of meat eaten by group members who did not hunt (nonhunters and cheaters) should be less than the average amount of meat eaten by the hunters (Packer and Ruttan 1988; Boesch in press b).

The fact that chimpanzees hunt small mammals for meat has been known for the last 30 years (Goodall 1963), and the growing body of evidence from studies of wild chimpanzees all over Africa—including Côte d'Ivoire, Liberia, Tanzania, Senegal, Guinea, Uganda, and Zaire—indicates that this may be a general behavior (Boesch and Boesch 1989; Anderson et al. 1983; Goodall 1986; Uehara et al. 1992; McGrew et al. 1978; Sugiyama 1981; Wrangham 1991; Mwanza et al. 1992). At present, detailed data on hunting behavior are available from three populations: Gombe Stream National Park, Tanzania; Mahale Mountains National Park, Tanzania; and Taï National Park, Ivory Coast.

In an attempt to answer questions on how cooperation in hunting develops, a comparative study of Gombe and Taï chimpanzees seemed essential. In agreement with Jane Goodall, I observed the Gombe chimpanzees for 19 weeks over a two-year period. To assure that the differences appearing in reviews of the relevant literature were real and did not arise as a result of different methods, definitions, or observers, I used the same observation method at both sites. Both at Gombe and at Taï, the chimpanzees have been habituated to human observers for at least a decade and can be followed individually from rising to the night nest. The standard method is to follow target individuals in all their daily activities; the individuals are minimally affected by a presence of the human observer. For the present study, I personally followed the chimpanzees at both sites making myself as inconspicuous as possible by wearing dull gray clothes, being very silent, and remaining behind the chimpanzees to avoid influencing their interactions with prey.

Gombe Stream National Park is situated on the shore of Lake Tanganyika and can be described as open woodland (93% of the park), with gallery forests at the base of the valleys (Collins and McGrew 1988; Goodall 1986), whereas Taï National Park is a homogeneous tropical rain forest (Boesch and Boesch 1989). When the comparative data for this study were collected in 1990, the chimpanzee community at Gombe included seven adult males (the most frequent hunters), eight adult females, and 17 subadults, whereas the chimpanzee community at Taï included seven adult males, 15 adult females, and 36 subadults (Boesch in press a).

Interaction of Chimpanzees and Red Colobus Monkeys

If two populations are to be compared, the observations must be made under the same conditions, as the conditions may directly affect the results of observation. For example, problems may have occurred in this case because the chimpanzees are habituated to human observers, while the red colobus monkeys are not and, thus, could well be influenced by a human presence. Although many descriptions exist of the colobus monkey chasing away chimpanzees, Gombe's red colobus monkeys have been described as generally afraid of the chimpanzees, retreating when chimpanzees approach (Busse 1978; Teleki 1981; Wrangham and van Bergmann-Riss 1990). At Taï, however, colobus monkeys clearly are afraid of chimpanzees, with rare exceptions, while chimpanzees are not afraid of the colobus monkeys (Boesch and Boesch 1989). During my stay at Gombe, it quickly became obvious that the red colobus monkeys were extremely afraid of human observers. Moreover, humans observers actually influenced the interactions between chimpanzees and red colobus in 71.4% of the cases in which the chimpanzees heard the red colobus monkeys (Boesch in press a). In the early days of her study, Jane Goodall (1986) noted that when she saw chimpanzees and red colobus from a distance through her binoculars, the colobus remained fearless despite the presence of the chimpanzees. Supporting that observation, I suggest that the red colobus monkeys' fear of chimpanzees proposed by previous observers is in fact a fear of the human observers following target chimpanzees. At Taï, situations in which humans cause red colobus to react—for example, by alarm calls, confusion, and/or flight—occurred only in 3% of the encounters between chimpanzees and colobus monkeys (Boesch in press a). The reason the colobus monkeys do not react to humans at Taï is certainly a result of the trees being taller. These tall trees diminish visibility of the ground and make discreet humans less conspicuous to the monkeys.

In the following analysis of the interactions between chimpanzees and red colobus at Gombe and at Taï, I shall compare only the encounters in which I could be reasonably sure that the observer(s) did not affect the colobus. These encounters include mainly my own observations, made by following the chimpanzees alone and trying, by hiding under branches or tree trunks, to be as inconspicuous as possible to the red colobus.

Reactions of Colobus Monkeys to Chimpanzees

When red colobus monkeys see chimpanzees coming, the red colobus are much more aggressive at Gombe than at Taï (see table 2) (panic + flight/no reaction versus threat + mobbing: $\chi^2 = 33.8$, df = l, $p < .001$). In the

Gombe woodlands, usually the red colobus would immediately advance in the trees toward the approaching chimpanzees and start to make aggressive *tschick* calls toward them. If the chimpanzees started to climb up toward the monkeys, as if testing their reactions, the adult monkeys would immediately come down the trees towards the chimpanzees, while the females with infants and the juveniles remained behind. In addition, the colobus monkeys would often attack the chimpanzees, even if the chimpanzees moved no farther. Moreover, Gombe colobus are cooperative when they face chimpanzees; generally two or more attack, with others present as a backup guard. If the trees were shorter, less than 12 meters in height, the colobus would chase and pursue one or more chimpanzees on the ground in spite of the presence of other chimpanzees. In five of the 16 close encounters that I observed without being spotted, the colobus adults pursued and attacked adult chimpanzees on the ground for about 20 meters without other chimpanzee group members providing support. The risk of wounds resulting from those attacks was considerable: Frodo showed cuts on the arm or the ears three times out of the seven times that I observed him being mobbed by colobus monkeys. On one occasion, Frodo received two deep cuts in the forearm that were still wide open 10 days later.

Thus, the red colobus monkeys at Gombe seem to pursue a *harassing strategy,* a strategy in which the colobus threaten chimpanzees each time they see them in trees, whether the chimpanzees are interested in colobus or not. The colobus may anticipate a possible threat and prevent it by threatening first. In addition, if chimpanzees show interest in the monkeys and are less than 10 meters away, the colobus mob them immediately with lots of threatening calls and some actual attacks. Even if the chimpanzees are not interested in them, a lone monkey may attack very suddenly without any previous warning. Once I observed the alpha male chimpanzee Wilkie

Table 2
Reactions of red colobus monkeys to the visual presence of Taï and Gombe chimpanzees in encounters without human influences (Boesch in press a).

Colobus reaction	Taï (%)	Gombe (%)
Panic	0	2
Flight/None	93	42
Threat	0	13
Mobbing	7	42
Total encounters	71	45

Notes: Flight and no reaction are combined since they are regularly observed in the same group or shown by the same individual during the encounter. *Threat* occurs when colobus monkeys advance in trees toward chimpanzees with aggressive calls. *Mobbing* is an attack against the chimpanzee with physical contact.

moving alone under a group of colobus that were very low to the ground in vine tangle. Suddenly, an adult male colobus jumped on Wilkie's back. Wilkie ran away silently after brushing the colobus monkey off his back.

At Taï, red colobus monkeys most frequently climb high in the emergent trees that make up the highest layer of the tropical rain forest (between 40 to 60 meters in height). The colobus climb into this layer to keep a great distance between themselves and the chimpanzees. Aggressive reactions to the chimpanzees' presence are less common at Taï as compared with Gombe.

Why do the colobus monkeys harass their potential predators more vigorously at Gombe than at Taï? At Gombe, the forest is very low in height and very irregular, with patches of forest intermingling with patches of vine tangles that have no trees at all. At a certain period of the year, these latter vine tangles are very rich in young leaves and fruits, and colobus monkeys as well as chimpanzees forage in these very low patches. If chimpanzees were not afraid of colobus, all age and sex classes might sneak silently under colobus groups and attempt a capture. This would significantly increase the colobus' losses to predation by the chimpanzees. Instead, because the colobus monkeys must prevent such loss, the colobus constantly harass all age and sex classes of their predators who, in turn, hesitate before attacking. Twice, while following chimpanzees in low forest, I saw chimpanzees attacked when they were not even aware of the colobus' presence.

At Gombe, even attacks on the not-yet-dangerous subadult chimpanzees can be productive from the colobus monkey's point of view. Wounds can be serious: Prof sustained an attack as an infant, suffered many wounds, lost part of a toe, and was thought to be dying. Victims may respect the colobus even more: Prof, for example, started to hunt very late, only when fully adult (Goodall pers. com.). Thus, harassment by colobus monkeys probably increases the chimpanzees' fear (especially in subadults and adult females) and that fear could cause adult male chimpanzees to start hunting later and, also, to fear the adult male colobus. This system of intimidation seemed to work even on the especially courageous Frodo, the only chimpanzee I observed to be prepared to brave mobbing adult colobus. However, I never saw him try to capture a colobus adult. He always aimed his attacks at the young. At Taï, on the other hand, the forest is taller and continuous, so that a consistently greater distance separates the two species, and colobus have no need to harass the chimpanzees.

Such a fear of adult colobus monkeys might account for the puzzling observations that whenever the Gombe chimpanzees kill an adult colobus, the meat seems not to appeal to them as much as meat of younger colobus. Regularly, chimpanzees leave much of the adult colobus meat untouched. Of the three adults killed out of 26 prey, only one adult was completely eaten. Of the two others, only the viscera and one arm was eaten, and this pattern has been observed regularly by others.

Chimpanzee Reactions to the Colobus Monkeys

From August to November, when Taï chimpanzees hear red colobus monkeys calling, they are more inclined to move toward the monkeys than are Gombe chimpanzees (no interest versus detour, test, and hunt; $\chi^2 = 8.19$, df = 1, $p < 0.01$) (see table 3). In comparison, during the rest of the year from December through July, Taï chimpanzees are less attracted to the red colobus monkeys than are Gombe chimpanzees who hunt more or less throughout the year (see Boesch in press a). Thus, the period from August to November constitutes a *hunting season* for Taï chimpanzees, a characteristic specific to that population.

Once the chimpanzees are positioned below the colobus monkeys, the reactions of the two populations of chimpanzees differ: Taï chimpanzees more frequently start to hunt (detour and test versus hunt; $\chi^2 = 14.66$, df = 1, $p < .001$); Gombe chimpanzees either just look at the colobus or test the reaction of the colobus by displaying, by looking for different trees to climb, or by climbing up a few meters. Reaching the level of the colobus is our criterion for scoring an interaction as a hunt (Boesch and Boesch 1989). This difference between chimpanzee populations is not affected by the hunting season of the Taï chimpanzees. Throughout the year, once they have begun to move toward the colobus, the Taï chimpanzees remain more keen to hunt than Gombe chimpanzees. At that stage of interaction, the major behavioral difference is that the Gombe chimpanzees are more afraid of the aggressive reactions of the colobus than their counterparts in the Taï forest and, although the Gombe chimpanzees are interested in the meat the monkeys represent, they often simply dare not hunt them.

When testing the Gombe red colobus, the chimpanzees evaluate the monkeys reactions without coming close enough to be attacked. If the adult colobus stand their ground or even attack the chimpanzees, the chimpanzees

Table 3
Reactions of Taï and Gombe chimpanzees to red colobus monkeys in encounters without human influences (Boesch in press a).

Chimpanzee reaction	Taï (%)	Gombe (%)
No interest	43	63
Detour	22	20
Test	1	10
Hunt	33	7
Total detections	168	83

Notes: Detour occurs when the chimpanzees move towards the colobus monkeys. *Test* occurs when, in addition, the chimpanzees charge or approach the colobus monkeys.

usually begin to look for some other members of the monkey group. Then the male colobus follow the chimpanzees through the trees. If the chimpanzees find some easily attainable infants, they rush toward that prey. However, whenever the colobus monkeys face them, all Gombe chimpanzees stop their approach and retreat to a safe distance. The only exception during my stay was Frodo, the keenest and bravest male hunter at the time. I saw him continue to approach and, rushing and zigzagging through a group of aggressive adult colobus monkeys, pushing them aside if need be, he tried to dash toward the youngest colobus. Once I saw the chimpanzees testing a group of colobus monkeys for 45 minutes in a small woodland at the limits of the grassland and, despite the fact that 12 of the colobus were cornered in two trees within the savanna with only one tree bridging them back to the forest, none of the 15 chimpanzees dared to climb up and attack. On the contrary, two adult male colobus monkeys chased one chimpanzee around on the ground.

When Taï male chimpanzee hunters face a mobbing group of colobus monkeys, they always respond by both trying to repel the group as a whole and trying to capture one individual, either by seizing it or by pulling it by the tail or by a limb toward the ground. If the colobus monkeys apply too much pressure, the chimpanzee hunters will retreat for some meters; but in most cases the hunt will go on against the same colobus group. Only if the colobus mob a group of subadult chimpanzees will the hunt be interrupted.

Chimpanzee Strategies for Hunting Red Colobus Monkeys

When chimpanzees hunt for red colobus monkeys, Taï chimpanzees hunt more in groups ($\chi^2 = 29.94$, $p < .001$) and collaborate more when in the groups ($\chi^2 = 24.39$, $p < .001$) compared to Gombe chimpanzees (see table 4). When Gombe chimpanzees hunt in groups, they tend to hunt in an uncoordinated manner, each chasing a different target, usually in different directions. Gombe chimpanzees could be described as performing *simultaneous solitary hunts* (81% of the hunts), whereas such hunts in Taï chimpanzees are less frequent (26% of the hunts) ($\chi^2 = 41.8$, $p < .001$). Goodall (pers. com.) proposed that the Gombe chimpanzees profit from the disorder created in the colobus monkeys' defense by the presence of other chimpanzee hunters. More specifically, an individual—during my stay usually Frodo—would face the colobus group while the other chimpanzees would place themselves low in adjacent trees where the colobus were or might come. Only after the initial hunter succeeds in making some colobus run away will other chimpanzees attack whichever infant colobus happens to move near his tree.

Table 4
Gombe and Taï chimpanzee strategies for hunting red colobus monkeys (Boesch in press a).

	Total hunts	Solitary hunts	Group Hunts			
			Similarity	Synchrony	Concordance	Collaboration
Gombe:						
1973–75	45	26	8	5	3	3
1990–92	17	4	4	3	4	2
Total	62	30	12	8	7	5
%	—	48	19	13	11	8
Taï:						
1984–86	78	6	5	9	9	49
%	—	8	6	11	11	63

Notes: Similarity: all hunters concentrate similar actions on the same prey but without any spatial or time relation between them. *Synchrony:* hunters concentrate similar actions on the same prey and try to relate in time to the others' actions. *Coordination:* hunters concentrate similar actions on the same prey and try to relate in time and space to the others' actions. *Collaboration:* hunters perform complementary actions, all directed towards the same prey (Boesch and Boesch 1989). Observations on Gombe hunts from the 1973–75 period (from the long-term Gombe predation files) and my observations from 1990–1992 are presented separately.

The important point in comparing the two populations is not that one is more sophisticated than the other. I clearly saw Gombe chimpanzees collaborate in the way Taï chimpanzees collaborate: Frodo slowly driving the colobus monkeys down the slope in a region of tall forest; Beethoven and Prof chasing the monkeys by climbing up under their line of retreat; Evered, looking up at their progression in the trees, running fast on the ground to get ahead of their advance, and climbing in full anticipation into a tree where they would arrive. At Taï and at Gombe, the oldest male was taking the most demanding role. The key point to understand here is why the Gombe chimpanzees hunt less frequently in groups when they clearly have all the necessary abilities to do so.

Should Gombe Chimpanzees Hunt More Frequently in Groups?

One would suppose that hunters would act together to achieve greater success. If they do not hunt in groups, we might reason that either they do not increase their success when doing so, or they have no need to increase their success. I shall now use cases where Taï and Gombe chimpanzees hunted for red colobus monkeys, whether or not human observers were influencing the interactions between the two species.

A comparison of the hunting success for lone hunters in the two chimpanzee populations gives the following results (Boesch in press b): For Gombe chimpanzees a lone hunter captures on average a prey of 1.6 kilograms within seven minutes of hunting (the rate of hunting success for lone hunters is 50%). For Taï chimpanzees a lone hunter captures on average a prey of 9.5 kilograms within 39 minutes of hunting (the rate of hunting success for lone hunters is 13%).

Lone hunting is profitable at both sites. According to my calculations, an individual chimpanzee must capture 1 kilogram of meat for every 25 minutes of hunting to compensate for energy expenditure. (Boesch in press b). Gombe chimpanzees achieve a capture five times more rapidly than Taï chimpanzees because of both a shorter hunting time and a much higher rate of hunting success. The high rate of success of lone Gombe hunters limits the advantage of cooperation: for group hunting to be worthwhile, two hunters have to double the lone hunter's rate of return per minute of hunting time. In other words, the Gombe hunters would have to achieve a capture about every three minutes of hunting and would have to be more or less successful each time they hunted, which seems a very difficult task indeed.

The other possibility for Gombe chimpanzees would be to capture prey larger than the infant colobus (estimated to weigh about 1 kilogram), but their fear of the colobus interferes. Even the bravest of the Gombe hunters, Frodo, never tried to capture an adult monkey but clearly concentrated on younger ones. The fear of adult monkeys seems to prevent Gombe chimpanzees from increasing the weight of their prey, leaving reduction in hunting time as the only benefit of hunting in groups. This, together with the high efficiency at Gombe of lone hunters on young colobus prey, tends to restrict the evolution of group hunting and seems to support the point of view that Gombe chimpanzees have no great need to hunt in groups.

On the other hand, Taï chimpanzees still have ample reasons for hunting in groups if they succeed in reducing the hunting time. The higher success rate of Gombe lone hunters over Taï lone hunters is decisively influenced by environmental conditions. The red colobus are arboreal monkeys that rarely descend to the ground (Clutton-Brook 1972; Galat and Galat-Luong 1985). The trees in Gombe National Park are in woodlands that make up 93% of the park (Collins and McGrew 1988) and are 10 to 15 meters in height. The low density of the woodland limits the possible escape routes for the red colobus. In contrast, trees in the Taï forest are about 20 to 30 meters in height at the canopy level, which is dominated by an additional layer of emergent trees that are 40 to 60 meters in height. Not only do Taï chimpanzees need more time to reach up to the colobus monkeys but the colobus also have more time to avoid the approaching chimpanzees. Thus, several factors combine to promote the evolution of group hunting in Taï chimpanzees.

Discussion

The fear chimpanzees show when facing adult male colobus monkeys has been reported from all study sites where both species occur including Gombe, Mahale and Kibale (Goodall 1986; Uehara et al. 1992; Ghiglieri 1984; Wrangham 1991). However, at Taï, although most subadult chimpanzees are also respectful of the mobbing colobus, this fear seems to disappear in adults. Such respect for the prey changes the approach of the predator: Taï chimpanzees, for example, look for a large group of adult colobus monkeys and then organize themselves to overcome the defenses of the colobus. Gombe chimpanzees test the colobus before hunting and, if they hunt, they look for subadult monkeys and take advantage of disorganization within the monkeys' defense or bypass the defense (Boesch and Boesch 1989). The difference in the levels of cooperation in hunting between the two populations seems to originate in the distance at which chimpanzees and colobus encounter each other, which is, in turn, influenced by the ecology of the habitats. A test of this hypothesis could be one in a habitat where both chimpanzees and colobus monkeys occur and where the height of the trees is similar to Taï trees, but where chimpanzees are afraid of colobus monkeys.

A lone hunter at Gombe expends five times less energy measured as the hunting time to achieve a capture than does a lone hunter at Taï. The ecological conditions prevailing within the small woodlands of Gombe allow the chimpanzees to hunt opportunistically with a net benefit despite their fear of the adult colobus. Thus, no selective pressure requires Gombe chimpanzees to collaborate more. However, when chimpanzees hunt red colobus in an environment with higher trees, as in the Taï rain forest, the distance between the prey and the hunter is greater and, accordingly, the time it takes the hunter to catch a prey increases. In a tropical rain forest, emergent trees rarely touch each other, and chimpanzees unlike the red colobus are prevented by their weight from jumping between the emergent trees. This condition further increases the hunting time. If a Gombe chimpanzee's hunting time increased to 39 minutes (the average time necessary for a lone hunter in the Taï rain forest to achieve a capture), then hunting with the Gombe technique would no longer provide a positive payoff. Thus, to get a positive payoff, Taï chimpanzees must capture larger colobus monkeys. Chimpanzees that live among tall trees with a fear of adult red colobus monkeys have two options: either hunt less often because it is not worthwhile, or overcome their fear and capture more adult colobus.

Once chimpanzees overcome their fear of adult red colobus, the monkeys are at much greater risk: At Gombe, for example, adult colobus are fairly safe from the chimpanzees (18% of 130 captured prey) (Goodall 1986). At Taï, on the other hand, 45% of 132 prey were adult colobus ($\chi^2 = 22.21$, $p < .001$). The red colobus in Taï seem aware of this greater risk, as individual Taï monkeys unlike individual Gombe monkeys never attack chimpanzees on their own. The higher proportion of adult colobus individuals captured at Taï leads to an increase in the losses to predation for the red colobus. This increase has led to two behavioral changes in the colobus. One, the colobus try to keep large distances between themselves and the chimpanzees: they never come face-to-face with the chimpanzees on their own initiative. Two, the Taï colobus are conspicuously quieter whenever they hear the chimpanzees. This quietness around chimpanzees has not been observed in Gombe red colobus monkeys (Boesch in press a). This quietness, in turn, increases the cost of hunting: it has forced chimpanzees to invest in searching for prey that is hard to find. Consequently, I reason that the hunting behavior of chimpanzees of the tall forests must become more intentional and planned as chances for opportunistic hunting become rarer. Chimpanzees facing the colobus monkeys that try to keep their distance must coordinate their hunting strategies better so that the colobus can be attacked from different directions. This strategy reduces the distance between the colobus and the chimpanzees and allows a capture. At, Taï hunters collaborate in their movements in the majority of hunts. The interaction between the chimpanzee and the red colobus at Taï seems to have led to behavioral changes in both the colobus and the chimpanzees.

Conclusion

To conclude, the low height of trees in Gombe Stream National Park forces the red colobus monkeys to harass the chimpanzees. This harassing strategy prevents the chimpanzees from attacking the colobus while foraging low in trees when the colobus are most vulnerable. Lone Gombe chimpanzees, despite their fear of the adult colobus, hunt infant colobus very successfully. In contrast, the tall height of trees in the Taï rain forest make the hunting so difficult for lone hunters that the Taï chimpanzees have had to overcome fear of the colobus and hunt, mainly in groups, for larger adult prey to make hunting profitable. Boesch (in press b) provides a more detailed discussion about the evolution and the stability of cooperation in chimpanzees. Thus, a difference in the single environmental factor of tree height may be the origin of all behavioral differences we observe between Taï and Gombe in the interactions of chimpanzees and red colobus monkeys.

References

Anderson, J., E. Williamson, and J. Carter. 1983. Chimpanzees of Sapo Forest, Liberia: Density, nests, tools and meat-eating. *Primates* 24:594–601.

Axelrod, R., and D. Dion. 1988. The further evolution of cooperation. *Science* 242:1385–90.

Axelrod, R., and W.D. Hamilton. 1981. The evolution of cooperation. *Science* 211:1390–6.

Boesch, C. In Press a. Chimpanzees-red colobus monkeys: A predator-prey system. *Anim. Behav.*

Boesch, C. In Press b. Cooperation in hunting in wild chimpanzees. *Anim. Behav.*

Boesch, C., and H. Boesch. 1989. Hunting behavior of wild chimpanzees in the Taï National Park. *Am. J. Phys. Anthro.* 78:547–73.

Busse, C.D. 1978a. Chimpanzee predation as a possible factor in the evolution of red colobus monkey social organization. *Evolution* 31:907–11.

Busse, C.D. 1978b. Do chimpanzees hunt cooperatively? *Am. Nat.* 112:767–70.

Clutton-Brock, T.H. 1972. Feeding and ranging behaviour of the red colobus monkey. Ph.d. diss., Cambridge University.

Collins, D.A., and W.C. McGrew. 1988. Habitats of three groups of chimpanzees *(Pan troglodytes)* in western Tanzania compared. *J. Human Evol.* 17:553–74.

Estes, R.D., and J. Goddard. 1967. Prey selection and hunting behavior of the African wild dog. *J. Wildl. Manag.* 31:52–70.

Galat, G., and A. Galat-Luong. 1985. La communauté de primates diutnes de la forêt de Taï, Côte d'Ivoire. *Rev. Ecol.* (Terre Vie) 40:3–32.

Ghiglieri, M.P. 1984. *The Chimpanzees of the Kibale Forest.* New York: Columbia Univ. Press.

Goodall, J. 1963. Feeding behaviour of wild chimpanzees: A preliminary report. *Symp. Zool. Soc. Lond.* 10:39–48.

———. 1968. Behaviour of free-living chimpanzees of the Gombe Stream area. *Anim. Behav. Monogr.* 1:163–311.

———. 1986. *The Chimpanzees of Gombe: Patterns of Behavior.* Cambridge, Mass.: Belknap Press.

Hamilton, W.D. 1964. The genetical theory of social behaviour (I and II). *J. Theor. Biol.* 7:1–32.

Isaac, G. 1978. The food sharing behavior of protohuman hominids. *Sci. Am.* 238:90–108.

Johanson, D., and M. Edey. 1981. *Lucy: The Beginnings of Humankind.* New York: Simon and Schuster.

Kawanaka, K. 1982. Further studies on predation by chimpanzees of the Mahale Mountains. *Primates* 23:364–84.

Kruuk, H. 1972. *The Spotted Hyena.* Chicago: Univ. of Chicago Press.

Leakey, R. 1980. *The Making of Mankind.* London: Book Club Associates.

Maynard-Smith, J. 1982. *Evolution and the Theory of Games.* Cambridge, England: Cambridge Univ. Press.

McGrew, W.C., C.E.G. Tutin, P.J. Baldwin, M.J. Sharman, and A. Whiten. 1978. Primates preying upon vertebrates: New records from West Africa. *Carnivores* 1:41–5.

Mech, D.L. 1970. *The Wolf.* New York: Natural History Press.

Mesterton-Gibbons, M., and L.A. Dugatkin. 1992. Cooperation among unrelated individuals: Evolutionary factors. *Q. Rev. Biol.* 67(3):267–81.
Mwanza, N., J. Yamagiwa, T. Maruhashi, and T. Yumoto. 1992. Animal eating and tool-use by chimpanzees in the Kahuzi-Biega National Park, Zaire. *Abst. 14th Cong. Intern. Primatol. Soc.* pp. 153.
Nishida, T., S. Uehara, and R. Nyondo. 1983. Predatory behavior among wild chimpanzees of the Mahale Mountains. *Primates* 20:1–20.
Packer, C., and L. Ruttan. 1988. The evolution of cooperative hunting. *Am. Nat.* 132(2):159—98.
Schaller, G.B. 1972. *The Serengeti Lion.* Chicago: Univ. of Chicago Press.
Strum, S.C. 1981. Processes and products of change: Baboon predatory behavior at Gilgil, Kenya. In R.S.O. Harding and G. Teleki, eds., *Omnivorous Primates: Gathering and Hunting in Human Evolution,* pp. 255–302. New York: Columbia Univ. Press.
Sugiyama, Y. 1981. Observations on the population dynamics and behavior of wild chimpanzees at Bossou, Guinea, 1979–1980. *Primates* 22:435–44.
Takahata, Y., T. Hasegawa, and T. Nishida. 1984. Chimpanzee predation in the Mahale Mountains from August 1979 to May 1982. *Internat. J. Primatol.* 5:213–33.
Teleki, G. 1973. *The Predatory Behavior of Wild Chimpanzees.* Brunswick: Bucknell Univ. Press.
Teleki, G. 1981. The omnivorous diet and eclectic feeding habits of chimpanzees in Gombe National Park, Tanzania. In R.S.O. Harding and G. Teleki, eds., *Omnivorous Primates: Gathering and Hunting in Human Evolution* pp. 303–43. New York: Columbia Univ. Press.
Trivers, R.L. 1971. The evolution of reciprocal altruism. *Q. Rev. Biol.* 46:35–57.
Uehara, S., T. Nishida, M. Hamai, T. Hasegawwa, H. Hayaki, M. Huffman, K. Kawanaka, S. Kobayashi, J. Mitani, Y. Takahata, H. Takasaki, and T. Tsukahara. 1992. Characteristics of predation by the chimpanzees in the Mahale Mountains National Park, Tanzania. In T. Nishida, W.C. McGrew, P. Marler, M. Pickford and F.B.M. de Waal, eds., *The Symposium Proceedings of the 13th Congress of the International Primatological Society* Basel: Karger AG.
Wrangham, R.W. 1975. The behavioural ecology of chimpanzees in the Gombe Stream National Park. Ph.D. diss., Cambridge University.
———. 1982. Mutualism, kinship and social evolution. In King's College Sociobiology Group, eds. *Current Problems in Sociobiology,* pp. 269–89. Cambridge, England: Cambridge Univ. Press.
———. 1991. Explaining diversity: Ecology and history in lives of chimpanzees. Paper presented at symposium, Understanding Chimpanzees: Diversity and Survival. 11–5 December, Chicago Academy of Sciences.
Wrangham, R.W., and E. van Bergmann-Riss. 1990. Rates of predation on mammals by Gombe chimpanzees. *Folia Primatol.* 31:157–70.

Wrestling lesson.
Photograph courtesy of J. Goodall.

Comparative Locomotor Behavior of Chimpanzees and Bonobos

Species and Habitat Differences

Diane M. Doran and Kevin D. Hunt

Introduction

Bonobos and chimpanzees differ in their locomotor anatomy, with bonobos having more curved phalanges (finger bones), a lower intermembral index and, at least for males, a longer and narrower scapula (shoulder blade) (Susman 1979, Jungers and Susman 1984: Doran 1993a). In addition, bonobos *(Pan paniscus)* weigh less than *Pan troglodytes troglodytes,* but do not differ in body size from *P. t. schweinfurthii* (Jungers and Susman 1984). The sole data available on body weight for *P. t. verus* places it closer in size to *P. t. schweinfurthii* than to *P. t. troglodytes* (Jungers and Susman 1984).

Before field studies were conducted, researchers engaged in considerable debate about the behavioral significance of these morphological differences. Some researchers predicted that adult bonobos would be more *suspensory,* that is, to locomote or hang under branches, than chimpanzees (Coolidge 1933; Frechkop 1935; Roberts 1974; Susman 1979; Susman et al. 1980), whereas others researchers predicted that there should be no positional behavior difference between the two species since they considered the bonobo to be an allometrically scaled, or proportional, version of the chimpanzee (Corruccini and McHenry 1979; McHenry and Corruccini 1981). However, preliminary behavioral data on bonobo positional behavior suggested that bonobos were not only more suspensory but also more likely to engage in leaping and diving than chimpanzees (Susman et al 1980; Susman 1984; Hunt 1991).

Recently, Doran (1993a) demonstrated that these anatomical features are linked with behavioral differences, particularly with the propensity of bonobos to engage in more arboreal travel than chimpanzees. Her study revealed that bonobos sometimes travel more than one kilometer in the trees before to descending to the ground, whereas chimpanzees travel on the ground between feeding sites and resting sites. In addition, during arboreal locomotion, bonobos use more quadrupedalism, particularly more palmigrade quadrupedalism, and less climbing and scrambling than chimpanzees; male bonobos use more suspensory behavior than chimpanzees.

Doran's (1993a) conclusions were based on a single study of one population of lowland rain forest chimpanzees and one population of bonobos. The chimpanzee subspecies studied was chosen because its habitat was similar to the habitat of bonobos. However, to understand whether the behavioral differences between chimpanzees and bonobos reflect accurately interspecific variation (related to interspecific morphological differences), it is essential to understand the diversity of chimpanzee positional behavior.

The three subspecies of chimpanzees (*P. t. schweinfurthii*, *P. t. troglodytes* and *P. t. verus*) range across Africa from Senegal to Tanzania (Teleki 1989). Throughout their range, chimpanzees occupy a wide variety of habitats, from dry savanna-woodlands to lowland evergreen rain forests. Bonobos, on the other hand, are restricted in their distribution to the lowland rain forest of the Congo basin in equatorial Zaire. Therefore, one might expect that, if positional behavior is determined primarily by habitat rather than by morphology, there may be greater positional behavior differences between some chimpanzee populations than between lowland rain forest dwelling chimpanzees and bonobos.

The locomotor behavior of both eastern and western chimpanzees is well documented (Doran 1992a, 1993a, 1993b; Hunt 1991, 1992a, 1992b). However, the data are not directly comparable because both the behavioral categories and the emphases of these studies differ.

The first goal of this study is to examine the variation that occurs in chimpanzee locomotor behavior both within one subspecies at two different sites and between two different subspecies. The second goal is to reconsider the issue of locomotor differences in chimpanzees and bonobos. On the assumption that locomotor behavior is influenced more by morphological than by habitat differences, we make the following three predictions: one, the locomotor behaviors of one subspecies of chimpanzees (*P. t. schweinfurthii*) at two different sites, in this case Mahale and Gombe, should show great similarity; two, results of a comparison of two subspecies of chimpanzees (*P. t. schweinfurthii* and *P. t. verus*) should show less similarity than the comparison of a single subspecies at different sites; three, a comparison of chimpanzees and bonobos should show the greatest difference when compared to the two situations just predicted, and the data should further document the differences of those two species.

Methods

This study compares the work of two researchers who studied chimpanzees and bonobos at a total of four sites. Hunt studied one chimpanzee subspecies, *P. t. schweinfurthii,* at two sites, Mahale Mountains and Gombe Stream National Park, both in Tanzania. Doran studied a second subspecies of chimpanzee, *P. t. verus,* in the Taï National Park, Ivory Coast, and the bonobo, *P. paniscus,* at Lomako Forest, Zaire.

Comparison of methods

Chimpanzees at Mahale, Gombe and Taï, were studied using instantaneous sampling of focal animals (Altmann, 1974). The sampling interval at Mahale and Gombe was two minutes. The sampling interval at Taï was one minute, although 15-minute sampling intervals were used to estimate the height of the arboreal substrate. There is no reason to expect sampling biases due to different sampling techniques (Altmann, 1974). At each interval, both researchers collected information on locomotor activity, arboreal substrate, and estimated height of substrate.

All locomotor or postural activity was grouped into one of five categories (Susman 1984; Doran 1993b). These categories are: *quadrupedalism,* a mode of locomotion that employs all four limbs in a definable gait on a horizontal or a diagonal substrate and includes knuckle-walking; tripedalism; palmigrade quadrupedalism (where hands are not clearly visible); crutch walking; and running; *quadrumanous climbing and scrambling,* a mode of locomotion using hands and feet in varying combinations during unpatterned, diverse gaits that always occur above substrate and include quadrumanous vertical climbing, scrambling, bridging, tree-swaying and pull-ups; *suspensory behavior,* a positional behavior in which the body's trunk is vertical and suspended below substrate, with the weight borne by forelimbs and includes arm swings, dropping, riding, and crashing foliage to the ground; *bipedalism,* a mode of locomotion in which the body's weight is borne on the hind limbs with the body's trunk vertical and includes bipedalism and aided bipedalism; *leaping and diving,* a mode of locomotion that includes leaping, diving and hopping.

The arboreal substrate was grouped into one of five categories (Susman et al. 1980): *trunk,* a stout, primary member of the tree; *bough,* secondary elements that range from 15–20 centimeters in diameter; *branch,* tertiary supports that range from 2–15 centimeters in diameter; *liana,* vines of any diameter; *foliage,* stems less than two centimeters in diameter and leaves.

Categories of locomotor behavior and arboreal substrate used in sampling bonobo behavior are the same as those used in sampling chimpanzee behavior. Data for bonobos on locomotor behavior and the use of the substrate are based on continuous, focal animal, locomotor-bout sampling (Fleagle 1976). The data include the proportion of all bouts spent in each locomotor activity weighted by the distance traveled per bout. This sampling method was used instead of instantaneous sampling because the mean length of time spent following a focal animal was so short (mean = 18 minutes, range = 2–102 minutes) and, thus, instantaneous sampling could not provide reliable information. For each locomotor bout, the following information was recorded: locomotor activity, substrate type, estimated height, and related other activity.

Data obtained from concurrent, instantaneous sampling and from continuous, locomotor-bout sampling of locomotor behavior yield similar results (Doran 1992b). Thus, we reasoned the data obtained from instantaneous sampling of chimpanzees and from locomotor-bout sampling of bonobos are directly comparable.

Comparison of habitats

Habitats vary considerably across sites. Gombe Stream National Park is described as a thicket woodland, or a semideciduous forest, with an annual rainfall of 1,495 millimeters (Collins and McGrew 1988). Mahale Mountains is a closed forest, or woodland, with an annual rainfall of 1,870 millimeters (Collins and McGrew 1988). Neither Gombe nor Mahale has much primary forest. The Taï forest is a lowland evergreen rain forest with an annual rainfall of 1,800 millimeters (Boesch and Boesch 1983). The Lomako forest is a primary lowland rain forest with an annual rainfall of 1,900 millimeters (Malenky 1990).

Comparison of data

Direct comparisons can be drawn because similar sampling methods were used, because all categories of locomotor activity and all categories of the arboreal substrate were the same; and because data on social and feeding behavior were collected similarly at the four study sites.

However, other differences at the study sites complicate interpretation of results. First, no measure was made to compare how the habitats differed structurally from each other. Second, the level of habituation at the study sites was uneven: Gombe and Taï chimpanzees were completely habituated; Mahale chimpanzees were slightly less well habituated; and the bonobos were the least habituated. Third, at Mahale, Taï, and Gombe the mean length of observation time per month was longer than at Lomako (Mahale—71.4 hours, Taï—61.4 hours, Gombe—43.3 hours, and Lomako—22 hours).

Statistical analyses

G-tests of independence (R x C contingency tables) were used to determine whether the frequency of locomotor behavior of one species (or subspecies) differed from that of the other. Data were pooled across same-sex individuals at each site. Statistical methods follow Sokal and Rohlf (1981), and significance values follow Rohlf and Sokal (1981). Unless stated otherwise, all G-test values have been corrected by the Williams correction factor to obtain a better approximation of the χ^2 distribution. All tests are two-tailed and performed on raw data.

Results

Comparison of the Two Populations of *P. t. schweinfurthii*

Despite significant differences in their habitats, the two populations of Mahale and Gombe chimpanzees differed little from each other in their locomotor activities. They did not differ significantly in overall locomotor behavior (see table 1), in arboreal locomotor activity (see table 2), or in their use of arboreal substrates (see table 3). Chimpanzees at both sites spent roughly 90% of their locomotor time in knuckle-walking quadrupedalism and traveled on the ground between feeding sites and resting sites.

Table 1

Percent of overall locomotor time spent in each locomotor category of activity. Overall locomotion includes combined arboreal and terrestrial locomotion.

	Taï *(P.t.v.)*		Mahale *(P.t.s.)*		Gombe *(P.t.s.)*	
	Male	Female	Male	Female	Male	Female
Quadrupedalism	86.6	85.6	93.6	91.3	96.5	89.5
Quadrumanous climb/scramble	11.1	10.9	5.1	7.7	3.5	8.9
Suspensory	1.1	1.4	0.8	0.9	—	0.5
Bipedal	1.2	1.2	0.3	0.2	—	—
Leap	—	0.6	0.2	—	—	—
N	732	685	922	560	257	190
G (males vs. females)	5.9 NS		5.6 NS		8.9 *	

Notes:

Mahale male vs. Gombe male overall locomotion	6.0	NS
Mahale female vs. Gombe female overall locomotion	2.5	NS
Mahale + Gombe males vs. Taï males	34.1***	
Mahale + Gombe females vs. Taï females	13.6***	
Taï female vs. Gombe female	3.7	NS

* = p < .05: *** = p < .001; NS = Not significant

P.t.v. = *Pan troglodytes verus*
P.t.s. = *Pan troglodytes schweinfurthii*

Table 2
Percent of arboreal locomotor time spent in each locomotor activity.

	Taï *(P.t.v.)*		Mahale + Gombe *(P.t.s.)*	Lomako *(P.p.)*	
	Male	Female		Male	Female
Quadrupedalism	11.7	30.3	31.2	26.1	44.4
Quadrumanous climb/scramble	76.7	59.8	58.8	57.9	42.8
Suspensory	5.8	7.4	6.8	10.0	7.8
Bipedal	5.8	0.8	2.6	1.1	1.9
Leap	—	1.6	0.5	4.8	3.1
N	103	122	192	993	468
G	17.7***			50.6***	

Notes:

Intrasubspecific comparison of *P. t. schweinfurthii:*	
Mahale male vs. Mahale female arboreal locomotion	2.5 NS
Gombe male vs. Gombe female arboreal locomotion	1.9 NS
Mahale male vs. Gombe male arboreal locomotion	2.5 NS
Mahale female vs. Gombe female arboreal locomotion	3.5 NS
Intersubspecific comparison *(P. t. schweinfurthii* vs. *P. t. verus):*	
Mahale + Gombe vs. Taï male arboreal locomotion	15.3***
Mahale + Gombe vs. Taï female arboreal locomotion	2.2 NS
Interspecific comparison:	
Mahale + Gombe vs. Lomako male arboreal locomotion	16.1***
Mahale + Gombe vs. Lomako female arboreal locomotion	18.6***
Taï males vs. Lomako males	33.2***
Taï females vs. Lomako females	11.8*

a. Gombe and Mahale males and females are combined because there is neither a significant sex difference at either site nor any intersite variation in the results.

P.t.v. = *Pan troglodytes verus*
P.t.s. = *Pan troglodytes schweinfurthii*
P.p. = *Pan panciscus*

* = $p > .05$; *** = $p < .001$; NS = not significant

The major difference between the chimpanzees at the two sites was that Gombe chimpanzees were more arboreal than Mahale chimpanzees (see table 4). However, once above ground, there was no difference in height use at the two sites (see table 5). Mahale and Gombe chimpanzees spent half of their arboreal time at heights of less than 5 meters and the other half at heights of between 5 to 20 meters.

There was no sex difference in either arboreal locomotor behavior or use of substrate for either Mahale or Gombe chimpanzees, or in overall locomotor behavior for Mahale chimpanzees (see tables 2 and 3). A marginally significant sex difference in overall locomotor behavior occurred at Gombe, with females climbing more and using less quadrupedalism than males (see table 2). This difference was directly related to the increased time

Table 3
Percent of arboreal locomotor time spent on each substrate[a].

	Taï *(P.t.v.)*	Mahale + Gombe[b] *(P.t.s.)*	Lomako *(P.p.)* Male	Lomako *(P.p.)* Female
Trunk	34.7	15.0	24.7	15.5
Bough	8.4	28.5	12.4	19.6
Branch	20.9	44.9	32.3	40.0
Liana	14.2	—	14.1	9.8
Foilage	21.8	11.6	16.5	15.0
N	225	207	963 bouts	453 bouts
			33.9***	

Notes:

Intrasubspecific comparison of *P. t. schweinfurthii:*
Mahale vs. Gombe 7.5 NS

Intersubspecific comparison:
Mahale + Gombe vs. Taï 109.2***

Interspecific comparisons:
Taï vs. Lomako males 20.2***
Taï vs. Lomako females 60.8***
Mahale + Gombe vs. Lomako males 95.0***
Mahale + Gombe vs. Lomako females 39.6***

a. Substrates are as follows: trunk = stout, primary members of the tree; bough = secondary elements that range from 15–20 centimeters in diameter; branch = tertiary supports that range from 2–15 centimeters in diameter; liana; and foliage. b. Gombe and Mahale males and females are combined because there is neither a significant sex difference at either site nor any intersite variation in the results.

P.t.v. = *Pan troglodytes verus*
P.t.s. = *Pan troglodytes schweinfurthii*
P.p. = *Pan panciscus*

*** = $p < .001$; NS = not significant

spent above the ground and increased frequency of climbing into and out of trees by some females; once in the trees there was no sex difference in locomotor activity (see table 2).

Variation between *P. t. schweinfurthii* and *P. t. verus*

Taï chimpanzees *(P. t. verus)* were similar to Mahale and Gombe chimpanzees *(P. t. schweinfurthii)* in that they travel on the ground between feeding sites and resting sites and they spend nearly 90% of their locomotor time in knuckle-walking quadrupedalism (Doran 1993b). However, there were differences in the locomotor behavior of the two subspecies of chimpanzees. Taï chimpanzees differed from Mahale and Gombe chimpanzees

Table 4
Degree of terrestriality: percent of time spent on the ground versus above ground.

	Taï (*P.t.v.*)		Mahale (*P.t.s*)		Gombe (*P.t.s.*)	
	Male	Female	Male	Female	Male	Female
Ground	51.1	35.1	67.1	52.2	62.6	31.6
Above Ground	48.9	64.8	32.9	47.8	37.4	68.4
N	5975	6543	6124	4175	1543	1234
G (Yates)	324.1***		231.3***		267.3***	

Notes:

G (Mahale vs. Gombe males) 11.02***
G (Mahale vs. Gombe females) 165.0***

G (Mahale vs. Taï) males 324.0***
G (Gombe vs. Taï) males 66.2***
G (Mahale vs. Taï) females 304.3***
G (Gombe vs. Taï) females 5.7**

P.t.v. = *Pan troglodytes verus*
P.t.s. = *Pan troglodytes schweinfurthii*

** = p < .01; *** = p < .001

Table 5
Height use: above ground heights only.

	Taï (*P.t.v.*)		Mahale (*P.t.s*)		Gombe (*P.t.s.*)	
	Male	Female	Male	Female	Male	Female
0–5m	12.6	15.7	51.0	47.1	52.6	51.1
5–20m	36.7	32.0	49.0	52.9	47.3	48.9
>20m	50.7	52.3	—	—	—	—
N	199	300	1764	1740	509	748
	15 msp		2 msp		2 msp	

Notes:

Sex differences:
Taï G = 1.6 NS
Mahale G (Yates) = 5.2*
Gombe G (Yates) = 0.2 NS

Mahale males vs. Gombe males G (Yates) = 0.4 NS
Mahale males vs. Gombe males G (Yates) = 0.4 NS
Taï (male + female) vs. Gombe (male + female) G = 812.6***
Taï (male + female) vs. Mahale male G = 951.7***
Taï (male + female) vs. Mahale female G = 933.1***

msp = minute sampling points

P.t.v. = *Pan troglodytes verus*
P.t.s. = *Pan troglodytes schweinfurthii*

* = p < .05; *** = p < .001; NS = not significant

in their overall locomotor behavior, with Taï chimpanzees climbing and scrambling more frequently, and using less quadrupedalism than Mahale and Gombe chimpanzees (see table 1). During arboreal locomotion, Taï chimpanzees, unlike Mahale and Gombe chimpanzees, showed a significant sex difference. Taï females were similar to Mahale and Gombe chimpanzees in their arboreal locomotor behavior, whereas Taï males engaged in less palmigrade (as opposed to knuckle-walking) quadrupedalism and considerably more climbing and tree-swaying than either Taï females or combined Mahale and Gombe males and females (see table 6).

Table 6
A comparison of Taï and Mahale arboreal locomotion.

	Taï *(P.t.v.)*		Mahale *(P.t.s.)*	
	Male	Female	Male	Female
Quadrupedalism				
Knuckle-walk	7.8	16.4	8.4	1.6
Palmigrade quadrupedalism	2.9	12.3	25.3	23.4
Quadrupedalism	1.0	1.6	—	—
Subtotal: Quadrupedalism	11.7	30.3	.7	5.0
Quadrumanous climb/scramble				
Quadrumanous climb	60.2	52.4	51.8	53.1
Scramble	1.9	4.9	—	1.6
Bridge	3.9	—	1.2	10.9
Tree-sway	10.7	2.5	1.2	—
Subtotal: Quadrumanous climb/scramble	76.7	59.8	54.2	65.6
Suspensory				
Arm-swing	3.9	6.6	8.4	7.8
Drop	1.9	0.8	—	—
Subtotal: Suspensory	5.8	7.4	8.4	7.8
Bipedal				
Bipedalism	1.9	—	2.4	1.6
Aided bipedalism	3.9	0.8	—	—
Subtotal: Bipedalism	5.8	0.8	2.4	1.6
Leap				
Leap	—	1.6	1.2	—
Dive	—	—	—	—
Subtotal: Leap	—	1.6	1.2	—
N—one-minute sampling points	103	122	83	64
G (Williams)	30.9***		11.8 ns	

Notes:
G (Taï vs. Mahale males) 33.4***
G (Taï vs. Mahale females) 30.4***

P.t.v. = *Pan troglodytes verus*
P.t.s. = *Pan troglodytes schweinfurthii*

*** = $p < .001$; NS = not significant

There was a subspecific difference between subspecies in the type of quadrupedalism used (see table 7). When in the trees, Taï males used relatively little quadrupedalism, but what quadrupedalism they did use was most often knuckle-walking quadrupedalism, rather than palmigrade quadrupedalism. Mahale and Gombe chimpanzees, on the other hand, used palmigrade quadrupedalism more frequently than knuckle-walking arboreal quadrupedalism (see table 7). Taï females, although somewhat intermediate, were more like Taï males than Mahale and Gombe chimpanzees in this regard.

Taï chimpanzees also differed from Mahale and Gombe chimpanzees in their use of substrate. Taï chimpanzees used more trunks, foliage, and liana and fewer boughs and branches than Mahale and Gombe chimpanzees (see table 3).

Female chimpanzees were more arboreal than males at all three sites (see table 4). In addition, Taï male chimpanzees were more arboreal than either Gombe or Mahale chimpanzees (see table 4). However, Gombe females spent more time arboreally than either Taï or Mahale chimpanzees (see table 4). There was a difference between subspecies in time spent at different heights: Taï chimpanzees spent considerably more time at heights greater than 20 meters than either Mahale or Gombe chimpanzees (see table 4).

Table 7
Palmigrade versus knuckle-walking quadrupedalism in arboreal locomotion.

	Taï *(P.t.v.)*		Mahale *(P.t.s.)*		Lomako *(P.p.)*	
	Male	Female	Male	Female	Male	Female
Knuckle-walking	72.7	57.1	23.3	6.25	17.0	13.0
Palmigrade	27.3	42.9	76.7	93.75	83.0	87.0
N (arboreal quadrupedalism)	11	35	30	16	176	121
N (arboreal locomotion)	103	122	92	64	963	453
G (Yates)	0.33 NS		2.6 NS		0.77 NS	

Notes:

G (Yates) Mahale vs. Taï males	7.2***
G (Yates) Mahale vs. Taï females	11.1***
G (Yates) Mahale vs. Lomako males	0.5 ns
G (Yates) Mahale vs. Lomako females	0.2 ns
G (Yates) Taï vs. Lomako males	12.7***
G (Yates) Taï vs. Lomako females	24.1***

P.t.v. = *Pan troglodytes verus*
P.t.s. = *Pan troglodytes schweinfurthii*
P.p. = *Pan panciscus*

*** = $p < .001$; NS = not significant

Comparison of Two Species: Bonobos and Chimpanzees

Because bonobos could not be followed consistently on the ground, comparisons between bonobos and chimpanzees were restricted to arboreal observations. Both male and female bonobos differ significantly from chimpanzees in their arboreal locomotor behavior (see table 2). Male bonobos were more suspensory and engaged in more leaping than chimpanzees. Female bonobos used more quadrupedalism and less climbing and scrambling than chimpanzees. Bonobos differed significantly in the use of substrate from Taï, Mahale, and Gombe chimpanzees, using each substrate with a frequency that was intermediate between that of Taï chimpanzees on one hand and that of the Mahale and Gombe chimpanzees on the other hand (see table 3).

Discussion

Comparison of Two Populations

As might have been predicted, the data from the two populations (two sites) of the *P. t. schweinfurthii* subspecies of chimpanzee were remarkably similar. The data indicated no significant difference in arboreal locomotor behavior, in overall locomotor behavior, or use of substrate at the two sites. The only difference noted was in the degree of arboreality, with Gombe chimpanzees spending more time in the trees than Mahale chimpanzees, in spite of a more open habitat at Gombe. There is some indication that Gombe chimpanzees spend more time feeding than Mahale chimpanzees (Hunt 1989). Thus, Gombe chimpanzees are more arboreal because feeding tends to be an arboreal activity.

Comparisons of two subspecies

The results of the comparison of the two subspecies, *P. t. schweinfurthii* and *P. t. verus*, were similar as well. Both subspecies are knuckle-walking quadrupeds that travel on the ground between feeding sites and resting sites (Doran 1993a, Hunt 1992a). Taï chimpanzees *(P. t. verus)* differed from Mahale and Gombe chimpanzees *(P. t. schweinfurthii)* primarily in the time spent above ground (see table 8). At each site the chimpanzee females were more arboreal than males. However, Taï males were very arboreal compared with Mahale and Gombe males and, in fact, were more arboreal than Mahale females. In addition, Taï males and females spent the majority of above-ground time at heights greater than 20 meters, whereas Gombe and Mahale chimpanzees were limited by their habitat to lower heights.

Table 8
Relative degree of arboreality of male and female chimpanzees at three sites.

	Taï (*P.t.v.*)		Gombe (*P.t.s.*)		Mahale (*P.p.*)	
	Male	Female	Male	Female	Male	Female
Percent of time spent above ground	48.9	64.8	37.4	68.4	32.9	47.8
Height above ground (in meters) used most often	>20	>20	0–5	0–5	0–5	5–20
Percent of locomotion spent in climbing	11.1	10.9	3.5	8.9	5.1	7.7
Percent of locomotion spent in quadrupedalism	86.6	85.6	96.5	89.5	93.6	91.3

Notes:
Differences in frequency of overall locomotion:

Gombe female vs. Taï female	3.7 NS
Gombe female + Taï female vs. Taï male	4.9 NS
Gombe female + Taï female + Taï male vs. Mahale female	13.2**
Mahale male vs. Gombe male	6.0 NS
Mahale male + Gombe male vs. Mahale female	7.0 NS
Mahale male + Gombe male + Mahale female vs. Taï male	34.1***

P.t.v. = *Pan troglodytes verus*
P.t.s. = *Pan troglodytes schweinfurthii*

** = $p < .01$; *** = $p < .001$

Taï chimpanzees differed from Mahale and Gombe chimpanzees in the frequencies of overall locomotor activities performed. However, this difference was seen between male and female chimpanzees at Gombe and is related to the amount of time actually spent above ground as well as the height above ground. In fact, when the chimpanzees are ranked by degree of arboreality, there were no significant differences in the frequency of overall locomotion among Gombe females and Taï males and females (see table 8). The more arboreal group (Gombe females, Taï males and Taï females) climb more and use less quadrupedalism than the more terrestrial group (Gombe males, Mahale males and Mahale females). Thus, there was not a clear difference between the two subspecies in the frequency of overall locomotion. Instead, differences in the frequencies of locomotor behavior were related to how arboreal the chimpanzees were. That Taï chimpanzees were so arboreal is not surprising. Taï chimpanzees inhabit a lowland rain forest in which many food trees are taller than 30 meters. What is perplexing is why Gombe females were so arboreal, even in comparison with Gombe males.

There was no clear difference between subspecies in the frequency of arboreal locomotor behavior. In general, the chimpanzees at all three sites (Mahale, Gombe and Taï) were rather similar to each other, with the exception of the males at Taï. Taï males engaged in more frequent climbing, tree-swaying and bipedalism, and less frequent quadrupedalism than either Taï females or Mahale and Gombe male and female chimpanzees.

Why do Taï males move differently in the trees as compared to all other chimpanzees studied? One possible explanation relates to the fact that Taï males may be larger in body size than the other chimpanzees. It is known that, in a given arboreal habitat, a larger animal sees fewer discontinuities in the arboreal "highway", and so that animal leaps and bridges less frequently and climbs more frequently than smaller animals (Fleagle and Mittermeier 1980). These differences in locomotor behavior also seemed to distinguish Taï males from other chimpanzees studied. Thus, because Taï males engage in more frequent climbing, tree-swaying, and bipedalism and less frequent quadrupedalism, it may be that the Taï male is larger than Gombe and Mahale chimpanzees, and that there is no difference in body size between the Taï female and the Gombe and Mahale male.

At present, this hypothesis cannot be supported or refuted because data on body size and weight is not available for the Taï chimpanzees. If Taï female chimpanzees are similar in size to Mahale and Gombe males, it would help to explain the similarity in behavior of female Taï chimpanzees with Mahale and Gombe male chimpanzees. However, that similarity in size leaves unanswered the question of the similarity of Taï female and Mahale and Gombe female chimpanzee arboreal locomotor behavior, and why, considering the distinct sex difference in arboreal behavior at Taï, there is no sex difference in arboreal locomotor behavior at Mahale and Gombe in spite of sex differences in male and female body size.

Regardless of the explanation as to why Taï male chimpanzees differ from other chimpanzees, it is clear that there is a range of locomotor behavior among chimpanzees, largely based on the degree of arboreality. It would be interesting to test whether body size is the primary reason for that difference. This could be done by determining whether *P. t. troglodytes*, the heaviest of the three chimpanzees subspecies, is more like Taï males or more like Taï females and Mahale and Gombe males and females in its locomotor behavior.

There was a difference between subspecies in the type of arboreal quadrupedalism used. Taï males, in particular, used little palmigrade quadrupedalism when moving arboreally, whereas Mahale and Gombe chimpanzees used more palmigrade quadrupedalism and less knuckle-walking quadrupedalism when moving in the trees. This difference was previously ascribed to a behavioral difference between species (Doran 1993a), with bonobos using more palmigrade quadrupedalism and less knuckle-walking quadrupedalism than Taï chimpanzees. But because Mahale and Gombe chimpanzees, and to a lesser extent, Taï females all engaged in relatively frequent palmigrade quadrupedalism, this behavioral difference can no longer be considered a difference between the species.

Comparisons of the Species

The original question posed in this study was whether a broader sampling of chimpanzee locomotor behavior would support the differences between the two species cited previously: specifically, whether bonobos are more arboreal than chimpanzees; whether male bonobos are more suspensory than chimpanzees; and whether bonobos engage in more quadrupedalism, particularly more palmigrade quadrupedalism, than chimpanzees. The results of this study indicate that all Mahale, Gombe and Taï chimpanzees engage in terrestrial travel exclusively between feeding sites and resting sites, whereas previously reported data (Doran 1993) indicated that bonobos travel substantial distances arboreally. In addition, male bonobos were shown to be more suspensory than chimpanzees, a behavior that is clearly correlated with the male bonobo's longer, narrower scapula as compared with that of chimpanzees. The third issue, the bonobo's greater use of palmigrade quadrupedalism as opposed to knuckle-walking quadrupedalism and increased quadrupedalism during arboreal locomotion is not supported. Taï male chimpanzees differ from all other groups in that they engage in very little quadrupedalism when in the trees. Perhaps the most surprising result of this study is the distinctiveness of locomotor behavior of Taï male chimpanzees compared to the surprising similarity of Taï females and male and female chimpanzees at Mahale and Gombe.

Summary

Considerable debate has occurred about the behavioral significance of the morphological features that distinguish bonobos from chimpanzees. A recent study indicated that these anatomical features are linked to specific behavioral differences of the two species (Doran, 1993). However, these conclusions were based on a single study of one population of lowland rain forest chimpanzees when, in fact, many chimpanzees populations representing three subspecies occupy a wide range of habitats across Africa. This study examines the variation in locomotor behavior that occurs both within one chimpanzee subspecies at two different sites and between two different chimpanzee subspecies before turning to the issue of locomotor differences between chimpanzees and bonobos. Results indicate that there is considerably less variation in locomotor behavior between sites and between subspecies than there is between the two species. Bonobos are more suspensory and engage in more arboreal travel than chimpanzees.

Acknowledgments

We thank the Chicago Academy of Sciences and especially Paul Heltne and Linda Marquardt for the invitation to contribute to this volume. The comments of Richard Wrangham greatly improved an earlier version of this manuscript. The field studies were made possible by the governments of Ivory Coast, Tanzania, and Zaire, the Tanzanian Scientific Research Council, the Serengeti Wildlife Institute, the Ministry of Scientific Research and the Station d'Ecologie Tropical (Ivory Coast),and the Institut de Recherché Scientifique (Zaire). In addition we gratefully acknowledge T. Nishida and the staff of the Mahale Mountains Wildlife Research Center, the village of Mugambo, Tanzania, Jane Goodall and the staff of the Gombe Stream Research Center, Christophe and Hedwige Boesch, Gregoire Nohon, Randall Susman and the staff of the Lomako Forest Pygmy Chimpanzee Project. Generous financial support was provided by the Louis B. Leakey Foundation, the Wenner Gren Foundation, the National Science Foundation, the Margaret Wray French Fund, Sigma Xi, and a Harvard Postdoctoral Fellowship.

References

Altmann, J. 1974. Observational study of behavior: sampling methods. *Behav.* 49:227–67.

Boesch, C., and H. Boesch. 1983. Optimisation of nut-cracking with natural hammers by wild chimpanzees. *Behav.* 34:265–86.

Collins, D.A., and W.C. McGrew. 1988. Habitats of three groups of chimpanzees (*Pan troglodytes*) in western Tanzania compared. *J. Human Evol.* 17:553–74.

Coolidge, H.J. 1933. *Pan paniscus*: pygmy chimpanzee from the south of the Congo river. *Am. J. Phys. Anthro.* 18:1–59.

Corruccini, R.S. and H.M. McHenry. 1979. Morphological affinities of *Pan paniscus* and human evolution. *Science* 204:1341–3.

Doran, D.M. 1992a. The ontogeny of chimpanzees and pygmy chimpanzee locomotor behavior: a case study of paedomorphism and its behavioral correlates. *J. Human Evol.* 23:139–57.

———. 1992b. A comparison of instantaneous and locomotor bout sampling methods: a case study of adult male chimpanzee locomotor behavior and substrate use. *Am. J. Phys. Anthro.* 89:85–99.

———. 1993a. The comparative locomotor behavior of chimpanzees and bonobos: the influence of morphology on locomotion. *Am. J. Phys. Anthro.* 91:83–98.

———. 1993b. Sex differences in adult chimpanzee positional behavior: the influence of body size on locomotion and posture. *Am. J. Phys. Anthro.* 91:99–115.

Fleagle, J.G. 1976. Locomotion and posture of the Malayan Siamung and implications for hominoid evolution. *Folia Primatol.* 26:245–69.

Fleagle, J.G., and R.A. Mittermeier. 1980. Locomotor behavior, body size, and comparative ecology of seven Surinam monkeys. *Am. J. Phys. Anthro.* 52:301–14.

Frechkop, S. 1935. Notes sur les mammiferes. A propos du chimpanzé de la rive gauche du Congo. *Mus. R. Hist. Nat. Belg. Bull.* 11:1–41.

Hunt, K.D. 1989. Positional behavior in *Pan troglodytes* at the Mahale Mountains and the Gombe Stream National Parks, Tanzania. Ph.D. diss. University of Michigan, Ann Arbor.

———. 1991. Positional behavior in the Hominoidea. *Internat. J. Primatol.* 12(2):95–118.

———. 1992a. Positional behavior of *Pan troglodytes* in the Mahale Mountains and Gombe Stream National Parks, Tanzania. *Am. J. Phys. Anthro.* 87:83–105.

———. 1992b. Social rank and body size as determinants of positional behavior in *Pan troglodytes. Primates* 33(3):347–57.

Jungers, W.L. and R.L. Susman. 1984. Body size and skeletal allometry in African apes. In R.L. Susman, ed., *The Pygmy Chimpanzee,* pp. 131–77. New York: Plenum Press.

Malenky, R.K. 1990. Ecological factors affecting food choice and social organization in *Pan paniscus.* Ph.D. diss. State University of New York at Stony Brook.

McHenry, H.M. and R.S. Corruccini 1981. *Pan paniscus* and human evolution. *Am. J. Phys. Anthro.* 54:355–67.

Roberts, D. 1974. Structure and function of the primate scapula. In F.A. Jenjins, ed., *Primate Locomotion,* pp. 171–200. New York: Academic Press.

Rohlf, F.J., and R.R. Sokal. 1981. *Statistical Tables.* New York: W.H. Freeman and Co.

Sokal, R.R. and F.J. Rohlf. 1981. *Biometry.* New York: W.H. Freeman and Co.

Susman, R.L. 1979. Comparative and functional morphology of hominoid fingers. *Am. J. Phys. Anthro.* 50:215–36.

Susman, R.L. 1984. The locomotor behavior of *Pan paniscus* in the Lomako Forest. In R.L. Susman, ed., *The Pygmy Chimpanzee,* pp. 369–93. New York: Plenum Press

Susman, R.L., N.L. Badrian, and A.J. Badrian. 1980. Locomotor behavior of *Pan paniscus* in Zaire. *Am. J. Phys. Anthro.* 53:69–80.

Teleki, G. 1989. Population status of wild chimpanzees *(Pan troglodytes)* and threats to survival. In P.G. Heltne and L.A. Marquardt, eds., *Understanding Chimpanzees,* pp. 312–53. Cambridge, Mass.: Harvard Univ. Press.

Tuttle, R.H. 1967. Knuckle-walking and the evolution of hominoid hands. *Am. J. Phys. Anthro.* 26:171–206.

Comparative Analyses of Nest Building Behavior in Bonobos and Chimpanzees

Barbara Fruth and Gottfried Hohmann

Introduction

One trait thought to separate nonhuman primates from other mammals is cognitive ability (Harcourt 1988). This ability can be expressed in a variety of ways. One way, which we will discuss, is the modification and/or utilization of physical objects or tools. Although reports on tool use and object manipulation are available for a large number of species, many of those reports concern captive animals, and corresponding reports from the field are limited in many cases to a few species performing single tasks of goal-oriented manipulation of objects. The only species known to utilize and manufacture a number of different tools both in captivity and in the wild is the chimpanzee *(Pan troglodytes)*. Moreover, the tools used by different chimpanzee populations vary in both number and quality (McGrew 1992).

Traditionally, nest building has been treated separately from tool use (Beck 1980; Tuttle 1986). However, this separation has been disputed (Galdikas 1982; McGrew 1992) although some of the arguments appear arbitrary. Whatever the case may be, nest building is certainly the most pervasive form of object manipulation among the Pongidae. Nest building differs from other forms of tool use in several ways: it is common practice in all Pongidae species while, at the same time, a practice that is absent in all other simian primates; it occurs daily, performed by all mature males and females with similar frequencies; and it is characterized by specific combinations of different objects.

Nest building shows variation both between and within species: while gorillas often build their nests close to or at the forest floor, orangutans, chimpanzees, and bonobos build their nests almost exclusively within trees

(Schaller 1963; Galdikas 1982; Goodall 1962; Kano 1992). Furthermore, this species-specific preference in nesting sites persists in areas where chimpanzees and gorillas are sympatric, that is, where gorillas and chimpanzees occur in the same range, strongly suggesting that this preference may not be exclusively related to environmental conditions. Another possible species-specific difference in nest building concerns the habit of improving nests by adding soft twigs and leaves, a behavior widely reported for chimpanzees (Goodall 1962) and bonobos (Kano 1992), but apparently absent in mountain gorillas (Schaller 1963).

Within one species, nest-building behavior seems to be comparatively uniform, and variations have been mostly attributed to ecological differences. While comparison of nests for two distinct chimpanzee populations revealed significant variation within and between populations, all structural differences were most likely related to environmental factors such as seasonality, predator pressure, and available vegetation, demonstrating adaptive responses to changes in environmental conditions (Baldwin et al. 1981). At Gombe, chimpanzees use oil palms for nest building (Goodall 1968). According to Wrangham (1975), this habit seems merely to reflect seasonal

Table 1
Habitat and nest characteristics of chimpanzees.

Country		Senegal		Republic of Guinea		Liberia	Ivory Coast		Gabon	
Study site		Mt. Assirik				Sapo	Taï		Lopé	
Author		Baldwin et al. 1981	1982	Nissen 1931[a]	de Bouron-onville 1967[a]	Anderson 1983	Boesch 1978	Fruth 1990[b]	Tutin and Fernan-dez 1983[c]	Wroge-mann 1992
Habitat		SW	SW	SW	—	SW	RF	RF	RF	RF
Nests	N	252	4,478	100	184	67	146	154	1,741	523
Height in meters:										
	average	—	—	11.5	—	—	—	23.2	8.7	11.7
	median	11	—	—	—	12	—	20	—	10
	range	0–44	5–22	4–31.5	2–24	6–20	3.5–15	5–45	2–32	2–45
	%	100	94	100	100	81	82	100	100	100
DBH in centimeters:										
	average	—	—	—	—	—	—	42	—	34.6
	median	39–45	—	—	—	52	—	33	—	25
	range	—	—	—	—	6–25	—	6–168	—	5–400
	%	—	—	—	—	93	—	100	—	—

Notes: a. study site not specified. b. four different study sites. c. nationwide. SW = savanna-woodland (dry); RF = rain forest; GWF = mosaic of grassland-woodland and forest; % = percentage of the nests the range is referring to; DBH = diameter breast high. * (Tuttle 1986).

variation in available materials. However, the fact that palm trees are used for nest building at Gombe but not at other study sites suggests that culture might be a factor in variation (McGrew 1985).

These examples show different factors involved in variations (Reynolds and Reynolds 1965). Accordingly, identification of the factors responsible for differences and similarities requires several kinds of comparison. By incorporating data from additional chimpanzee populations and from other closely related species into a comprehensive analysis, we may be able to identify the links between environmental conditions and nest building, to weigh the magnitude of single factors as, for example, hunting pressure on particular aspects of this behavior (nest height), and to select those features that most likely represent local culture.

We had precisely such a scheme in mind when we started to work on this chapter. Available data on nest building for the chimpanzee *(Pan troglodytes)* include all three subspecies and a wide range of habitats occupied by this species (see table 1) (Tuttle 1986). Moreover, the nest building behavior of the bonobo *(Pan paniscus),* has been investigated at four study sites (see table 2) and, thus, provides sufficient data for comparative analyses. However, during an initial survey of the literature, it became evident that

Equatorial Guinea		Uganda			Tanzania		Zaire	
		Budongo	Ngogo	Kanyawara	Gombe	Ugalla	Lake Kivu	Virunga
Jones and Sabater Pi 1971[a]	Baldwin et al. 1981[a]	Reynolds and Reynolds 1965	Ghiglieri 1984[a]	Ghiglieri 1984[b]	Goodall 1968	Itani 1979	Rahm* 1967	Sept 1992
RF	RF	RF	RF	RF	GWF	GWF	RF	GWF
195	195	259	372	63	384	491	?	101
—	—	—	12.2	10.3	—	19	—	13.5
10	10	—	—	—	—	—	—	—
0–20	0–40	3–45	2–35	5–23	12–24	5–40	10–25	—
96	—	100	100	100	57	100	100	—
—	—	—	—	—	—	—	—	—
—	13–19	—	—	—	—	—	—	—
—	—	—	—	—	—	—	—	—
—	—	—	—	—	—	—	—	—

Table 2
Habitat and nest characteristics of bonobos.

Country		Zaire				
Study Site		Lomako		Lake Tumba	Yalosidi	Wamba
Author		Badrian and Badrian 1977[a]	Fruth and Hohmann 1977[b]	Horn 1980	Kano 1983	Kano 1992
Habitat		RF	RF	RF	G/RF	RF
Number of nests		174	1,155	107	2,380	3,353
Height in meters:						
	average	—	16.6	—	—	—
	median	—	16	—	—	13[c]
	range	5– >35	3–50	4–28	0–50	0– >40
	%	100	100	100	100	100
DBH in centimeters:						
	average	—	23.8	—	—	—
	median	—	18	—	—	—
	range	—	2–137	1–36	>0–>40	—
	%	—	100	100	100	—

Notes: a. unspecified community. b. eyengo community. c. night nests only. RF = rain forest; G/RF = mosaic of grassland and rain forest. % = percentage of the nests the range is referring to.

the majority of data sets for the two *Pan* species were highly incommensurate, that is, lacking a common basis for comparison. For example, in many studies, data on nest building were collected for only short periods of time (less than one year) and, with very few exceptions, the data were collected from abandoned, anonymous nests of unknown age. After selecting only studies containing compatible information, the comparative data boiled down to a few structural characteristics of nests and nest trees. Because most of the studies have reported on nests and not on nest-building behavior, data on the social context or the manner of use are almost nonexistent.

Consequently, we had to abandon our first scheme and create a more realistic scheme. These comparisons now involve only a small number of studies that provide the most comprehensive information on some of the major aspects of nest-building behavior. Many categories of data we would have liked to compare were available from only a few sites or from only one site but not the others. Therefore, in spite of better intentions, our comparisons of nest building in the two *Pan* species remain fragmentary and preliminary.

This chapter presents data on the following three topics: single nests, considering morphological and ecological characteristics of nests; nest groups, considering the number of nests per group and the changes in party size during day and night; and the ethology of nest building, considering ontogeny, sex differences, context, and social significance.

Methods

The data for comparison appear in table 3. Whenever possible, the data are presented in their original form and physical dimension. For reasons of compatibility some data was modified or rearranged. For example, absolute values were transformed into percentages and measurements of structural parameters of nests and nest trees using different scales such as feet, yards, or meters were standardized. If not specified otherwise, data from Lomako was derived from our 14-month study of bonobos conducted between August 1990 and June 1992 (Fruth and Hohmann 1993). The Lomako study site is described in detail by Badrian and Badrian (1977, 1984), White (1988), and Malenky and Stiles (1991).

Table 3

Structural features of nests for chimpanzees and bonobos.

Species		*Pan Troglodytes*				*Pan Paniscus*		
Country		Senegal	Liberia	Gabon	Equatorial Guinea	Zaire		
Study Site		Mt. Assirik	Sapo	Lopé	a.	Lomako	Yalosidi	Wamba
Nest	N	252	67	523	195	1,156	2,380	3,353
Uncovered	%	75.0	21.0	38.2	17.0	50.4	—	—
Covered	%	25.0	79.0	61.8	83.0	49.6	—	—
Nests/Tree	N	2	?	+	1	+	+	+
Trees/Nest	max	—	—	2	—	6	5	6
Int. Nests	%	—	—	8.4	—	37.1	14.6	32.8
Ground	%	—	—	—	—	—	+	+
ID species	N	—	—	86	—	80	—	—
Used	N	—	—	45	—	24	100	108
UNT	N	—	—	321	—	476	2,142	—
TTT	%	—	—	61.2	—	75.6	63.9	72.0
Nest Groups	N	83	58	66	127	—	—	—
Nests/Group	range	1–18	1–10	1–26	1–12	1–24	—	—
Nests/Group	median	4	1	1	2	7	—	—
NN-Nest	N	224	—	284	33	—	—	—
NN-Distance	m	4	—	6	4	—	—	—

Note: a. study site not specified. Nest N = total sample size; Uncovered = percent of nests not covered by upper vegetation; Int. nests = percent of integrated nests; ID species = total of identified species in the sample area; Used = number of species used for nest construction; UNT = number of individual trees used for nesting from which the rate of used species was calculated; TTT = percent of all UNT representing the 10 tree species most often used for nesting; Nest Groups = number of obscured nest groups; NN-Nest = total number of nests occurring in nest groups for which NN-distance was computed; NN-Distance = mean distance between next neighbors for all nests in a given group.

Results: Single Nests

Type of Construction

The technique of nest building has been described for chimpanzees by Bolwig (1959) and Goodall (1968) and for bonobos by Horn (1980) and Kano (1979). Despite individual, age, or species variations, all nests combine plant materials into three nest components: a solid foundation, or frame; a central mattress; and a lining made of additional leafs and twigs (McGrew 1992).

From the available descriptions of nest building by chimpanzees, we have information from six sites on the median number of nests observed in one tree (see table 3). However, with the exception of data from Lopé (Wrogemann 1992), little has been mentioned about whether nests incorporate parts of more than one tree. At Gombe, for example, integrated nests constructed of materials from more than one tree seemed to be the exception rather than the rule (Goodall 1962).

In contrast, data about bonobo populations at Yalosidi, Wamba, and Lomako have demonstrated that nests commonly incorporate parts of more than one tree (see table 3) (Kano 1983, 1992; Fruth and Hohmann 1993). At Lomako, almost 37% of all bonobo nests consisted of parts of two or more trees, with a maximum of six trees involved in a single nest. The only available source for comparison is the study on chimpanzees conducted at Lopé, Gabon (Wrogemann 1992). At Lopé, less than 10% of the total sample (N = 523) of nests incorporated parts of more than one tree, and the maximum number of trees constituting a nest was two.

Height of Nests

Tables 1 and 2 present data on nest height above forest floor from 23 studies of chimpanzees and bonobos. All values on nest height indicate a large range of variability within populations. Height varied most (3 to 45 meters) at Budongo (Reynolds and Reynolds 1965) and least (3.5 to 15 meters) at Taï (Boesch 1978). There is no evidence for a correlation between the height of nests and the type of habitat (savanna-woodland versus rain forest) or a correlation between the height of nests and their affiliation to forests (West versus Central Africa). However, the variation found within a given population was always greater than that found between populations. This is true for both chimpanzees and bonobos.

Diameter of Nest Trees

Trees used for nest construction had a minimum diameter of five centimeters for chimpanzees at Lopé (Wrogemann 1992) and one centimeter for bonobos at Lake Tumba. The values for the two populations of bonobo fit well within the range given for chimpanzees (see tables 1 and 2). As mentioned before, bonobos often combine trunks of several different trees into their nests. This technique may enable them to use saplings, which could never support a nest alone. Consequently, at least in bonobos, the diameter of trees does not seem to be a limiting factor, although height of trees and the height of the nest remains important.

Seasonal Differences

Speculation about the reasons for differences between populations or between sites has often overshadowed the magnitude of variation within one population or one site. However, studies across seasons demonstrate variations in characteristics such as the height and the position of nests. Both chimpanzees (Baldwin et al. 1981; Wrogemann 1992) and bonobos (Fruth and Hohmann unpubl. data) build their nests higher during the rainy season. In Equatorial Guinea, Senegal, and Gabon, chimpanzees build fewer nests with covers during the rainy season than during dry months (Baldwin et al. 1981; Wrogemann 1992). Thus, for chimpanzees, seasonal variation appear to be consistent across habitats. At Lomako, the ratio of uncovered to covered bonobo nests remained steady throughout the year (47% versus 49%; N = 995), with a tendency to cover more in the wet season.

Day and Night Nests

The fact that chimpanzees and bonobos build nests not only at night but also during the day is well established (Nissen 1931; Goodall 1968; Kano 1983). Descriptions of day nests of chimpanzees have been published by Goodall (1962) and Hiraiwa-Hasegawa (1989), but the data reported permits only general comparisons with night nests. More detailed analyses have been conducted for bonobos at Wamba (Kano 1992) and at Lomako. At Lomako, it became apparent that day nests differed in four respects from night nests, time of construction, duration of use, mode of construction, and nest sites.

Time of construction: on average, bonobos needed 4.2 minutes to build a night nest, or roosting site, (M = 4, range 1–7 minutes, N = 35) but less than 1 minute to build a day nest (M = 0, range 0–5 minutes, N = 105). These patterns are significantly different (U-test; $p < .001$).

Duration of use: in the mean, day nests were used for 31.3 minutes (M = 25, range 2–120 minutes, N = 134), while the estimated time spent in night nests was 10 to 12 hours.

Mode of construction: out of 110 day nests, 96% incorporated material from only a single tree, but 41% of all night nests (N = 595) combined parts of several trees. Hence, the number of trees incorporated in a day nest differed significantly from the number used in a night nest ($\chi^2=57.7$, $p < .001$).

Nest sites: day nests were built considerably higher in the trees ($x=20.4$ meters, $SD=8.5$, range 5–50 meters, $N = 102$) than night nests ($x=15.4$ meters, $SD=4.8$, range 3–35 meters, $N = 595$) ($t=8.47$, $df=695$, $p <.001$). These data on differences between bonobo day and night nests at Lomako correspond with the results published by Kano at Wamba (1992).

Another important difference seems to exist concerning the use of feeding trees for nest building: corresponding to our data from Lomako, Kano (1983) observed that of 19 day nests, 11 were built within a tree previously used for feeding.

Most of these features may be compared for both species when a comparable set of detailed data becomes available for chimpanzees.

At Gombe and Mahale, the rate of day-nest construction by chimpanzees seems to be very low (Goodall 1962, Hiraiwa-Hasegewa 1989). However, chimpanzees at Kibale and bonobos at Lomako and Wamba apparently build day nests frequently (Wrangham pers. com.; Kano 1983). Therefore, the frequencies of day-nest construction appears to be habitat-specific rather than species-specific.

Nesting Site Selection

Chimpanzees and bonobos spend half their lives or more in nests. Therefore, it is reasonable to assume that nesting sites are carefully chosen. Various studies have hinted about preferences regarding nesting sites, but those studies often refer only generally to the size of the trees or the position of the nests within trees (Reynolds and Reynolds 1965; Jones and Sabater Pi 1971; Badrian and Badrian 1977; Horn 1980).

The studies providing more systematic data on nest building and the location of nesting sites suggest the following: first, nest groups are not randomly distributed but located in particular areas of the home range. At Lomako, traces of old nests indicate that some nesting sites are reused by generations of bonobos (Fruth and Hohmann in press) Studies of chimpanzees at Sapo, Liberia (Anderson et al. 1983); Equatorial Guinea (Baldwin et al. 1981); and Ishasha, Zaire (Sept 1992) and of bonobos at Yalosidi (Kano 1983) and Lomako have demonstrated that, if available, primary forest and gallery forest are the two most preferred habitat types for nest building. At Mount Assirik, Senegal, nests were found in equal proportion in woodland and gallery forest (Baldwin et al. 1981), and at Ugalla, Tanzania, most chimpanzee nests were found in dry open woodland but not in gallery forest (Itani 1979). With the exception of Ugalla, the data indicate that chimpanzees and bonobos preferred to build their nests in areas of dense, high vegetation.

Second, chimpanzees and bonobos tend to select particular tree species for nest building. For example, when nesting in the grassland, chimpanzees at Mount Assirik choose two species *(Spondias mombin* and *Adansonia digitata)* as a nesting tree (Baldwin et al. 1981), and chimpanzees at Ishasha, Zaire, have been reported to prefer *Cynometra alexandrii* as a nesting tree (Sept 1992). Kano (1983, 1992) found that bonobos at Yalosidi and Wamba show a high preference for a single species *(Leonardoxa romii),* accounting for 45% of all trees chosen for nest building at Yalosidi and 34.5% at Wamba.

Third, chimpanzees and bonobos rarely build their night nests within trees offering ripe fruit or other food but rather in neighboring trees (Goodall 1962; Kano 1992; Fruth and Hohmann unpubl. data). Although analyses of our data from Lomako are still in progress, it appears that feeding trees are used for night nests at other times of the year, such as when the tree is not in fruit. Hence, the state of a tree may be another criterion for selection of nesting sites.

Conclusions regarding the use of selection criteria (height of nesting trees) for particular trees by chimpanzees and bonobos must take into account the abundance and distribution of trees of different height. This information has only recently been sought. Earlier reports, while pointing to possible selection criteria, do not have the full range of data to be conclusive.

More extensive quantitative data on the selection of nesting trees have been collected in independent studies, one on chimpanzees at Lopé, Gabon (Wrogemann 1992) and the other on bonobos at Lomako. In both studies, evaluation of choices for nesting trees were based on the relationship between the used nesting trees (UNT) and the potential nesting trees (PNT). At Lopé, the characteristics of PNT were obtained by the following sampling method: At 100-meter intervals along several paths, the height of the 10 nearest trees over 5 centimeters DBH (diameter at breast height) was estimated. The heights were arranged into 5-meter classes. At Lomako, data were taken from 10 forest plots, each of 400 square meters, containing nesting sites. The height of all trees with a minimum diameter of 2 centimeters and a minimum height of 5 meters were measured. In order to obtain data on the selection of particular species as preferred nesting sites, 15 plots randomly placed along standardized transects were compared to 17 plots containing nesting sites.

The results of these studies can be summarized as follows. Bonobos at Lomako built their nests in tall trees more often than expected. The majority of nests (82%) were located in trees of the middle forest layer (11–30 meters above the forest floor); in contrast, only 12% of the potential nesting trees were this high. The difference between potential and used nesting trees in this forest layer, then, is highly significant (U-test; $p < .001$). Chimpanzees at Lopé select UNT that have a height distribution quite similar to PNT (see figure 1.) In general, nests at Lopé are constructed slightly lower.

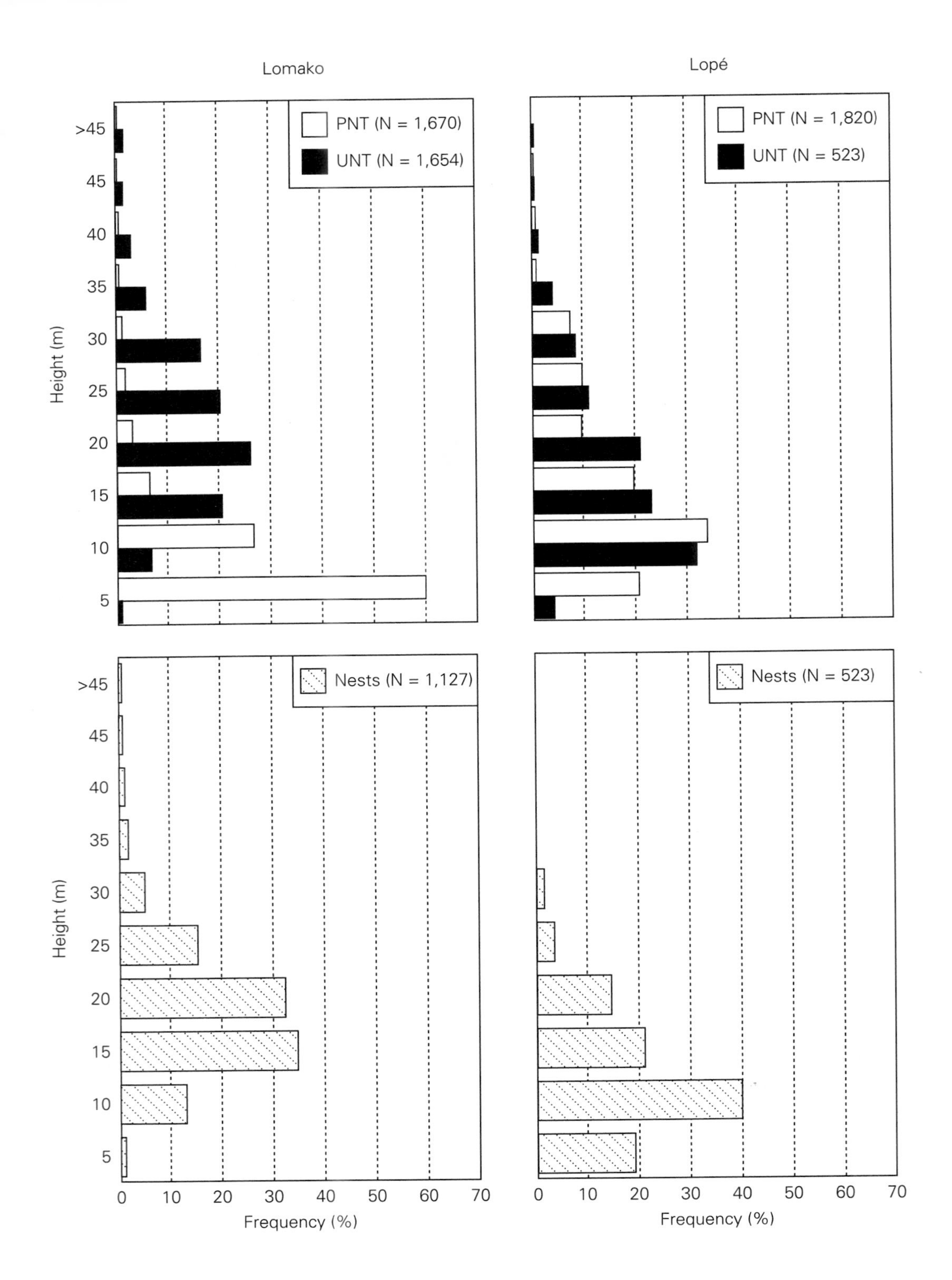
Lomako
Lopé
PNT (N = 1,670)
UNT (N = 1,654)
PNT (N = 1,820)
UNT (N = 523)
Nests (N = 1,127)
Nests (N = 523)
Height (m)
>45
45
40
35
30
25
20
15
10
5
0
10
20
30
40
50
60
70
Frequency (%)

◀ **Figure 1**
Top left and right: Height of potential nesting trees (PNT) and used nesting trees (UNT). Bottom left and right: Frequency of nest heights for bonobos at Lomako (left) and chimpanzees at Lopé (right).

At Lomako, the species of 83% of 476 UNT have been identified, compared to 1052 PNT of which 91% have been identified into 80 species. Of these 80 species of PNT, bonobos have used 24 species as nesting trees. Of the 10 tree species most frequently chosen at Lomako, five were used with much higher frequencies than expected by chance. In contrast, at Lopé, the species of 63% of 321 UNT have been identified, compared with 1092 PNT of which 77% have been identified into 86 species. Of the 86 species of PNT, 56 have been chosen for nest building. Chimpanzees at Lopé focused their nest building efforts on seven out of the 10 most frequently used species.

Overall, bonobos appear to be more selective about tree height than chimpanzees. This difference, however, might be caused by the high frequency of the shortest class of PNT at Lomako (an artifact of including trees of DBH down to 2 centimeters) and the high frequency of bonobo day nests, which are constructed considerably higher than bonobo night nests.

Groups of Nests

The term *nest group* is restricted to clusters of nests thought to be built by different individuals at the same time. When construction was not directly observed, a nest group was inferred when the nests were close together and were similar in the degree of decomposition. Most of these studies dealt with single nests or clusters of nests of unknown age.

Comparing the ranges of the number of nests per group from different studies (see table 3), it would appear that chimpanzees and bonobos form sleeping groups of similar size. However, differences become evident when comparing the median number of nests per group for bonobos (M = 7) and for chimpanzees (M = 1–4). Data collected for chimpanzees at Lopé (Wrogemann 1992) and for bonobos at Lomako strongly support this difference in the size of nest groups. At Lopé, single nests accounted for more than 53% of all chimpanzee nest groups whereas at Lomako, more than 96% of all bonobo nest groups consisted of two or more nests, with most groups consisting of 2 to 13 nests (see figure 2).

Further, at Lomako, nest groups of bonobos are always larger than daytime travel parties (see figure 3). Data collected during a third field trip (1993) verify the tendency for fusion of parties at night. In contrast, data collected by Wrangham and Smuts (1980) at Gombe show that chimpanzees do not gather to form large sleeping groups. However, observations from chimpanzees who live in a rain forest are required to decide whether this variation in the size of sleeping groups is related to environmental

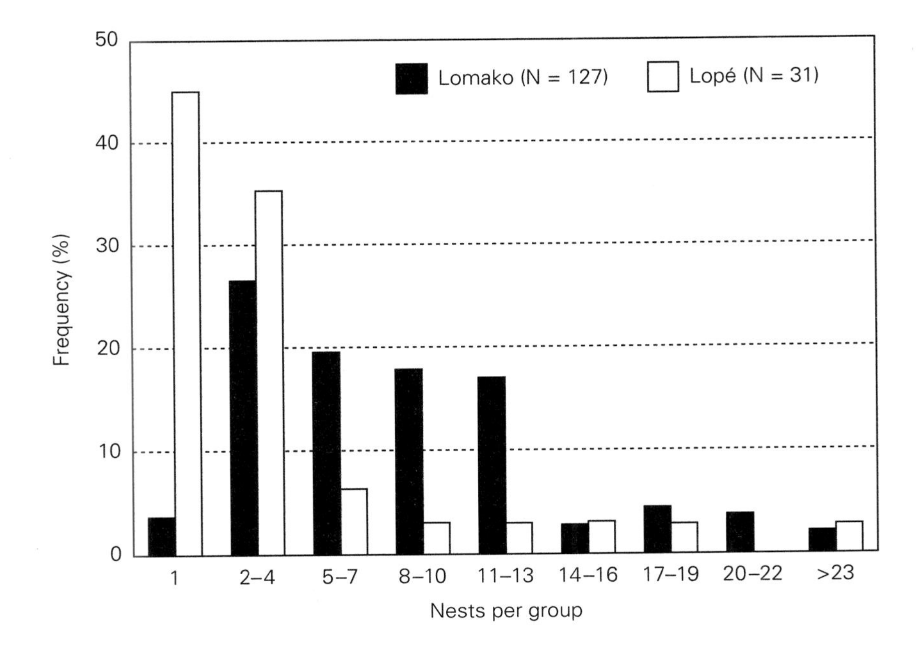

Figure 2
Comparison of nest group size (= number of nests per group) between bonobos at Lomako and chimpanzees at Lopé.

conditions or represents a species-specific pattern. If this disparity in grouping patterns can be confirmed, this would be another striking difference between the two *Pan* species.

Ethology

Ontogeny

Chimpanzees and bonobos build nests every day, and individuals of all age-sex groups participate. Despite the regular and frequent occurrence of nest building, very little is known about the ontogeny of this behavior. Captive chimpanzees separated from mothers soon after birth make only incomplete nests (Bernstein 1961). On the other hand, wild-born immature chimpanzees captured and subsequently raised by human beings have demonstrated the regular occurrence of nest-building activities (Reynolds 1967). Observations suggest that chimpanzees acquire nest-building skills through learning (Bernstein 1961; Baldwin et al. 1981; Goodall 1962), but this proposal remains to be demonstrated by systematic research.

At Lomako, immature bonobos frequently make day nests not by standing in the center of the prospective nest (as adults do), but by bending and folding leaves and twigs in front of them. Moreover, day-nest construction by infants was rarely related to nest-building activities of mature individuals; instead day-nest building by infants occurred when other party

Figure 3
Mean party size of bonobos at Lomako during the day (white bars, N = 247) and at night (black bars, N = 156).

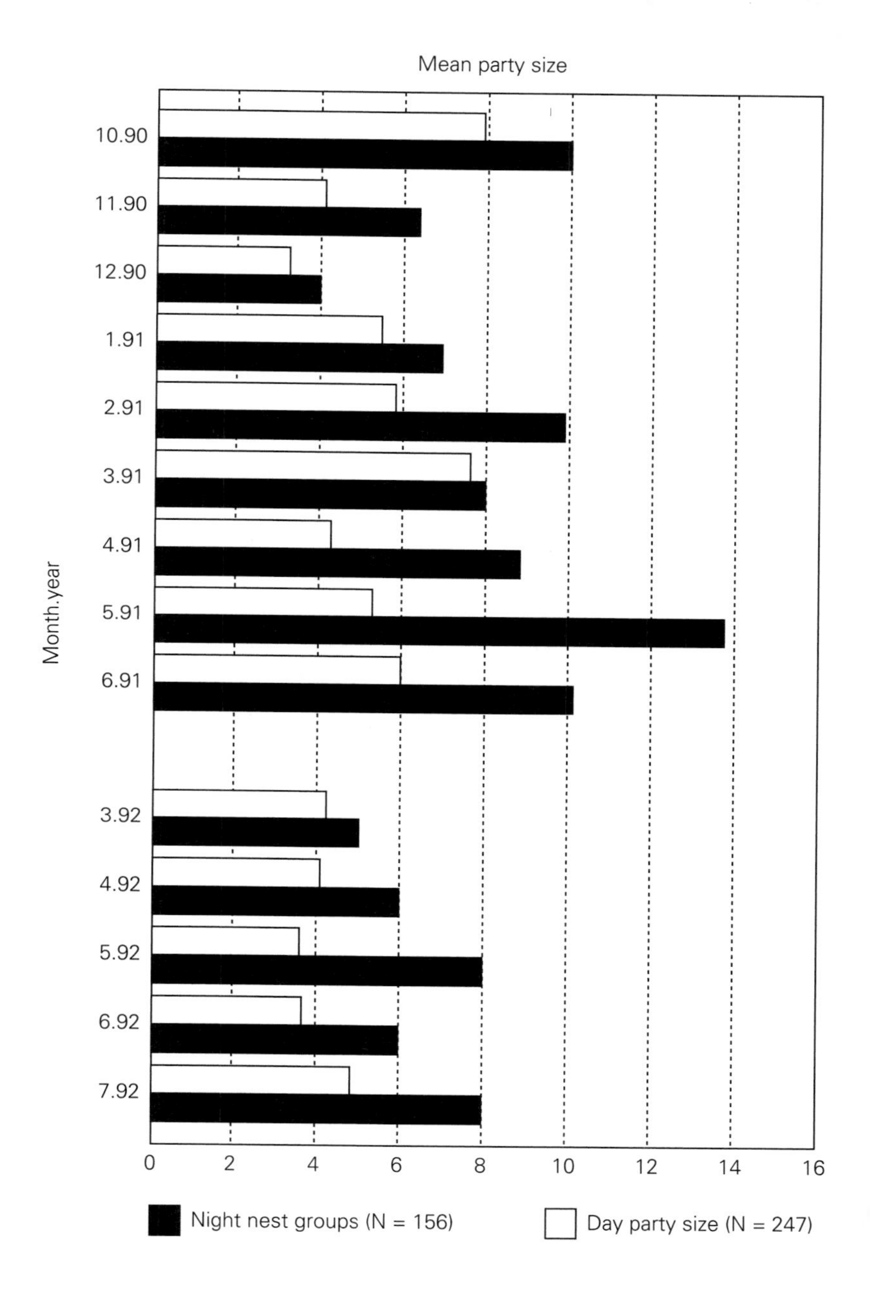

members were feeding or resting. With few exceptions, infants used their day nests not for rest but for solitary or social play. However, observations from Gombe (Goodall 1962) and Lomako show that an infant may participate in the building of a night nest with their mother.

In chimpanzees, the first attempts at nest building were seen at the age of 8 months at Gombe (Goodall 1968) and at 12 to 14 months at other sites (Plooji 1984; Hiraiwa-Hasegawa 1989). Goodall's (1968) description of nest building refers to attempts of an infant to build a ground nest out of twigs and grass blades. In Goodall's example, the infant built the nest not under, but on top of, its body. At Mahale, Hiraiwa-Hasegawa (1989) observed that, initially, infants did not break or bend branches but made nests out of a collection of loose material. Later, the quality and quantity of nest building steadily increased, and the frequency of day-nest construction reached a peak at the age of three (Hiraiwa-Hasegawa 1989). Thus, at Mahale, nest building becomes an important part of the daily behavior long before the infant is weaned.

At Lomako, older infants often approached their mothers while the mothers were resting in a day nest with the obvious intention of making physical contact. Usually, infants did not jump into the nest immediately but stopped at the fringe, making begging gestures (pout face) and related vocalizations (whimpering). In response, mothers usually permitted physical contact. However, the contact very often occurred outside the nest. Although quantitative data are still lacking, the prohibition of offspring from the maternal nest may be part of the weaning process, encouraging infants to build and occupy their own nests. Another example of nest building in response to weaning stress has been provided by Kummer (1991). This event involves a five-year-old female and its mother. When the mother put down the daughter who had been clinging to her body, the infant, "comforted herself by building a flimsy nest of twigs and sucking her toe" (Kummer 1991:80). It is tempting to conclude that, in this case, the nest may have served as a surrogate.

Sex-specific Differences

To investigate the possibility of sex differences in nest-building behavior of bonobos, only nests of mature individuals have been used. Generally, females started the daily process of nest building, while males began later. Considering all nests built by identified adults, males' nests were at an average height above the forest floor of 14.6 meters, significantly lower than the average height of 19.4 meters for female nests (t-test, $p < .01$). When day nests were analyzed separately from night nests, this difference became even more apparent. In addition, both the time required for nest building as well as the duration of use differed: males built day nests faster than females (0.1 minute versus 0.7 minute) and used the day nests for shorter periods of time (25 minutes versus 35 minutes). Finally, females built day nests more often than males (2.88 versus 0.71 per 10 hours of observation). Similar differences have been found by Hiraiwa-Hasegawa (1989) for chimpanzees at Mahale (females vs. males: 0.08 versus 0.01 per 10 hours). Unfortunately, these are the only data on sex differences in nest building of chimpanzees; a comparison between the chimpanzees and bonobos is not yet possible.

Nothing has been published on sex differences in nest building from other wild populations of bonobos, but a recent study of a captive group confirmed our field observations (Berle 1993). Not withstanding the low degree of sexual dimorphism in bonobos, the differences mentioned here seem to be related to environmental rather than social factors. Various explanations are possible: if the quality of potential nesting trees at a given site differs, whoever makes the first choice (females) can occupy the best places. The choice made by males, however, may represent a compromise between two distinct interests: the proximity to particular females and the quality of the nesting site. Another possible interpretation of the males' choice of lower nest sites is that guarding females against ground predators is part of their social task.

Social Context

The function of nests within a social context remains largely unexplored. The major function traditionally ascribed to nests is that they facilitate sleep and rest. This is certainly true for nests that individuals regularly occupy at dusk and, with few exceptions, continue to occupy until the next morning. However, this may not be the limit of nest function. As we know from our own species, beds may serve not only sleep, but a number of other, and often vital, functions. Very little is known about the nocturnal activity of the two *Pan* species, and that part of their life remains, literally, in the dark. What we do know points to important activity. Chimpanzees at Gombe and bonobos at Wamba have been observed to mate in night nests (Goodall 1968; Kano 1992), and chimpanzees are known to travel, and thus leave and, sometimes, reoccupy nests at night (Goodall 1968).

Some additional information on the social context of nest building is available from studies of animals that construct and occupy nests during the day. At Lomako, bonobos used day nests not only for rest but also for a number of other activities. Observations of 176 day nests revealed that only 53% of nest owners used their nests solely for rest. Other activities recorded were feeding (10%), mutual grooming (5%), and social play (3%). In the remaining cases (29%), the nests were occupied by mother-infant pairs, and at least one individual (mostly the infant) performed activities other than rest.

According to Reynolds and Reynolds (1965), unhabituated chimpanzees may build nests to hide from human observers; similar observations have been made for orangutans (Harrison 1962) and gorillas (Reynolds and Reynolds 1965). At Lomako we have observed bonobo nests serving as refuges during interactions between members of the Eyengo-community

(Fruth and Hohmann 1993). These cases (N = 13) occurred when females or immature individuals feeding on highly preferred food were approached by other adult male or female party members. Following the building and occupation of a nest by the first individual, the approaching individual made no further attempt to contact or displace the nest owner, and the nest owner consumed food undisturbed for extended periods of time. Only one case did not involve food. In that case, an adult male escaped a charging display of another adult male by climbing into a tree and building a rudimentary nest. In response, the charging male stopped at the base of the nesting tree and moved away, leaving a dragged branch which he used for display at the base of the nesting tree.

Several observations indicate the social significance of nests as a means of recognizing the presence of conspecifics (Wrangham 1975; Goodall et al. 1978). When, for example, adult chimpanzees encountered fresh nests at the periphery of their home range, they inspected the nests visually and olfactorily and then displayed at the nests destroying them partly or completely. Such instances of redirected aggression closely resemble the violent acts of human beings during conflict and warfare (Eibl-Eibesfeldt 1989). The examples given escape a quantitative approach but supply important anecdotal clues about the social significance of nest building. Neither the legendary goose Martina (Lorenz 1965) nor the Japanese macaque female Imo (Kawamura 1959) would fit the median of the normal distribution but both have lent tremendous input into our understanding of phylogenetic ingenuity (Sommer 1985).

Summary

This chapter deals with the ecological and behavioral differences in nest building of chimpanzees and bonobos. Data from 23 independent field studies have been reviewed. The major findings are organized into three subject categories: single nests, nest groups, and ethology.

Most of the structural characteristics of single nests were similar for both species. For example, chimpanzees and bonobos preferred to build their nests in areas of dense, high vegetation; favored particular tree species; and rarely built night nests in trees offering ripe fruit. Furthermore, nests of both species showed seasonal variations, appearing higher in trees during the rainy season. However, bonobos more often build integrated nests and include trees with small diameters than do chimpanzees. Data from bonobos showed that, as compared with night nests, day nests were higher, less sophisticated in structure, required less time for construction, and were used for shorter periods of time. Overall, structural characteristics of nests and nesting trees for both species show large variation but not necessarily at the level of species specificity.

Comparisons between nest groups of chimpanzees and of bonobos revealed striking differences. While chimpanzees seemed to prefer to rest solitarily or in small groups, bonobos showed a tendency to gather at night, forming sleeping clusters consistently larger than daytime parties. If this pattern is not related to environmental conditions, which remains to be seen, it will represent a significant difference between the two *Pan* species.

Finally, regarding ethology, several points can be made. In both infant chimpanzees and infant bonobos, nest building becomes a daily routine long before weaning. Nesting female bonobos showed a spatial intolerance toward their dependent offspring keeping infants out of the nest, which appears to be part of weaning. Bonobos revealed a number of sex-specific differences: males constructed nests lower than females, males built day nests less frequently and faster, and males used their nests for shorter periods of time. These differences between males and females seem to be related to social rather than to environmental conditions.

Both bonobos and chimpanzees are known to build nests not only for rest but also to serve a number of other functions. Bonobos, for example, have been observed to build nests that served as refuges when potential or imminent conflicts arose. Nests are tools that the apes make to alter their environment and, sometimes, the behavior of conspecifics. Moreover, nests provide tools for humans who want to know more about the lives of their nearest relatives.

Acknowledgments

The authors would like to thank I. Eibl-Eibesfeldt, G. Neuweiler, and D. Ploog for continuous support, encouragement, and advice. We are grateful to P. Heltne and L. Marquardt for the invitation to contribute to this book and to R. Wrangham and E. Altman for comments on this manuscript. Special thanks are due to D. Wrogemann, who gave us permission to include the data from her dissertation. For inspiring discussions, we would like to thank W.C. McGrew and V. Sommer. Fieldwork in Zaire would have been impossible without the help of numerous institutions: the Institut de Recherché Scientifique (Kinshasa), the Centre de Recherché en Sciences Naturelles (Lwiro), the German Embassy (Kinshasa), the GTZ (Kinshasa), and the Catholic Missions at Kinshasa, Bamanya, Boende, and Befale. The help of F. and L. Christiaans, H. Dettmann, J. Jans, C. Kühn, P. Laschan, R. Malenky, E. Ott, N. Thompson-Handler, and E. Young is gratefully acknowledged. For assistance in the field, we would like to thank B. Bolesa,

J.P. Bontamba-Lokuli, M. Ikala-Lokuli, and G. Lomboto. For assistance at our institute, we thank D. Krull, D. Leippert, M. Pötke, and C. Schulte. Financial support was provided by the German Academic Exchange Service, the German Research Council, and the Max Planck Society. Thanks to C. Bonoberts for correcting the English text.

References

Anderson, J.R., E.A. Williamson, and J. Carter. 1983. Chimpanzees of Sapo forest, Liberia: Density, nests, tools and meat-eating. *Primates* 24:594–601.

Badrian, A., and N. Badrian. 1977. Pygmy chimpanzees. *Oryx* 13:463–8.

———. 1984. Social organization of *Pan paniscus* in the Lomako forest, Zaire. In R.L. Susman, ed., *The Pygmy Chimpanzee,* pp. 325–46. New York and London: Plenum.

Baldwin, P.J., J. Sabater Pi, W.C. McGrew, and C.E. Tutin. 1981. Comparisons of nests made by different populations of chimpanzees *(Pan troglodytes). Primates* 22:474–86.

Beck, B.B. 1980. *Animal Tool Behavior.* New York: Garland STPM.

Berle, A. 1993. Nestbauverhalten von Bonobos in Zootierhaltung. Diplom. thesis, Ludwig-Maximilians: University of Munich.

Bernstein, I.S. 1961. Response to nesting materials of wild born and captive born chimpanzees. *Anim. Behav.* 10(1–2):1–6.

Boesch, C. 1978. Nouvelles observations sur les chimpanzés de la fóret de Taï. *La Terre et la Vie* 32:195–201.

Bolwig, N. 1959. A study of the nests built by mountain gorilla and chimpanzee. *South Afr. J. of Sci.* 286–91.

de Bournonville, D. 1967. Contribution à l'étude du chimpanzé en République de Guinée. *Bull. Instit. Fond. Afr. Noire* 29A:1188–1269.

Eibl-Eibesfeldt, I. 1989. *Human Ethology.* New York: Aldine de Gruyter.

Fruth, B. 1990. Nussknackplätze, Nester und Populationsdichte von Schimpansen: Untersuchungen zu regionalen Differenzen im Süd-Westen der Elfenbeinküste. Master's thesis, Ludwig-Maximilians: University of Munich.

Fruth, B., and G. Hohmann. 1993. Ecological and behavioral aspects of nest building in wild bonobos *(Pan paniscus). Ethology.* 94:113–26.

Fruth, B. and G. Hohmann. In Press. Nests: Living artifacts of recent apes? *Curr. Anthro.*

Galdikas, B.M.F. 1982. Orangutan tool use at Tanjung Putin Reserve, Central Indonesian Borneo (Kalimantan Tengah). *J. Human Evol.* 10:19–33.

Ghiglieri, M.P. 1984a. *The Chimpanzees of Kibale Forest: a Field Study of Ecology and Social Structure.* New York: Columbia Univ. Press.

———. 1984b. Feeding ecology and sociality of chimpanzees in Kibale forest, Uganda. In P.S. Rodmann and Y.G.H. Cant, eds., *Adaptation for Foraging in Nonhuman Primates,* pp. 161–94. New York: Columbia Univ. Press.

Goodall, J. 1962. Nest-building behavior in the free ranging chimpanzee. *Ann. N.Y. Acad. Sci.* 102:455–568.

———. 1968. The behaviour of free-living chimpanzees in the Gombe Stream Reserve. *Anim. Behav. Monogr.* 1:163–311.

Goodall, J., A. Bandora, E. Bergmann, C. Busse, M. Hilali, E. Mpongo, A. Pearce, and D. Riss. 1978. Intercommunity interactions of the chimpanzee population of the Gombe National Park. In D.A. Hamburg and E.R. McCown, eds., *The Great Apes,* pp. 13–53. Menlo Park, Calif.: Benjamin/Cummings.

Harcourt, A.H. 1988. Alliances in contests and social intelligence. In R. Byrne and A. Whiten, eds., *Machiavellian Intelligence,* pp. 132–52. Oxford, G.B.: Clarendon.

Harrisson, B. 1962. *Orangutan.* London: Collins.

Hiraiwa-Hasegawa, M. 1989. Sex differences in the behavioral development of chimpanzees at Mahale. In P.G. Heltne and L.A. Marquardt, eds., *Understanding Chimpanzees,* pp. 104–15. Cambridge, Mass.: Harvard Univ. Press.

Horn, A.D. 1980. Some observations on the ecology of the bonobo chimpanzee *(Pan paniscus,* Schwarz 1929) near Lake Tumba, Zaire. *Folia Primatol.* 34:145–69.

Itani, J. 1979. Distribution and adaptation of chimpanzees in an arid area. In D.A. Hamburg and E.R. McCown, eds., *The Great Apes,* pp. 54–71. Menlo Park, Calif.: Benjamin/Cummings.

Jones, C., and J. Sabater Pi. 1971. Comparative ecology of *Gorilla gorilla* (Savage and Wyman) and *Pan troglodytes* (Blumenbach) in Rio Muni, West Africa. *Bibl. Primatol.* 13:1–96.

Kano, T. 1979. A pilot study of the ecology of pygmy chimpanzees, *Pan paniscus.* In D.A. Hamburg and E.R. McCown, eds., *The Great Apes,* pp. 123–35. Menlo Park, Calif.: Benjamin/Cummings.

———. 1983. An ecological study of the pygmy chimpanzees *(Pan paniscus)* of Yalosidi, Rep. of Zaire. *Internat. J. Primatol.* 4:1–31.

———. 1992. *The Last Ape.* Palo Alto, Calif.: Stanford Univ. Press.

Kawamura, S. 1959. The process of sub-culture propogation among Japanese macaques. *Primates* 2:43–60

Kummer, H. 1991. Evolutionary transformations of possessive behavior. In F.W. Rudmin ed., *To Have Possessions: A Handbook on Ownership and Property* (Special Issue). *J. Social Behav. and Personal.* 6(6):75–83.

Lorenz, K. 1965. *Uber tierisches und menschliches Verhalten. Aus dem Werdegang der Verhalten slehre I and II.* Munich: Piper.

Malenky, R.K., and E.W. Stiles. 1991. Distribution of terrestrial herbaceous vegetation and its consumption by *Pan paniscus* in the Lomako Forest, Zaire. *Am. J. Primatol.* 23(3):153–69.

McGrew, W.C. 1985. The chimpanzee and the oil palm: Patterns of culture. *Social Bio. and Human Affairs* 50:7–23.

———. 1992. *Chimpanzee Material Culture: Implications for Human Evolution.* Cambridge, England: Cambridge Univ. Press.

Nissen, H.W. 1931. A field study of the chimpanzee. Observations of chimpanzee behavior and environment in Western French Guinea. *Comp. Psychol. Monogr.* 8:1–122.

Plooji, F.X. 1984. *The Behavioral Development of Free-Living Chimpanzee Babies and Infants.* Norwood N.J.: Ablex Publishing.

Reynolds, V. 1967. *The Apes.* London: Cassell.

Reynolds, V., and F. Reynolds. 1965. Chimpanzees of the Budongo forest. In de Vore, ed., *Primate Behavior,* pp. 368–424. New York: Holt, Rinehart, and Winston.

Schaller, G.B. 1963. *The Mountain Gorilla.* Chicago: Univ. of Chicago Press.

Sept, J.M. 1992. Was there no place like home? A new perspective on early hominid archaeological sites from the mapping of chimpanzee nests. *Current Anthro.* 33:187–207.

Tutin, C.E., and M. Fernandez. 1983. *Recensement des Gorilles et des Chimpanzés du Gabon.* Franceville, Gabon: Centre International de Recherches Médicales de Franceville.

Tuttle, R.H. 1986. *Apes of the World.* New Jersey: Noyes.

White, F.J. 1988. Party composition and dynamics in *Pan paniscus. Internat. J. Primatol.* 9:179–93.

Wrangham, R. 1975. The behavioral ecology of chimpanzees in Gombe National Park, Tanzania. Ph.D. diss., Cambridge University.

Wrangham, R., and B.E. Smuts. 1980. Sex differences in the behavioral ecology of chimpanzees in the Gombe Stream National Park, Tanzania. *J. Reprod. Fert. (Suppl.)* 28:13–31.

Wrogemann, D. 1992. Wild Chimpanzees in Lopé, Gabon: Census Method and Habitat Use. Ph.D diss., Bremen University, Germany.

Diversity of Medicinal Plant Use by Chimpanzees in the Wild

Michael A. Huffman and Richard W. Wrangham

Introduction

The presence of secondary compounds, many of which may have medicinal value and/or may be toxic, is known to influence the kinds of plants that primates select as food (Glander 1982). The major focus of interest about plant secondary compounds in the diet of primates has been on how and why primates cope with their presence (Glander 1975; Hladik 1977a, 1977b; Janzen 1978; McKey 1978; Milton 1979; Oates et al. 1977; Oates et al. 1980; Wrangham and Waterman 1981).

Janzen (1978) first suggested the possibility that the incidental ingestion of secondary compounds may help combat pathogens and parasites. Pathogens and parasites can cause a variety of diseases, affecting the overall behavior and reproductive fitness of an individual (Hart 1990; Holmes and Zohar 1990). Therefore, the need to counteract such diseases should be great. However, it is difficult to distinguish the relative nutritional values and the relative medicinal values of plant foods, containing active secondary compounds, that occur in the diet of primates (Glander 1982; Janzen 1978; Phillips-Conroy 1986).

Various sources of evidence have been used to suggest that certain plant foods found to contain toxins are selected for their medicinal value. This evidence includes restriction of plant use to areas associated with high risk of parasitic infection (Phillips-Conroy 1986), a habit of ingestion that does not contribute important nutritional benefits (Wrangham and Nishida 1983; Huffman and Seifu 1989), apparent sickness of the individual at the time of ingestion (Huffman and Seifu 1989), association of use with seasonal periods of demonstrated, high risk from parasitic infection (Huffman 1991; Huffman et al. 1992; Kawabata and Nishida 1991), and the low-frequency

intake of plant species that are not a regular part of the diet (Wrangham and Nishida 1983; Huffman and Seifu 1989; Wrangham and Goodall 1989).

In recent years, a growing body of new evidence supporting the possibility of self-medication in animals has given a burst of momentum to a field of study that we call *zoopharmacognosy* (Rodriguez and Wrangham 1992, 1993). Chimpanzees have provided more evidence of self-medication than other primates (Rodriguez and Wrangham 1992).

In this chapter, we will discuss the current state of research on this topic with respect to chimpanzees. We will also review related new findings and observations from a number of multidisciplinary investigations. The use of plants for possible medicinal value, in particular pith chewing and swallowing of leaves whole in the Mahale, Gombe, and Kibale populations of chimpanzees in Tanzania and Uganda is discussed and compared.

Comparison of the Study Areas

Three main study sites and six groups are the focus of this discussion: Mahale (M group and the now extinct K group), Gombe (Kasakela, Kahama), and Kibale (Kanywara, Ngogo). They are located at approximately the same longitude (30° E) along the Rift Valley, at roughly 6°17', 4°40', and 0°32' south latitudes respectively (see figure 1).

Figure 1

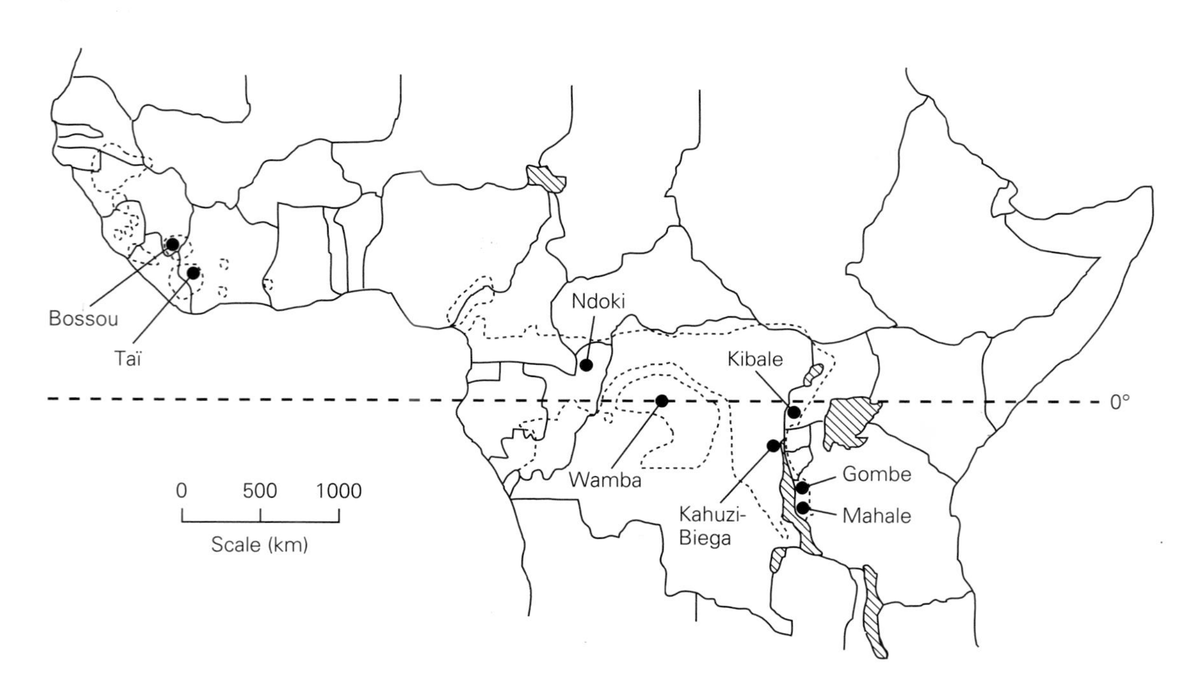

Mahale and Gombe, situated about 150 kilometers apart on the eastern shore of Lake Tanganyika, share a common climate, influenced by weather from the lake and from the mountainous terrain, ranging from 772 meters to 1,500–2,000 meters above sea level. Chimpanzees are supported mainly by the semideciduous gallery forests between 780 and 1,300 meters above sea level. At both sites, the year is divided into two distinct seasons, the rainy season and the dry season. The rainy season lasts from mid-October to mid-May with an average annual rainfall of approximately 1,800 millimeters for both sites (Collins and McGrew 1988; Goodall 1986; Moore 1992; Nishida 1990).

Kibale, on the other hand, is located inland at an altitude of 1,500–1,700 meters above sea level. The terrain is hilly and includes elements of lowland tropical rain forest, montane rain forest, and mixed deciduous forest. The year is divided into four distinct seasons: two rainy seasons from March to May and from September to November with relatively drier periods in between. The average annual rainfall of 1,500–1,800 millimeters is similar to that of Mahale and Gombe (Wing and Buss 1970; Struhsaker 1975; Ghiglieri 1984).

Kibale has a cooler, more equable year-round climate, as compared to the more distinct differences between the dry and rainy seasons at Mahale and Gombe. These differences in moisture and temperature should have some effect on the timing of reinfection, on the relative degree or severity of parasitic infection, and on other factors that may be responsible for some of the variation between sites in the medicinal use of plants.

Comparing Mahale and Gombe, Gombe and Kibale, and Kibale and Mahale shows that variability in diet between sites is great. Comparing the number of in-common food items actually eaten at two sites versus the number of in-common food items available at the same two sites shows an overlap of 36% (104 of 286) for Mahale and Gombe (Nishida et al. 1983), 23% (17 of 75) for Gombe and Kibale, and 54% (32 of 54) for Kibale and Mahale (Wrangham and Isabirye-Basuta in prep.). While a larger total number of in-common food species are found at Mahale and Gombe, the Kibale and Mahale chimpanzees were found to share a larger total number of in-common food items in their diets.

Medicinal Use of Plants by Chimpanzees

Thus far, 13 plant species from 10 genera and eight families have been identified as possibly being used by Mahale, Gombe, and Kibale chimpanzees for their medicinal value. Table 1 shows these species with a partial listing of the known ethnomedicinal and/or pharmacological properties of plant parts ingested and the study sites at which they can be found.

Table 1

Plant pith and leaves ingested possibly for their medicinal value by chimpanzees at Mahale, Gombe, and Kibale.

Method of ingestion / Family / Genus and species	Ethnomedicinal information on plant part ingested	Pharmacology of plant part ingested	Plant part ingested		
			Mahale	Gombe	Kibale
Pith Chewing:					
Compositae					
Vernonia amygdalina (Del.) #*V. colorata* (Willd.) Drake	anthelminthic, epidermal infections, malarial fever, colic (1)	antischistosomal, leshmanicidic, antiplasmodial, antitumor, antibiotic (1, 2, 3, 4, 5, 6)	+ P*	+# P	—
Leaf Swallowing:					
Commelinaceae					
Commelina diffusa? Burm. f.	eye wash, toxic to livestock if overeaten (7)	ND	+ L**	—	—
Aneilema aequinoctiale (P. Beauv.) Loudon	ND	ND	—	—	+ L**
Compositae					
Aspilia mossambicensis (Oliv.)	anthelminthic, topical antiseptic, febrifuge, stomach upset, epidermal infections, ringworm galactagogue, menstrual cramps (8, 9, 10)	antifungal, anti-nematodal, antibiotic, uterotary stimulatory effect, antibacterial, antihepatotoxic (5, 8, 10, 11, 12, 13)	+ L**	+	—
A. pluriseta (O. Hoffm.) Wild	epidermal infections, topical antiseptic (8, 11)	antifungal, anti-nematodal, antibiotic uterotary stimulatory effect, antibacterial, antihepatotoxic (8, 10, 11, 12, 13)	—	+ L**	—
A. rudis Oliv. and Hiern	ND	ND	—	+ L**	—

(Table continues on next page)

The bitter piths of two plants are chewed, and the leaves of 11 plants are swallowed whole without being chewed. Other ingested items of possible medicinal value to be discussed are bark, cambium, live and dry wood, termite mound clay, and a saponin-rich berry.

Method of ingestion Family Genus and species	Ethnomedicinal information on plant part ingested	Pharmacology of plant part ingested	Plant part ingested		
			Mahale	Gombe	Kibale
Leaf Swallowing continued:					
Malvaceae					
Hibiscus aponeurus Sprague and Hutch	ND	ND	+ L	+ L**	—
Melastomataceae					
Melastomastrum capitatum Vahl A. and R. Fernandes	stomach upset (14)	ND	+ L**	—	—
Moraceae					
Ficus exasperata Vahl	colic, cough, component of arrow poison poisonous to livestock, for kidney pain (7, 15)	antinematodal, insecticidal (5, 7, 16)	+ L** F	+ L? F	+ L? F
Rubiaceae					
Rubia cordifolia L.	locally cultivated in garden for stomach ailments (9)	rubiatrol, bioactive anthraquinones, cyclic hexapeptide (14, 17)	—	—	+ L**
Ulmaceae					
Trema orientalis (L) Blume	anthelminthic, emetic, jaundice (7, 15)	insecticidal, high tannin alkaloid content, detoxify alkaloids? (5, 18)	+ L**	+	+
Verbenaceae					
Lippia plicata Baker	stomach upset, menstrual cramps (19)	some/no biocidal activity detected (5, 19)	+ L**	+ P	—

Notes: + = species recorded; — = species not recorded; * = of suspected medicinal values; ** = leaf swallowed whole; ? = leaf swallowing has not been observed. Plant parts ingested: L = leaf, F = fruit, P = pith. ND = no data.

Sources: 1. Huffman and Seifu (1989) 2. Jisaka et al. (1992b) 3. Jisaka et al. (1993) 4. Kirby and Allen (unpub. data) 5. Ohigashi et al. (1991) 6. Timmon-David (unpub. data) 7. Abbiw (1990) 8. Rodriguez et al. (1985) 9. Kokwaro (1987) 10. Page et al. (1992) 11. Wrangham and Goodall (1989) 12. Lwande et al. (1985) 13. Yang et al. (1986) 14. M.S. Kalunde (pers. com.) 15. Terashima et al. (1991) 16. Rodriguez et al. (unpub. data) 17. Itokawa et al. (1983) 18. Oates et al. (1977) 19. Takasaki and Hunt (1987).

Bitter-Pith Chewing

Nishida and Uehara (1983) list the pith of *Vernonia amygdalina* (Del.) as a food item ingested by apparently healthy members of M group and members of the former K group. The hypothesis that this species has medicinal value for chimpanzees began with detailed observations of an ailing female's use of the plant at Mahale in November of 1987. The female meticulously removed the leaves and outer bark from several young shoots and chewed on the exposed pith, sucking out the extremely bitter juice. Within 24 hours, she had fully recovered from a lack of appetite, malaise, and constipation (Huffman and Seifu 1989).

A similar incident was observed in December of 1991 and, in this case, parasite levels dropped noticeably within 20 hours after an adult female chewed *V. amygdalina* pith (Huffman et al. in press). In both cases, the rate of recovery within 20–24 hours was comparable to that of the Tongwe local human inhabitants who use this plant as a treatment for similar symptoms. Among many African peoples, this plant is prescribed treatment for stomachaches and intestinal parasites (Burkill 1985; Dalziel 1937; Watt and Breyer-Brandwijk 1962).

Detailed information on the use of *V. amygdalina* by chimpanzees prior to these observations is limited. At Gombe, a chimpanzee was observed to ingest the equally bitter pith of *V. colorata* (Willd.) Drake (Wrangham 1975). Whether the individual was ill at the time is not known. *V. colorata* and *V. amygdalina* are very closely related species, not distinguished from each other by traditional African healers with regard to medicinal properties and folk classification (Burkill 1985, and see table 1).

Phytochemical analysis of *V. amygdalina* collected from Mahale revealed the presence of two major classes of bioactive compounds: the sesquiterpene lactones and the newly discovered group, steroid glucosides (Ohigashi et al. 1991; Jisaka et al. 1992a). The compounds from both groups are present in varying concentrations in the leaf, bark, root, and pith (Jisaka et al. 1992b).

The sesquiterpene lactones present in *V. amygdalina,* in *V. colorata,* and in a number of other *Vernonia* species, are well-known compounds with demonstrated anthelmintic, antiamoebic, antitumor, and antibiotic properties (Asaka et al. 1977; Gasquet et al. 1985; Jisaka et al. 1992b; Jisaka et al. 1993; Kupchan et al. 1969; Toubiana and Gaudemer 1967). *In vitro* tests on the antischistosomal activity of the pith's most abundant steroid glucoside (vernonioside B1 and its aglycones) and sesquiterpene lactones (vernodaline, vernolide, hydroxyvernolide, vernodalol) showed significant inhibition of movement of adult parasites and of the adult female's egg-laying capacity (Jisaka et al. 1992b). Preliminary tests on the antiparasitic activity of the sesquiterpene lactones were done by Dr. P. Timon-David of the Université D'Aix, Marseilles, France, and by Drs. G.C. Kirby, D. Allen, and associates of the University of London, England, who are colleagues in the CHIMPP Group (Chemo-ethology of Hominoid Interactions with Medicinal Plants and Parasites) These tests using *Leishmania donovani infantum* and a K1 multidrug-resistant strain of

Plasmodium falciparum (C.W. Wright, G.C Kirby, D. Allen, D.C. Warhurst, and J.D. Phillipson unpubl. obs.) have demonstrated significant antiparasitic activity.

Hypotheses

Given the abundance of cytotoxic and otherwise pharmacologically active secondary compounds in the pith, its low frequency of ingestion, the association of its use with illness, and its uncommonly bitter taste as compared to other chimpanzee food items, the nutritional value of *V. amygdalina* or *V. colorata* for chimpanzees is not likely to be significant when compared to their pharmacological effect.

Based on the behavioral, ethnomedicinal, and phytochemical evidence described above, Huffman hypothesized that chimpanzees use the *Vernonia* species for the control of parasite-related disease (Huffman and Seifu 1989; Huffman 1991; Huffman et al. 1992).

At Mahale, despite year-round availability, chimpanzees tend to use *Vernonia* species during the rainy season (Nishida and Uehara 1983; Huffman et al. 1990; Huffman et al. 1992). Also, during the rainy season, the number of individuals with parasitic infections tends to increase, as does the number of kinds of parasitic infections (Huffman et al. in prep.). Indeed, given the broad range of pharmacological activity found in *V. amygdalina,* the benefits of its use may be quite diverse (Huffman and Seifu 1989 and see table 1).

Leaf-Swallowing

Leaf-swallowing behavior was first reported in detail by Wrangham (1975, 1977) and then by Wrangham and Nishida (1983) from the perspective of its possible non-nutritional value for chimpanzees. The rough-surfaced leaves of *Aspilia mossambicensis* (Oliv.), *A. pluriseta* (O. Hoffm.) Wild, and *A. rudis* Oliv. and Hiern are usually selected one at a time and placed into the mouth, where they are not chewed. Instead, they are apparently rubbed with the tongue against the inside of the mouth for about five seconds and then swallowed. Leaves are typically swallowed within an hour of dawn and are found whole and undigested in the feces. At Mahale, however, the leaves are chewed when ingested in the afternoon (Nishida unpubl. observ.). In addition to this peculiar habit of ingestion, the presence of powerful bioactive compounds in the leaves of *A. mossambicensis* and *A. pluriseta* suggest the possible medicinal value of these plants for chimpanzees (Page et al. 1992; Rodriguez et al. 1985; Wrangham and Goodall 1989).

Eight other plant species, observed to be ingested in the same way at Mahale, Gombe, and Kibale have been identified (see table 1). It is the similar habit of ingesting whole leaves of these species that suggest they too are used for something other than their nutritional value. While detailed behavioral data regarding their use by chimpanzees is still limited, all of the following have been observed to be swallowed whole and in the early morning hours: *Lippia plicata* Baker (Takasaki and Hunt 1987), *Commelinacea* species *Commelina diffusa?* Burm. f., *Ficus exasperata* Vahl (Newton and Nishida 1990), *Trema orientalis* (L) Blume (K. Kawanaka, unpubl. observ.; first observer), *Aneilema aequinoctiale* (P. Beauv.) Loudon (Wrangham unpubl. data), and *Hibiscus aponeurus* Sprague and Hutch (Wrangham and Goodall 1989). As for *Rubia cordifolia* L. (Wrangham and Goodall 1989) and *Melastomastrum capitatum* Vahl A. and R. Fernandes (M.S. Kalunde pers. com.), their use has been confirmed only from the presence of whole leaves in the feces.

At five other sites across Africa recent evidence of whole-leaf swallowing behavior also has been found in three populations of two subspecies of chimpanzees (*Pan troglodytes schweinfurthii,* and *P.t. verus*), in one population of bonobos (*P. paniscus*), and one population of Eastern lowland gorillas (*Gorilla gorilla graueri*) (see figure 1). For chimpanzees, leaf swallowing has been observed at Bossou, Guinea, using *Ficus mucuso* Ficalho (M. Nakamura pers. com., first observ.) and *Polycephalium capitatum* (Baill.) (T. Matsuzawa pers. com.); at the Taï Forest National Park, Ivory Coast using *Manniophyton fulvum* Mull. Arg., *Tristemna coronatum* Benth., and *Dichaetanthera africana* (Hook. f.) Jacques-Félix (= *Sakersia africana)* (C. Boesch in prep.), and at the Kahuzi-Biega National Park, Zaire using *Commelina cecilae* C.B. Clarke, *Ipomea involucrata* P. Beauv., and *Lagenaria abyssinica* (Hook. f.) C. Jeffrey (J. Yamagiwa pers. com.). For bonobos, evidence comes from Wamba, Zaire for the use of *Manniophyton fulvum* and two unidentified fern species (S. Kuroda pers. com.). For Eastern lowland gorillas evidence comes from the Kahuzi-Biega National Park, Zaire for the use of *Commelina cesilae* (J. Yamagiwa pers. com.).

The literature contains clues to other possible benefits for chimpanzees from the consumption of medicinal plants. *C. diffusa?*, one of the species whose leaves are swallowed whole in the early morning hours, has been reported by Russo (1992) to be taken as a tea before breakfast by the Shuar of the Ecuadorian Amazon as a common treatment for headaches. Chimpanzees may be obtaining a similar effect.

Based on Nishida's observations of *F. exasperata* leaf swallowing at Mahale (Newton and Nishida 1990), Rodriguez et al. (unpubl. data) investigated this species in Kibale and isolated 5-methoxypsoralen, a well-known furanocoumarin of potent biological activity, from its leaves. Rodriguez and colleagues established that this compound showed effective nematicidal activity at a concentration of 13 micrograms per milliliters. Most abundant in young leaves according to observations by Barrera (1989), a quantity of 5-methoxypsoralen sufficient for nematicidal activity could be obtained by consuming 50–100 young leaves. However, it is not known whether chewing, versus swallowing the leaves whole, could alter the pharmacological activity of 5-methoxypsoralen.

The leaves of *F. exasperata* are often chewed at all three study sites and are normally considered to provide only nutritional value. However, in Mahale in 1991 a large quantity of these leaves was consumed over several days by an adult male suffering from an acute parasitic infection of nematodes (*Strongyloides fuelleborni, Trichuris trichura, Ternidens* species). Subsequently, there was a sharp drop in the male's parasite load, along with a visible improvement in his health (Huffman et al. in press). This suggests that *F. exasperata* leaves may have medicinal value. However, further field observations of this kind are necessary before coming to any definite conclusions.

Hypotheses

Based on the presence of thiarubrine A, a powerful antibiotic and anthelminthic diathiane polyine, found in the leaves of two *Aspilia* species (Rodriguez et al. 1985; Towers et al. 1985, Page et al. 1992), Wrangham and Goodall (1989) have suggested that chimpanzees use these plants for the expulsion of parasites. While this hypothesis has yet to be investigated in detail, it is supported by data from Mahale showing a seasonal peak in the use of *A. mossambicensis* in January and February (Wrangham and Nishida 1983) which coincides with an increase in the number of parasitic infections during the rainy season (Huffman et al. 1990; Kawabata and Nishida 1991). This hypothesis is also strengthened by evidence that many of the leaves swallowed have important pharmacological activity and are used as ethnomedicines (see table 1).

A second hypothesis for the use of *A. pluriseta* and *A. rudis,* Page et al. (1992) suggest that the uterostimulatory effect of kaurenoic acid and grandiflorenic acid, two diterpenes, traditionally used by humans as an abortifacient, or inducer of labor, may be exploited by female chimpanzees as a regulator of fertility. They support this hypothesis with observations at Gombe that show that females ingested the leaves of these two species at significantly higher rates than did males (Wrangham and Goodall 1990).

Currently, there is some suggestion of the effects of secondary compounds in plants as reproductive modulators in muriquis (Strier 1993), vervet monkeys (Garey et al. 1992), and howler monkeys (Glander 1992).

Page et al. (1992) have been able to confirm the presence of thiarubrine-A in the roots, but not in the leaves, of *Aspilia* species collected from Mahale and Gombe. On the other hand, Page (pers. com.) thinks that the antibacterial and antihepatotoxic activities of kaurenoic acid (Lwande et al. 1985; Yang et al. 1986) may be of general therapeutic value to males and nonreproductive individuals using these leaves.

Also swallowed whole both by chimpanzees in the Taï Forest and by bonobos at Wamba, the leaves of *Manniophyton fulvum* are used in parts of Zaire as a treatment for diarrhea in humans (D.N. Muanza pers. com.).

A third hypothesis for leaf swallowing is that chimpanzees select the leaves of certain species for physical rather than chemical properties. All the leaves swallowed by chimpanzees are bristly haired hispid or have an otherwise semirough surface, whereas most leaves in their habitats are smooth (glabrous) or not noticeably rough (Wrangham pers. observ.). Wrangham and Goodall (1989) argued against this hypothesis because some notably rough-surfaced leaves were not swallowed by chimpanzees. They cited in particular *F. exasperata,* which was not known to be swallowed at the time. But *F. exasperata* leaves have now been observed to be swallowed in Mahale (see table 1). As the list has increased from four species (Wrangham and Goodall 1989) to 13 species (see table 1) roughness has remained a constant property of swallowed leaves, with the exception of one species, *D. africana.* This suggests that roughness is more important than indicated earlier and it raises the possibility that leaf-swallowing behavior by chimpanzees is functionally similarity to the grass-swallowing behavior of carnivores, which has no apparent chemical significance.

The pharmacological activity characteristic of swallowed leaves suggest, however, that leaf roughness alone is an insufficient criterion for selection by chimpanzees. Furthermore, chimpanzees swallow the leaves of only certain species within a genus while ignoring those of other species of the same genus with apparently very similar leaf surfaces. Examples includes *Aspilia* (Wrangham and Goodall 1989) and *Aneilema* (Wrangham pers. observ.). In addition, the leaves of some species are swallowed only at certain times of day. Chemical differences seem more likely than physical differences to account for such selectivity, but data are needed to test this. At present, it is safest to assume that both physical and chemical properties influence selection of leaves for swallowing.

The physical properties of leaves may be important in allowing targeted delivery of drugs that would otherwise be either detoxified by the liver or inactivated by the acidity of the stomach. Rodriguez and Wrangham (1992) propose that the *Aspilia* leaf is strong enough to act as a capsule, enabling the antiparasitic thiarubrine A to survive passage through the stomach. This is the only hypothesis to suggest why leaves are swallowed rather than chewed. On the other hand, Newton and Nishida (1990) noted that chimpanzees appear to rub *Aspilia* leaves against the inside of the mouth and thereby allow drugs to be absorbed directly into the blood stream without going through the stomach. This hypothesis would be supported by evidence that drugs are located in appropriate sites on the leaf as, for example, in the leaf hairs of many Compositae species (Rodriguez pers. com.).

The rough surface of leaves makes them difficult to swallow. Chimpanzees normally fold the leaves several times with their tongues in the process of pulling the leaves into the mouth. This behavior was interpreted as a way to make swallowing easier (Wrangham and Goodall 1989).

Other Food Items with Possible Pharmacological Significance

Other food items ingested by chimpanzees with potentially interesting pharmacological significance include berries rich in secondary compounds, bitter leaves, inner bark (cambium), live and dry wood, and termite-mound clay (Ghiglieri 1984; Nishida and Uehara 1983; Wrangham 1975).

The bitter-tasting berries of *Phytolacca dodecandra* L Herit are an abundant and frequently ingested food item of the Kanyawara group in Kibale (Wrangham and Isabirye-Basuta in prep.). Interestingly, these berries are a concentrated source of at least four toxic triterpenoid saponins (lemmatoxin, lemmatoxin-C, oleanoglycotoxin-A, and phytolacca-dodecandra glycoside). Ingestion of about two grams of these saponins by mice and rats is fatal. Found in the highest concentration in the berry, these compounds are now being developed in the United States to control the levels of schistosome-carrying snails (Abbiw 1990; Kloos and McCullough 1987). Other known properties of these triterpinoid saponins include antiviral, antibacterial, antifertility, spermicidal and embryotoxic activities (Kloos and McCullough 1987).

The bitter leaves of *Thomandersia laurifolia* (T. Auders. ex Benth.) Baill. are occasionally ingested (chewed) by Western lowland gorillas *(Gorilla gorilla gorilla)* in the Ndoki forest of Northern Congo. The local, human inhabitants use these leaves as a treatment for parasites and fever (S. Kuroda unpubl. observ.).

Bark and wood are characteristically highly fibrous, heavily lignified, sometimes toxic, relatively indigestible, and poor in nutrients (Waterman 1984). Chimpanzees at Mahale ingest bark and wood of several plant species including *Brachystegia bussei* Harms, *Sesbania sesban* (L.) Merrill, *Pycnanthus angolensis* (Welw.) Warb., *Ficus sur* Forssk. (= *F. capensis* Thunb.) (Nishida and Uehara, 1983) and *Grewia platyclada* K. Schum (Huffman unpubl. data); chimpanzees at Gombe ingest wood and bark of species such as *B. bussei* and *Entada abyssinica* Steud. ex A. Rich. (Wrangham 1975); and the chimpanzees at Kibale ingest the species *Ficus natalensis* Hochst., *Markhamia platycalyx* (Bak.) Sprague, and *Pterygota mildbraedii* Engl. (Wrangham and Isabirye-Basuta in prep.).

The African ethnomedicine literature suggests a variety of possible medicinal uses for some of these. The bark of *P. angolensis* is used as a purgative, laxative, digestive tonic, emetic, reliever of toothaches, and poison antidote (Abbiw 1990). Bark strips of *G. platyclada* are sometimes chewed for the relief of stomachaches (Mohamedi Seifu Kalunde pers. com.). *E. abyssinica* is used for diarrhea and as an emetic (Abbiw 1990).

The consumption of termite-mound clay and geophagy in general has been suggested to function as an adsorbent of tannins and other toxins that accumulate in the stomach of chimpanzees (Hladik 1977a, 1977b).

Variation in Medicinal Plant Use Between Sites

Variation between sites in medicinal use of plants is largely due to availability of plant species. Of the species listed in table 1 under the heading of leaf swallowing, 54% (6 of 11) are found at only one of the three study sites. While the remaining five plant species are found in at least two of the sites, the leaves of each species are swallowed whole at only one of the sites; however, as investigations continue at these sites this may change. At the other sites, the plant is either ingested in a different manner or completely ignored (see table 1).

At Gombe, chimpanzees are reported to take the leaves of *Aspilia pluriseta* most frequently within an hour after leaving their sleeping nests. When they do, it is frequently the first food item of the day to be ingested. At Mahale, however, the leaves of *A. mossambicensis* are swallowed in the morning and chewed in the afternoon or evening (Wrangham and Nishida 1983; T. Nishida unpubl. observ.).

The leaves of *H. aponeurus* are swallowed at Gombe while they are chewed at Mahale. The leaves of *F. exasperata* are swallowed whole in the early morning and chewed in the afternoon at Mahale, while they are only chewed at Gombe and Kibale (Newton and Nishida 1990; Wrangham 1975; Wrangham and Isabiyre-Basuta in prep.). At Mahale, the leaves of *L. plicata* are swallowed whole, while at Gombe only the pith is ingested.

The leaves of *T. orientalis* are swallowed whole at Mahale, but despite the abundance of this species in Gombe and Kibale, there is no evidence that these chimpanzees use any part of this plant (Clutton-Brock and Gillette 1979; Wrangham and Isabiyre-Basuta in prep.).

Variation in the Use of Medicinal Plants Between Groups of a Single Site

A different picture appears when we look closely at the behavior of neighboring groups at the same study site. At Mahale, both K group and M group chimpanzees are known to have chewed the pith of *V. amygdalina.* Likewise, both K and M groups swallowed whole leaves of *A. mossambicensis* at all times of the day. At Gombe, the Kasakela and Kahama groups were both known to swallow whole leaves of *A. pluriseta* and *A. rudis* significantly more often in the early morning hours (Wrangham and Goodall 1989).

At Kibale, leaves of *R. cordifolia* and *A. aequinoctiale* are swallowed whole by both the Kanyawara and Ngogo group of chimpanzees, at similar rates.

The above data show that chimpanzees of neighboring groups use medicinal plants in a similar fashion, suggesting possible transmission of knowledge between groups. Because it is the female that transfers between chimpanzee unit groups (Goodall 1986; Nishida and Kawanaka 1972), the prime route for transmission of medicinal-plant-use behavior between groups is most likely by way of immigrant females.

Acquisition of Medicinal-Plant-Use Behavior

Perhaps one of the biggest puzzles is the way in which medicinal-plant-use behavior is acquired by chimpanzees. Both when chewing bitter pith and when swallowing abrasive bristly haired leaves, chimpanzees must counter their normal tendency to avoid food items that are toxic or difficult to swallow.

Leaf-swallowing behavior has now been found to occur widely among chimpanzees including at least two of the three chimpanzee subspecies, the bonobo, and the Eastern lowland gorillas—even where mutual contact does not occur with chimpanzees (see table 1 and figure 1). Assuming that the medicinal-use hypothesis proves correct, the lack of mutual contact suggests that there is mechanism common to the African apes for the recognition of plants that have medicinal value.

If such a mechanism were in operation, we could expect chimpanzees from different areas to select leaves with similar characteristics and/or pharmacological activity. That this occurs is supported by the coincidence of the use of species from the same or related genera at different sites including two species of *Vernonia* at Mahale and Gombe; three species of *Aspilia* at Mahale and Gombe; three species of *Commelina-Aneilema* at Mahale, Kibale, and Kahuzi-Biega; two species of *Ficus* at Mahale, Bossou, and Taï Forest; and two species of *Melastomastrum-Tristema* at Mahale and Taï Forest (see table 1).

One possible explanation for the convergence on particular plants is that, when ill, chimpanzees and the other great apes may innately associate certain tastes with pharmacological activity. Continued use might be based on a feed-back mechanism that tells the individual when the medicine is no longer necessary. Huffman and Seifu (1989) have argued that the strong, bitter taste of *V. amygdalina* may be an important factor in the selection of this plant by Mahale chimpanzees. Historically, herbalists have emphasized the importance of taste and smell in the evaluation of the properties of medicinal plants (Crellin and Philpott 1989), suggesting that humans rely on such cues to select these plants.

It is a novel possibility that chimpanzees become either less aversive to ingested toxins or seek them out as a result of being ill. The principle that physical condition may influence the degree of aversion to toxins appears reasonable because, in a similar way, toxin aversion has been reported to increase as a result of pregnancy in humans. In the case of human pregnancy increased toxin aversion is suggested to protect against the teratogens that would harm the fetus (Profet 1992). In the case of sick chimpanzees, decreased aversion to toxins might provide increased protection against pathogens. Supportive evidence comes from an observation from Gombe. Goodall noted that sick chimpanzees accepted antibiotics when eating bananas laced with tetracycline. However, when apparently well or recovering, they refused the medicine. More such evidence, ideally from experiments, will enable the hypothesis to be tested.

Even if the response of chimpanzees to strong tastes changes when they are ill, it is doubtful that they could discover medicinal plants by random sampling of the vegetation: they are conservative in their selection of food items (Nishida et al. 1983), live in habitats with hundreds of plant species, and are rarely ill. Observational learning is more likely to be important, allowing many individuals to acquire information through the experience of a few (Nishida 1987). On at least four occasions (twice with *V. amygdalina,* once with *A. mossambicensis,* and once with an unidentified Commelinaceae species), an infant was observed to copy the feeding behavior of its mother, but stopped short of actually ingesting the bitter-pith juice or the whole leaf (Huffman and Seifu 1989; Nishida unpubl. observ.; Takasaki unpubl. observ.; Nakamura video records). These observations and the pattern described for neighboring unit-groups to use the same plant species in the same manner suggests that observational learning is important.

The suggestion that the use of medicinal plants is a behavioral tradition includes a range of possibilities for how the behavior started and how individuals become predisposed to ingest medicinal plants. This behavioral tradition may have started as a result of ill, hungry chimpanzees trying new foods during periods of sickness, recovering their health, and associating their improved health with the new food. This mechanism is speculative, however, because it is difficult to establish animal preferences for initially aversive tastes (Rozin 1987). Once the behavior is established in a population, the young presumably learn about it by observing their elders. However, whether they discover the benefits for themselves or, alternatively, recognize similarity between their own feelings of discomfort and those of other unhealthy chimpanzees is an open question (Galef 1990). It will clearly be of considerable interest to find out to what extent chimpanzees are able to use social input to guide their choice of plant use when ill.

Summary

It has been proposed that chimpanzees use a number of plant species for their medicinal value. This chapter compares and discusses in detail the behavioral diversity and pharmacology involved in the use of medicinal plants by chimpanzees in the wild at the Mahale, Gombe, and Kibale sites. With regard to bitter-pith chewing and whole-leaf swallowing behaviors, 13 plant species have been identified as being used for their possible medicinal value at these sites. Based on behavior, plant pharmacology, and ethnomedical information, hypotheses concerning the medicinal value of these plants for chimpanzees include the control of parasites, treatment of gastrointestinal disorders, regulation of fertility and, possible, antibacterial or antihepatotoxic activity. Chimpanzees of neighboring groups at the Mahale, Gombe, and Kibale study sites tended to use the same or closely related plant species for bitter-pith chewing and for whole-leaf swallowing.

Similar whole-leaf swallowing of another eight plant species by chimpanzees (*Pan troglodytes schweinfurthii, P. t. verus*), bonobos (*Pan paniscus*), and Eastern lowland gorillas (*Gorilla gorilla graueri*) have recently been recorded at five other sites. Also great apes of isolated sites tended to use different species of the same or closely related genera having similar properties (bitter taste, roughly surfaced leaves) in the same manner. Based on these observations and assuming that the hypothesis of medicinal value proves correct, it seems likely that medicinal use of plants is a behavioral tradition carried between groups. The mechanism by which medicinal use of plants becomes established is unknown. One mechanism may be a changed response to toxins as a function of being ill. Social learning is also likely to be an important mechanism.

Endnote

The CHIMPP Group (Chemo-ethology of Hominoid Interactions with Medicinal Plants and Parasites) is an interdisciplinary (ethology, natural plant products chemistry, parasitology, pharmacognosy, and pharmacology), multinational research group coordinated by M.A. Huffman from Kyoto University (Huffman 1993).

Acknowledgments

We wish to give our sincerest gratitude to the Governments of Tanzania and Uganda for the long-term cooperation and commitment extended to us and our colleagues at Mahale, Gombe, and Kibale. In particular, we are thankful to the Tanzanian National Scientific Research Council, Tanzanian National Parks, Serengeti Wildlife Research Institute, Mahale Mountains Wildlife Research Centre, Gombe Stream Research Centre, the Government of Uganda's Forest Department, and Makerere University. We wish to give our deepest gratitude to all those colleagues and field assistants mentioned in the text for their generous contributions of unpublished data and personal communications. Without their advice, support, and understanding, this study would not have been possible. We would also like to sincerely thank (in alphabetical order) M. Aregullin, G. Balansard, C. Chapman, A. Clark, R. Faden, J. Goodall, G. Isabirye-Basuta, J. Itani, K. Koshimizu, T. Nishida, Y. Sugiyama, and J. West for their generous support and advice. J. Page, E. Rodriguez, and L. Turner kindly reviewed and commented on the manuscript. Research done by M.A. Huffman in 1987, 1989, and 1990–91 in the field was supported by a grant under the Monbusho International Scientific Research Program to T. Nishida (# 61043017, 62041021, 63043017, 03041046) and during the preparation of this manuscript in 1992 by a Post-doctoral Fellowship from the Japan Society for the Promotion of Science. R.W. Wrangham was supported by the National Science Foundation (BNS–8704458) and National Geographic Society (3603–87).

References

Abbiw, D.K. 1990. *Useful Plants of Ghana.* Kew, England: Intermediate Technology Publications and Royal Botanic Gardens.

Asaka Y., T. Kubota, and A.B. Kulkarni. 1977. Studies on a bitter principle from *Vernonia anthelminthica. Phytochem.* 16:1838–9.

Burkill, H.M. 1985. *The Useful Plants of West Tropical Africa.* 2d ed. Vol. 1. Kew, England: Royal Botanical Gardens.

Barrera, E., and E. Rodriguez. 1990. Studies on the African plant *Ficus exasperata* eaten by wild chimpanzees *Pan troglodytes schweinfurthii.* Society for the Advancement of Chicanos and Native Americans in Science (SACNAS), Phoenix, Arizona.

Clutton-Brock T.H., and J.B. Gilette. 1979. A survey of forest composition in the Gombe National Park, Tanzania. *African J. Ecol.* 17:131–58.

Collins, D.A., and W.C. McGrew. 1988. Habitats of three groups of chimpanzees *(Pan troglodytes)* in western Tanzania compared. *J. Human Evol.* 17:553–74.

Crellin, J.K. and J. Philpott. 1990. *Herbal Medicine Past and Present.* Vol 2 *A reference guide to medicinal plants.* Durham, N.C.: Duke Univ. Press.

Dalziel, J.M. 1937. The useful plants of west tropical Africa. In J. Hutchinson and J.M. Dalziel, eds., *Appendix to Flora of West Tropical Africa* London: Whitefriars Press.

Galef, B.G. 1990. The question of animal culture. *Human Nature* 3:157–78.

Garey, J., L. Markiewicz, and E. Gurpide. 1992. Estrogenic flowers, a stimulus for mating activity in female vervet monkeys. In XIVth Congress of the International Primatological Society Abstracts, (Strasbourg, France), pp. 210.

Gasquet, M., D. Bamba, A. Babadjamian, G. Balansard, P. Timon-David, and J. Metzger. 1985. Action amoebicide et anthelminthique du vernolide et de l'hydroxyvernolide isoles des feuilles de *Vernonia colorata* (Willd.) Drake. *European J. Medicine and Chem. Theory* 2:111–5.

Ghiglieri, M.P. 1984. *The Chimpanzees of Kibale Forest* New York: Columbia Univ. Press.

Githens, T.S. 1949. *Drug Plants of Africa.* Philadelphia: Univ. of Penn. Press.

Glander, K.E. 1975. Habitat description and resource utilization: A preliminary report on mantled howler monkey ecology. In. R.H. Tuttle, ed., *Socioecology and Psychology of Primates,* pp. 37–57. The Hague: Mouton Press.

———. 1982. The impact of plant secondary compounds on primate feeding behavior. *Yearbook of Phys. Anthro.* 25:1–18.

———. 1992. Plant selection and gender bias in howling monkey births. Paper presented at the meetings of the American Association for the Advancement of Science (AAAS). Chicago, Ill.

Goodall, J. 1986. *The Chimpanzees of Gombe: Patterns of Behavior.* Cambridge, Mass.: Belknap Press.

Hart, B.L. 1990. Behavioral adaptations to pathogens and parasites: Five strategies. *Neuroscience and Biobehavioral Reviews* 14:273–94.

Hladik, C.M. 1977a. Chimpanzees of Gabon and chimpanzees of Gombe: Some comparative data on the diet. In T.H. Clutton-Brock, ed., *Primate Ecology,* pp. 481–502. London: Academic Press.

———. 1977b. A comparative study of the feeding strategies of two sympatric species of leaf monkeys: *Presbytis senex* and *Presbytis entellus.* In T.H. Clutton-Brock, ed., *Primate Ecology,* pp. 324–53. London: Academic Press.

Holmes, J.C., and S. Zohar. 1990. Pathology and host behavior. In C.J. Barnard and J.M. Behnke, eds., *Parasitism and Host Behavior,* pp. 34–63. London: Taylor and Frances.

Huffman, M.A. 1993. An investigation of the use of medicinal plants by wild chimpanzees. Current status and future prospect. *Primate Research* 9(2):179–87.

Huffman, M.A., and M. Seifu. 1989. Observations on the illness and consumption of a possibly medicinal plant, *Vernonia amygdalina* (Del.), by a wild chimpanzee in the Mahale Mountains National Park, Tanzania. *Primates,* 30:51–63.

Huffman, M.A., K. Koshimizu, H. Ohigashi, and M. Kawanaka. 1992. Medicinal plant use by wild chimpanzees: A behavioral adaptation for parasite control? Paper presented at the meetings of the American Association for the Advancement of Science (AAAS). Chicago, Ill.

Huffman, M.A., S. Gotoh, D. Izutsu, K. Koshimizu, M.S. Kalunde. In Press. Further observations on the use of *Vernonia amygdalina* (Del.) by a wild chimpanzee, its possible effect on parasite load, and its phytochemistry. *African Study Monogr.*

Itogawa, H., K. Takeya, K. Mihara, N. Mori, T. Hamanaka, T. Sonobe, Y. Iitoka. 1983. Studies on the antitumor cyclic hexapatides, obtained from *Rubia radix. Chem. and Pharmacol. Bull.* 37(6):1670–2.

Janzen, D.H. 1978. Complications in interpreting the chemical defenses of trees against tropical arboreal plant-eating vertebrates. In G.G. Montgomery, ed., *The Ecology of Arboreal Folivores,* pp. 73–84. Washington, D.C.: Smithsonian Institute Press.

Jisaka, M., M. Kawanaka, H. Sugiyama, K. Takegawa, M.A. Huffman, H. Ohigashi, K. Koshimizu. 1992b. Antischistosomal activities of sesquiterpene lactones and steroid glucosides from *Vernonia amygdalina,* possibly used by wild chimpanzees against parasite-related diseases. *Biosci., Biotech. and Biochem.* 56(5): 845–6.

Jisaka, M., H. Ohigashi, K. Takegawa, M.A. Huffman, and K. Koshimizu. In Press. Antitumor and antimicrobial activities of bitter sesquiterpene lactones of *Vernonia amygdalina,* a possible medicinal plant used by wild chimpanzees. *Biosci., Biotech. and Biochem.*

Jisaka, M., H. Ohigashi, K. Takagaki, H. Nozaki, T. Tada, M. Hirota, R. Irie, M.A. Huffman, T. Nishida, M. Kaji, and K. Koshimizu. 1992a. Bitter steroid glucosides, vernoniosides A1, A2, and A3 and related B1 from a possible medicinal plant *Vernonia amygdalina,* used by wild chimpanzees. *Tetrahedron* 48:625–32.

Kawabata, M., and T. Nishida. 1991. A preliminary note on the intestinal parasites of wild chimpanzees of the Mahale Mountains, Tanzania. *Primates* 32:275–8.

Kloos, H., and F.S. McCullough. 1987. Plants with recognized molluscicidal activity. In K.E. Mott, ed., *Plant Molluscicides,* pp. 45–108. New York: John Wiley and Sons.

Kokwaro, J.O. 1987. Some common African herbal remedies for skin diseases: With special reference to Kenya. In A.J.M. Leeuwenberg, compiler, *Medicinal and Poisonous Plants of the Tropics,* pp. 44–69. Wageningen: Pudoc.

Kupchan, S.M., R.J. Hemingway, A. Karim, and D. Werner. 1969. Tumor Inhibitors XLVII. vernodalin and vernomygdin, two new cytotoxic sesquiterpene lactones from *Vernonia amygdalina* Del. *J. Organic Chem.* 34:3908–11.

Lwande, W., C. MacFoy, M. Okecj, F. Delle Monache, and G.B. Marino Bettolo. 1985. Research on African medicinal plants. Part 8: Kaurenoic acids from *Aspilia pluriseta. Fitoerapia* 56:126–8.

McKey, D. 1978. Soil, vegetation and seed-eating by black colobus monkeys. In G.G. Montgomery, ed., *The Ecology of Arboreal Folivores,* pp. 423–37. Washington, D.C.: Smithsonian Institute Press.

Milton, K. 1979. Factors influencing leaf choice by howler monkeys: A test for some hypotheses of food selection by generalist herbivores. *Amer. Naturalist* 114:362–78.

Moore, J. 1992. "Savanna" chimpanzees. In T. Nishida et al., eds., *Topics in Primatology.* Vol. 1, *Human Origins,* pp. 99–118. Tokyo: Univ. of Tokyo Press.

Newton, P.N., and T. Nishida. 1990. Possible buccal administration of herbal drugs by wild chimpanzees, *Pan troglodytes. Anim. Behav.* 39(4):798–801.

Nishida, T. 1987. Learning and cultural transmission in non-human primates. In B.B. Smuts et al., eds., *Primate Societies,* pp. 462–74. Chicago: Univ. of Chicago Press.

———. A quarter century of research in the Mahale Mountains: An overview. In T. Nishida, ed., *The Chimpanzees of the Mahale Mountains: Sexual and Life History Strategies,* pp. 3–35. Tokyo: Univ. of Tokyo Press.

Nishida, T., and K. Kawanaka. 1972. Inter-unit-group relationships among wild chimpanzees of the Mahale Mountains. *African Studies Monogr.* 7:131–69.

Nishida, T., and S. Uehara. 1983. Natural diet of chimpanzees (*Pan troglodytes schweinfurthii*): Long term record from the Mahale Mountains, Tanzania. *African Studies Monogr.* 3:109–30.

Nishida, T., R.W. Wrangham, J. Goodall, and S. Uehara. 1983. Local differences in plant-feeding habits of chimpanzees between the Mahale Mountains and Gombe National Park, Tanzania. *J. Human Evol.* 12:467–80.

Oates, J.F. 1977. The guereza and its food. In T.H. Clutton-Brock, ed., *Primate Ecology,* pp. 276–321, London: Academic Press.

Oates, J.F., T. Swain, and J. Zantovska. 1977. Secondary compounds and food selection by colobus monkeys. *Biochem. Syst. Ecol.* 5:317–21.

Oates, J.F., P.G. Waterman, and G.M. Choo. 1980. Food selection by the south Indian leaf-monkey *Presbytis johnii,* in relation to leaf chemistry. *Oecologia* 45:45–56.

Ohigashi, H., M. Jisaka, T. Takagaki, H. Nozaki, T. Tada, M.A. Huffman, T. Nishida, M. Kaji, and K. Koshimizu. 1991. Bitter principle and a related steroid glucoside from *Vernonia amygdalina,* a possible medicinal plant for wild chimpanzees. *Agric. and Biol. Chem.* 55: 1201–3.

Page, J.E., F.F. Balza, T. Nishida, G.H.N. Towers. 1992. Biologically active diterpenes from *Aspilia mossambicensis,* a chimpanzee medicinal plant. *Phytochem.* 31(10):3437–9.

Phillips-Conroy, J.E. 1986. Baboons, diet, and disease: Food plant selection and schistosomiasis. in D.M. Taub and F.A. King, eds., *Current Perspectives in Primates Social Dynamics,* pp. 287–304. New York: Van Nostrand Reinhold.

Profet, M. 1992. Pregnancy sickness as adaptation: a deterrent to material ingestion of teratogens. In J.H. Barkow, L. Cosmides and J. Tooby, eds., *The Adapted Mind: Evolutionary Psychology and the Generation of Culture,* pp. 327–65. Oxford, England: Oxford Univ. Press.

Rodriguez, E., M. Aregullin, T. Nishida, S. Uehara, R.W. Wrangham, Z. Abramowski, A. Finlayson, and G.H.N. Towers. 1985. Thiarubrin A, a bioactive constituent of *Aspilia* (Asteraceae) consumed by wild chimpanzees. *Experientia* 41:419–20.

Rodriguez, E. and R.W. Wrangham. 1992. Zoopharmacognosy: Medicinal plant use by wild apes and monkeys. Paper presented at the meetings of the American Association for the Advancement of Science (AAAS). Chicago.

———. In Press. Zoopharmacognosy: The use of medicinal plants by animals. In K.R. Downum, J.T. Romeo and H.A. Stafford, eds., *Recent Advances in Phytochemistry.* Vol. 27, pp. 89–106. New York: Plenum Press.

Rozin, P. 1987. Psychological perspectives on food preferences and avoidances. In M. Harris and E.B. Ross, eds. *Food and Evolution: Toward a Theory of Human Food Habits,* pp. 181–206. Philadelphia, Penn.: Temple Univ. Press.

Russo, E.B. 1992. Headache treatments by native peoples of the Ecuadorian Amazon: a preliminary cross-disciplinary assessment. *J. Ethnopharmacol.* 36:193–206.

Struhsaker, T. 1975. *The Red Colobus Monkey.* Chicago: Univ. of Chicago Press.

Strier, K. 1993. Menu for a monkey. *Nat. Hist.* Feb.

Takasaki, H., and K. Hunt. 1987. Further medicinal plant consumption in wild chimpanzees? *African Study Monogr.* 8:125–8.

Terashima, H., S. Kalala, N.M. Malasi. 1991. Ethnobotany of the Lega in the tropical rain forest of eastern Zaire: Part one, Zone de Mwenga. *African Study Monogr. (Suppl.)* 15:1–61.

Toubiana, R., and A. Gaudemer. 1967. Structure du vernolide, nouvel ester sesquiterpique isole de *Vernonia colorata. Tetrahedron* 14:1333–6.

Towers, G.H.N., Z. Abramowski, A.J. Finlayson, and A. Zucconi. 1985. Antibiotic properties of thiarubrine-A, a naturally occurring dithiacyclohedadiene polyine. *Planta Medica* 3:225–9.

Waterman, P.G. 1984. Food acquisition and processing as a function of plant chemistry. In D.J. Chivers, B.A. Wood, and A. Bilsborough, eds., *Food Acquisition and Processing in Primates,* pp. 177–211. New York: Plenum Press.

Watt, J.M., and M.G. Breyer-Brandwinjk. 1962. *The Medicinal and Poisonous Plants of Southern and East Africa.* Edinburgh: E. and S. Livingstone.

Wing, L.D., and I.O. Buss. 1970. Elephants and Forests. In *Wildlife Monographs,* no. 19. Washington D.C.: The Wildlife Society.

Wrangham, R.W. 1975. The behavioural ecology of chimpanzees in Gombe National Park, Tanzania. Ph.D. diss., Univ. of Cambridge, England.

———. 1977. Feeding behavior of chimpanzees in Gombe National Park, Tanzania. In T.H. Clutton-Brock, ed., *Primate Ecology,* pp. 504–38. London: Academic Press.

Wrangham, R.W., and P.G. Waterman. 1981. Condensed tannins in fruits eaten by chimpanzees. *Biotropica* 15:217–22.

Wrangham, R.W., and T. Nishida. 1983. *Aspilia* spp. leaves: A puzzle in the feeding behavior of wild chimpanzees. *Primates* 24:276–82.

Wrangham, R.W., and J. Goodall. 1989. Chimpanzee use of medicinal leaves. In P.G. Heltne and L.A. Marquardt, eds., *Understanding Chimpanzees,* pp. 22–37. Cambridge, Mass.: Harvard Univ. Press.

Yang, L.-L., K.-Y. Yen, C. Konno, Y. Oshima, H. Hikino. 1986. Antihepatotoxic principles of *Wedelia chinensis* herbs. *Planta Medica* 52:499–500.

Section 2

Social Relations

Flo's family

Overview—Diversity in Social Relations

W.C. McGrew

Understanding behavioral diversity in chimpanzees is a challenge on all fronts, especially in the area of social relations. As in the ecology of individual behavior, we confront a host of variables, many of them interrelated or embedded. For example, the availability of a kind of fruit eaten by chimpanzees is linked to rainfall, sunshine, temperature, soil nutrients, and so on. An individual chimpanzee may compete for these fruits with members of its group or with other chimpanzee groups or with a wide variety of other species. But beyond the complexities of such interactive or confounded variables, social relations pose further problems in the form of contingent strategies under the control of intentional players. For example, predators presumably do not keep score of the strengths or weaknesses of individual prey, but chimpanzee competitors apparently do keep such accounts on other competitors within the same group. Prey presumably do not change their responses according to the identity of an individual predator, but when chimpanzees decide whether or not to cooperate, they respond according to the individual identity of other group members. Thus, culture will always be tougher to tackle than nature.

In trying to explain diversity in social behavior, the key is systematic comparison. Ideally, all but one independent variable is controlled, so that the one remaining variable can then be manipulated, either experimentally or opportunistically. Thus, subjects within the same species, the same subspecies, the same population, the same group, or the same family can be compared on grounds of gender, age, rank, or some other variable. Then the same dependent variables are measured; for instance, play is recorded using the same protocols of sampling, definition, observability, frequency, duration, and so on. Ideal protocols are hard to achieve, especially under the quixotic conditions of fieldwork, but each chapter in this section makes an attempt in one or more ways.

Arguably, the most basic social variable is species, with the obvious comparison being made between the two congeners, *Pan paniscus* and *Pan troglodytes.* Hashimoto and Furuichi's chapter on sexual behavior and Muroyama and Sugiyama's chapter on grooming undertake this comparison starting with data on bonobos from Wamba and chimpanzees from Bossou. These results are generalized to other data for comparison between bonobos and chimpanzees, casting the net as widely as current knowledge permits.

Tutin's chapter takes comparison of species to the next level with a study of the two sympatric African apes, chimpanzee and gorilla. (Whether this is considered congeneric depends on one's choice of taxonomist.) Tutin focuses on the basics of female reproductive success: age at first birth, birth interval, and lifetime output of offspring.

None of the chapters in this section tackles the obvious comparison between apes and humans, though the phylogenetic distance continues to narrow and the molecular anthropological grounds for combining the African Pongidae and the Hominidae are increasingly strong. If chimpanzees and humans can exchange blood transfusions and Ameslan signs, then perhaps their social relations deserve more than analogical analyses too.

Another potent set of comparisons is between subspecies or geographical races. For example, *Pan troglodytes troglodytes* of Central Western Africa can be contrasted with its conspecifics to either side: *P. t. verus* in Western Africa and *P. t. schweinfurthii* in Eastern Africa. No chapter in this section explicitly tackles this three-way comparison, though all authors are careful to cite material from the relevant field sites. Detailed behavioral data from *P.t.t.* clearly lag behind those from *P.t.v.* (Bossou and Taï) and *P.t.s.* (Gombe, Kibale, and Mahale), and so whoever habituates the first *P.t.t.* population will greatly advance the study of diversity across subspecies.

In captivity, reliance on confusing morphological cues has long prevented comparisons of known subspecies, but recent genetic advances now allow this comparison. The subspecific variable appears in another way, as traditional export-import lines from Africa to consumer countries were not random. Chimpanzees in North America tend to originate in Liberia and Sierra Leone *(P.t.v.),* while chimpanzees in Europe tend to originate in Congo and Zaire *(P.t.t.).* Whether this genetic diversity is responsible for behavioral diversity in captivity remains to be seen. It is also an issue in conservation.

The next level of social organization for comparison is that of the population (however fragmented these may now be from formerly more continuous and extensive demes). The potential for making social behavior comparisons is much less than for making socioecological comparisons, given the necessity of habituated subjects. Currently, there are only two such field sites for bonobos (Lomako and Wamba) and five for chimpanzees (Bossou, Gombe, Kibale, Mahale, and Taï). More are obviously needed, but the required investment of time, effort, money, and enthusiasm is sobering, and it is not just grant-giving bodies that must be patient. In this section, the most direct comparison appears in Mitani's chapter on vocal diversity between Gombe and Mahale populations. Muroyama and Sugiyama also present data on grooming in different populations, as does Tutin on female reproductive success.

Comparisons of two or more groups within a population are straightforward in principle, but such comparisons are not common in practice. For *Pan troglodytes,* the two best-known examples—K Group and M Group at Mahale, and the Kasakela and Kahama communities at Gombe—are defunct because, in each case, the larger of the two groups outcompeted the other and drove it to extinction. The best current candidates for such comparison may be Kanyawara group and Ngogo group at Kibale. In addition, there are efforts to habituate neighbors underway at Gombe and Taï. For *Pan paniscus,* researchers at both major sites of study follow two groups in parallel: E1 and E2 at Wamba, and Hedons and Rangers at Lomako.

Comparisons between groups are more readily done in captivity, as exemplified by Baker and Smuts's chapter on the competitive social strategies of females at the Detroit and Arnhem Zoos. Each of these locations houses a single group. Facilities that maintain multiple groups, such as Bastrop or Yerkes, would allow even tighter comparisons between groups, as many nonsocial variables such as diet and climate could be controlled. Another largely ignored source of between-groups-but-within-population comparisons occurs where chimpanzees are being rehabilitated back into the wild. For example, Brewer and Carter's project in Gambia has three breeding groups on separate islands in Baboon Island National Park. Unfortunately, the potential for captive studies of bonobos remains severely limited because, worldwide, the captive population of bonobos remains fragmented into tiny groups, none of which approximates natural age-sex composition.

For within-group comparisons, the usual unit is the individual, as tempered by his or her various lifelong familial ties and longstanding alliances with nonclose kin. However, in some groups, matrilines may exercise notable influence on group affairs. In Gombe's Kasakela community, for example, two matrilineal kin groups compete in the group's core area. These two groups, descended from now-dead ancestors, are Flo's "F" family with eight living members and Melissa's "G" family with five living members. Such extended families probably lead very different social lives than do smaller subgroups, but this assumption remains to be studied. Patrilines may also be influential; DNA fingerprinting now offers the chance to pursue this covert, unstudied set of factors.

However, the simplest and most telling single variable of comparison is that of life in the wild versus captivity. Actually this variable occurs as a gradient rather than as a dichotomy, with large enclosures and islands providing the intermediate points between the extremes of cramped cage and African national park. Confinement, that is, inability to avoid companionship in the short term and inability to disperse in the longer term seems to be a crucial factor of social relations. Compared with life in the wild, daily life of chimpanzees in zoo or laboratory is a social hothouse. Of the higher primates, chimpanzees are among the least sociable, spending much time alone or accompanied only by dependent offspring. Causing chimpanzees to live constantly in troops leads to notable social adaptations, as Boehm's chapter comparing pacifying interventions at Gombe versus Arnhem shows specifically, and de Waal's chapter shows more generally for male-male bonds and female-female relations. Tutin's chapter also compares female reproductive success in captivity versus the wild, but the likely causal factors are nutritional and not social.

So, just as psychologists confront diversity in humans in terms of individual differences and sociocultural anthropologists confront diversity in humans in terms of cross-cultural variation, so do primatologists confront diversity in chimpanzees. The extent to which differences and similarities in social life stem from social traditions and group processes rather than ecological vicissitudes remains to be seen. All the chapters in this section tackle these issues in one way or another.

Social Role and Development of Noncopulatory Sexual Behavior of Wild Bonobos

Chie Hashimoto and Takeshi Furuichi

Introduction

Adult bonobos show a variety of noncopulatory sexual behaviors that are not seen in chimpanzees. These behaviors serve many social functions, such as reducing tension among individuals, terminating agonistic interactions, and forming new affiliate social relations (de Waal 1987; Furuichi 1989; Kano 1992; Kuroda 1980; Thompson-Handler et al. 1984). These behaviors help bonobos to form social groups whose members aggregate more closely than chimpanzees (Furuichi 1987; Kano 1982; Kitamura 1983; White 1988). To date, the development of noncopulatory sexual behaviors has not been reported in detail.

The present study shows that bonobos begin to engage in various types of sexual behavior at an early age. However, the sexual behaviors of immature chimpanzees and immature bonobos are relatively similar, as compared to the sexual behaviors of mature chimpanzees and bonobos (Goodall 1968; Hasegawa and Hiraiwa-Hasegawa 1983; Kitamura 1989; Plooij 1984). This pattern suggests that the differences in sexual behaviors of the two species appear during the process of maturation.

This chapter examines the type, frequency, and context of noncopulatory sexual behaviors of immature and mature bonobos. Comparisons made between chimpanzees and bonobos help illustrate the divergence of sexual behaviors in the genus *Pan.*

Study Group and Methods

The subject of this study is a group of wild bonobos *(Pan paniscus),* called E1 Group, living at Wamba, Zaire (Furuichi 1989; Kano 1992). During the study period, between November 1990 and February 1991, this group consisted of 30 individuals, including seven adult males (15 years of age and older), two adolescent males (eight and 10 years old), seven juvenile and infant males (younger than five years old), nine adult females (16 years of age and older), and five juvenile and infant females (younger than seven years old). Immature individuals in this study included juveniles and infants. Data for the two adolescent males were analyzed together with data for adult males. Except for one adult male who ranged alone throughout the study period, all members of the group were studied as subjects of this paper. The subject individuals usually ranged together while forming one large mixed party, though they sometimes ranged while forming separate parties (Furuichi 1987,1989).

The data for this study were obtained by two observation methods. First, each immature individual was observed for about 500 minutes using focal animal sampling. All sexual behaviors occurring within two meters of a focal animal were recorded. Second, all sexual behaviors of both mature and immature individuals at feeding sites were recorded. The presence of individuals six years old and older was checked by scan sampling at 10 minute intervals. Individuals younger than six years old were assumed to be present at the site when their mothers were present. While some amount of sexual behavior occurred with members of groups other than the E1, these instances are excluded from the present study.

In this study, sexual behavior included all interactions between individuals involving the contact of genitals. Sexual behavior was subdivided into two categories: *copulation* and *genital contact.*

Mounting with insertion between adult or adolescent males and adult females was classed as copulation. Although insertion or ejaculation was not always confirmed, those interactions involving normal soliciting behaviors for copulation and thrusting movements were recorded as copulation. Interactions that lacked apparent thrusting movement or insertion were recorded as genital contact (Furuichi 1992). Males mounting females when one or both individuals were immature was recorded as genital contact.

Genital contact behaviors between adults or adolescents included noncopulatory mounting, genito-genital rubbing, and rump-rump contact (Kano 1992; Kitamura 1989; Kuroda 1980; Thompson-Handler et al. 1984). Moreover, genital contacts involving immature individuals were classified into ventro-ventral, ventro-dorsal, and dorso-dorsal types with respect to the posture of the participants.

A series of sexual behaviors separated by intervals of less than one minute was counted as one incident of the behavior if no other interaction between the participants or with other individuals occurred during the intervals.

Results

Genital Contact Behavior of Immature Individuals

Immature individuals engaged in noncopulatory genital contact behaviors with individuals of most age classes. As shown in figure 1, genital contact between immature individuals began when individuals were younger than one year. The frequency of such behaviors did not change with age, but males engaged in these behaviors more frequently than did females (Mann-Whitney U-test, $N_1 = 7$, $N_2 = 5$, $U = 3$, $p < .01$, two-tailed).

When most members were feeding, immature individuals who had finished eating earlier than adults, began to play with one another. They continued playing while adults were resting after eating. Genital contact between immature individuals occurred mostly as part of such play.

The most frequent posture for genital contact between immature individuals was ventro-ventral, which occurred when they were wrestling on the ground, when they were holding each other while hanging from a branch, and when an older individual was carrying a younger one. Movement during genital thrusting was rather irregular. Rhythmical movement from left to right, like that in a genito-genital rubbing between

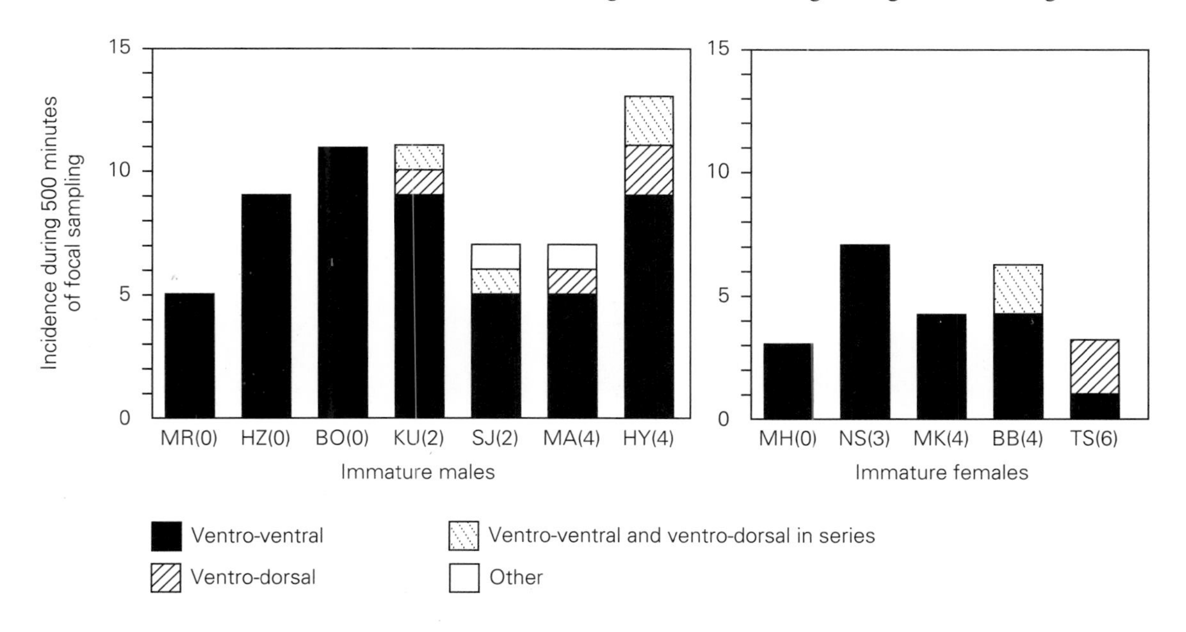

Figure 1
Incidence of genital contact between immature bonobos. Incidence includes cases of genital contact involving a focal individual and cases that occurred within two meters of a focal individual during 500 minutes of focal sampling. Figures in parentheses show age in years of each individual.

adult females, was not seen. Ventro-dorsal genital contact was observed for males from two years old and for females from four years old. Such contact occurred independently or in a sequence with ventro-ventral genital contact. Males served as both mounter and mountee, but females took part only as the mountee. Dorso-dorsal genital contact, like rump-rump contact between adult males, was not seen between immature individuals.

Both immature males and immature females engaged in genital contact with adult or adolescent males, although the frequency was low (see figure 2). In genital contacts involving three of the seven individuals younger than four years old, an adult male held an immature individual ventro-ventrally and shook its body with his leg or hand so that their genitals rubbed against each other. This behavior occurred in a playlike context, and penile erection was rarely observed in the adult male. On the other hand, most genital contact involving immature individuals from 4 years old and over occurred in a different context. When being attacked or threatened by an adult male,

Figure 2
Incidence of genital contact between immature bonobos and adult or adolescent males (upper figures) and between immature bonobos and adult females (lower figures).

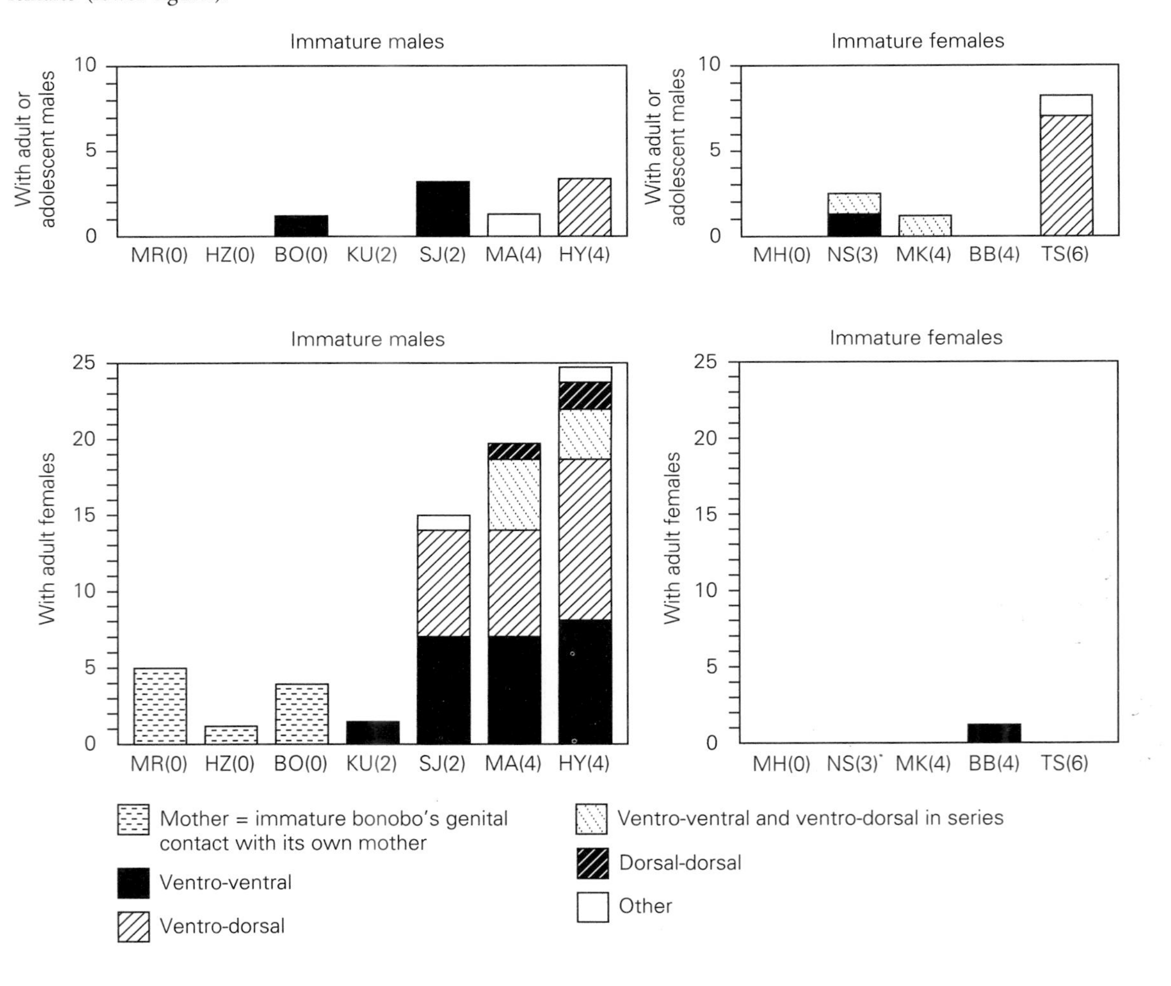

an immature individual showed presenting behavior, and the male mounted it. This kind of interaction resembled mounting between adults and seemed to serve the social function of resolving conflict.

Genital contact with adult females, on the other hand, occurred primarily with immature males (see figure 2). In all cases involving males younger than two years old, mothers held their sons ventro-ventrally and rubbed their genitals against their own genitals. It seemed that mothers performed this behavior to reduce their own emotional arousal, because this behavior occurred only during tense situations, such as when they entered the feeding sites or after they were involved in agonistic interactions.

Males two years old and over engaged in genital contact with adult females other than their mother, and the frequency increased with the male's age. When adults engaged in sexual behaviors, immature males approached them and rubbed their own genitals against those of the females, or inserted their penises, in a ventro-ventral, ventro-dorsal, or dorso-dorsal position. Some cases of genital contact between four-year-old males and adult females closely resembled copulation between adults and were accompanied by soliciting behaviors similar to those shown by adult males before copulation.

Figure 3 shows the reaction of immature individuals to sexual behavior between adult or adolescent individuals that occurred within two meters of the immature individuals. Some males who were two years of age and older and a female three years of age frequently joined the sexual behavior by mounting on the back or clinging to the chest of one of the adults or by performing genital contact with one of them. However, immature individuals rarely joined in when their own mother was involved. Finally, although they did join in, immature individuals did not disrupt the sexual behavior of adults during this study period; such interference was only rarely observed during other observation periods.

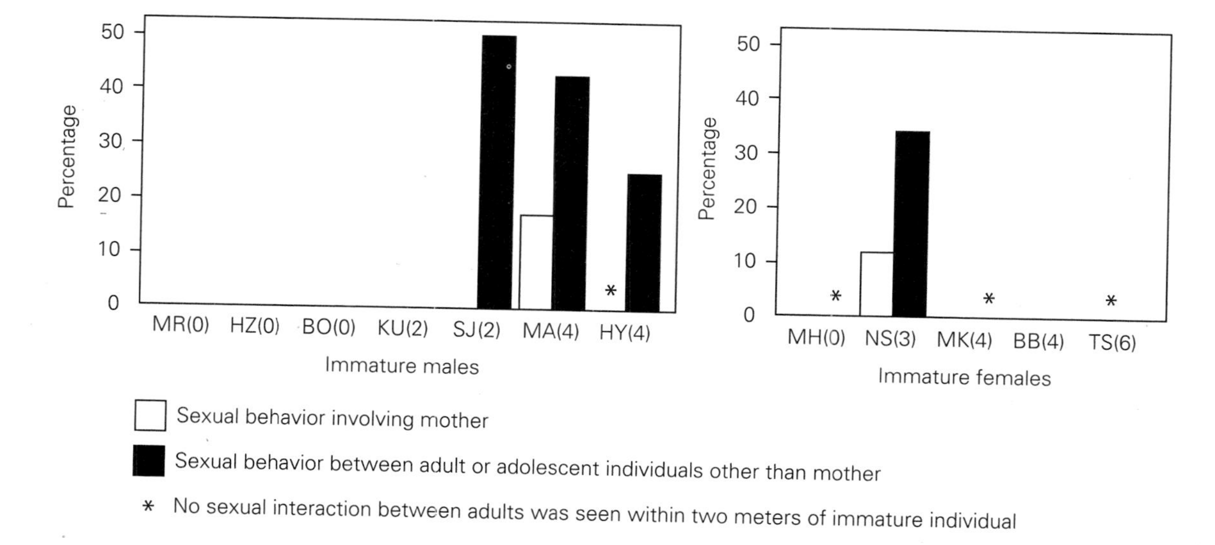

Figure 3
Percentage of occurrences when immature bonobos participated in sexual behavior between adults or adolescents within two meters of the immature. This figure shows only cases observed during focal sampling. * = no sexual interaction between adults was seen within two meters of the individual.

Changes in Sexual Behavior from Immature to Adult

The frequency of sexual behavior changed dramatically with age (see figure 4). For example, the frequency of sexual behavior of immature males increased with age. However, upon reaching adolescence, there was an abrupt drop in frequency of genital contact of adolescent males with adult females and with other immature individuals. This decline probably occurred because adolescent males tended to stay on the periphery of a mixed party, away from the core where most females and immature individuals congregated. In part, this peripheral position relates to a son's increased independence from his mother. In addition, adult males tended to be less tolerant of adolescent males than of younger males. Adults persistently attacked or threatened any adolescent male who tried to enter the party's core. Upon reaching adulthood, however, the frequency of sexual behavior of males increased again, because of the frequent genital contact between adult males. The frequency of copulation with adult females varied greatly from individual to individual, but remained lower than the frequency of genital contact between four-year-old males and adult females.

For females, a marked contrast in sexual behavior was seen between immature individuals born into the group and immigrant adults. Immature females showed very low frequency of sexual behaviors, although six-year-old TS began to engage in genital contact in patterns and contexts similar to those of adults. Females who were five to six years old tended to remain on the periphery of a party; they transferred to another group between the ages of seven and nine years (Furuichi 1989). All adult females, who were assumed to be immigrants from other groups, engaged in sexual behavior frequently. Most of their sexual behavior involved genital contact with other adult females. Frequency of copulation with adult or adolescent males and genital contact with immature males showed large individual variation because of differences in estrous state.

The types of genital contact between adults (including two adolescent males) differed from the types seen between immature individuals. Although immature individuals showed a wide variety of postures and movements during genital contact, the adults showed fixed patterns. Those included mounting, genito-genital rubbing, and rump-rump contact. Genital contact was most frequently observed between females (95 cases) and included genito-genital rubbing (91 cases) and mounting (four cases). Frequency of genital contact between males was about one third of that between females (36 cases) and included mounting (22 cases), rump-rump contact (13 cases), and only one case of genito-genital rubbing. Noncopulatory genital contact between males and females was restricted to only six cases of mounting, though rump-rump contact like that between males was also observed in a neighboring group.

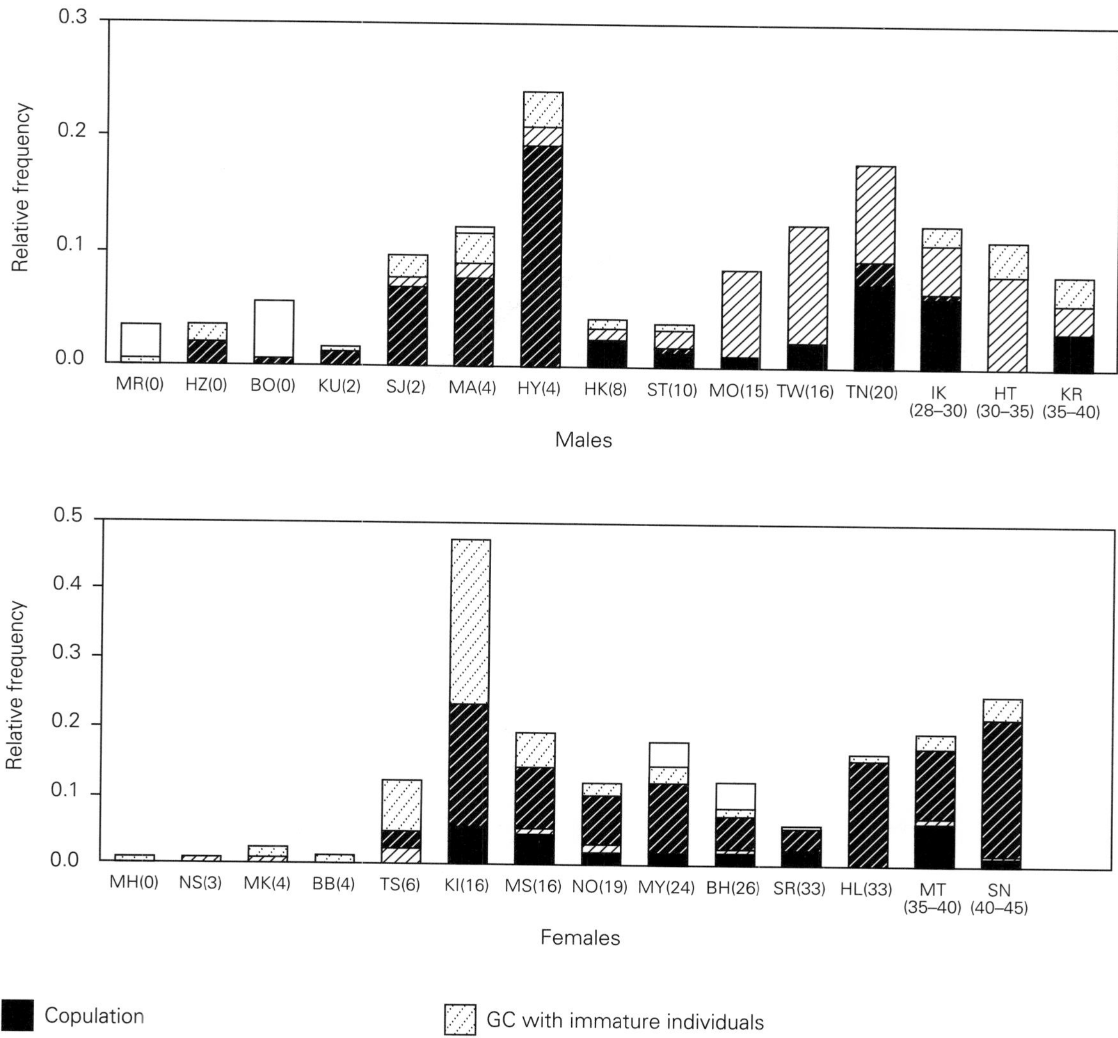

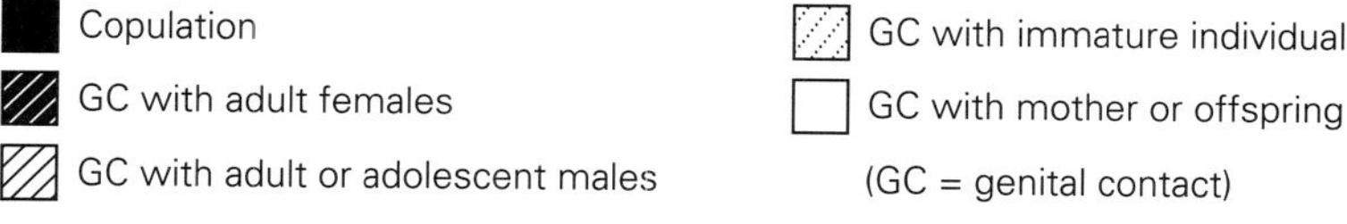

Figure 4

Frequency of sexual behavior involving each bonobo. Data for this figure were obtained by sampling all sexual behaviors that occurred at feeding sites. Relative frequency equals the number of cases of sexual behavior divided by total number of scanning observations at 10-minute intervals at which each individual was present. Ages of more than 15 years were estimated.

Social Functions of Genital Contact Between Adults

Although genital contact between immature individuals usually occurred as part of play, such contact between adults tended to occur in tense situations immediately after a party arrived at a feeding site (see figure 5). Of 137 cases of genital contact between adult individuals that were observed at feeding sites, 98 cases (71.5%) occurred after both participants had appeared at a feeding site but before the first scanning observation, that is, before 10 minutes had elapsed. By contrast, only 51 of 114 cases (44.7%) of genital contact between adult and immature individuals and five of 16 cases (31.3%) between immature individuals occurred before the first scanning observation. These differences were statistically significant for the adult-adult versus adult-immature comparison ($\chi^2 = 18.5$, df = 1, $p < .0001$), and for the adult-adult versus immature-immature comparison ($\chi^2 = 10.6$, df = 1, $p < 0.01$).

An analysis of social behaviors before and after the occurrence of genital contact suggests that genital contacts occurring between males and genital contacts occurring between females performed different social functions (see table 1). For example, genital contact between males occurred frequently following agonistic interactions between the participants (35.3%) or after display behavior by either of them (14.7%). However, once genital contact had occurred, neither male behaved agonistically toward the other or engaged in display behavior. Among the cases in which no social interaction was

Figure 5
Frequency of genital contacts per 10-minute sampling period after arrival of both participants at a feeding site. Data for this figure were obtained by sampling all occurrences of sexual behavior at feeding sites.

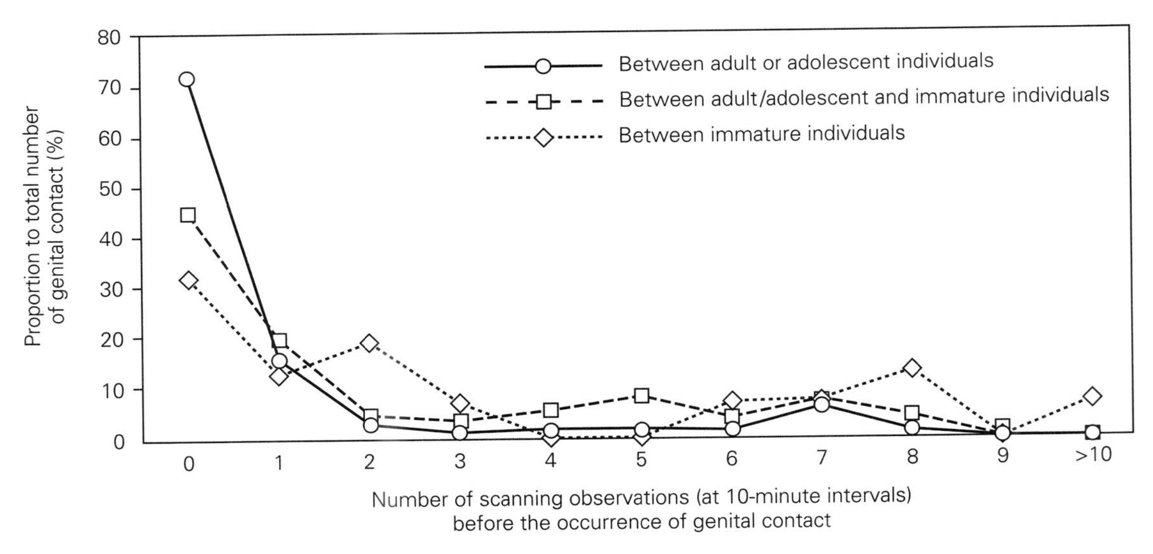

Because the scanning observations were begun when a first individual appeared at the feeding site, length of time from appearance of certain participants to occurrence of a first scanning observation varied from 0 to 10 minutes.

Table 1

Behavior of participants preceding or following genital contact.

	Male-male (%)	Female-female (%)	Male-female (%)
Behavior immediately preceding a genital contact:			
Agnostic behavior between participants[a]	35.3	0.0	60.0
Agnostic behavior with a third individual[a]	0.0	0.0	0.0
Display behavior by either participant[b]	14.7	0.0	0.0
Excited situation involving the whole party	0.0	7.7	20.0
Sexual behavior with a third individual	2.9	10.8	0.0
No social behavior observed	47.1	81.5	20.0
Approached from more than 1 meter	41.2	47.7	20.0
Stayed less than 1 meter apart	5.9	33.8	0.0
Number of observations	34	65	5[c]
Behavior immediately following a genital contact:			
Agnostic behavior between participants[a]	0.0	0.0	0.0
Agnostic behavior with a third individual[a]	6.3	0.0	0.0
Display behavior by either participant[b]	0.0	0.0	0.0
Excited situation involving the whole party	0.0	5.8	0.0
Sexual behavior with a third individual	3.1	18.4	0.0
No social behavior observed	90.6	75.8	100.0
Moved more than 1 meter apart	78.1	37.9	75.0
Stayed less than 1 meter apart	12.5	37.9	25.0
Number of observations	32	87	4[c]

Notes: Cases of genital contact for which preceding or following behavior was not confirmed are excluded. a. Includes both aggressive and submissive behaviors. b. Includes displays toward unspecified individuals. c. Results for the genital contact between males and females are not discussed in the text because of the small sample size.

observed before or after a genital contact, one male often approached the other immediately before the contact (41.2 of 47.1%) and the participants often moved apart more than one meter immediately following the contact (78.1 of 90.6%).

Genital contact between females sometimes occurred during excitement involving a whole party, with individuals showing agonistic behaviors, displaying, or emitting loud calls (7.7% of preceding behavior and 5.8% of following behavior). Such excitement was usually triggered by an agonistic interaction occurring in a feeding aggregation or by an intergroup encounter. However, no genital contact between females was immediately preceded or followed by agonistic or display behavior by the participants. Moreover, genital contact between females was fairly frequently preceded (10.8%) or followed (18.4%) by sexual interaction between one participant and a third individual. Genito-genital rubbing between females was seemingly linked to some extent with sexual arousal.

As with males, many cases of genital contact between females were immediately preceded by approach (47.7 of 81.5%) when no other social interaction had occurred prior to the genital contact. However, unlike between males, genital contact between females frequently occurred after the participants had already been feeding or resting within one meter of each other (33.8 of 81.5%). Also, unlike between males, females frequently remained within one meter after engaging in the genital contact (37.9 of 75.8%). Thus, while a major role of genital contact between males seemed to terminate agonistic interactions, genital contact between females seemed to initiate or maintain close proximity.

Discussion

We have little quantitative data on the sexual behavior of immature chimpanzees in the wild. However, several studies have reported that generally their sexual behavior is similar to that of bonobos. For example, immature chimpanzees begin to exhibit sexual behavior when they are younger than one year of age. They have partners of all age classes, and immature males frequently engage in copulation-like behavior with young adult females. Sexual behavior between immature individuals is mostly seen during play, and males engage in sexual behavior more frequently than do females (Goodall 1968; Hasegawa and Hiraiwa-Hasegawa 1983; Plooij 1984).

However, genital contact in the ventro-ventral position seems to be unique to immature bonobos, just as genito-genital rubbing is unique to adult female bonobos. Although immature chimpanzees sometimes show ventro-ventral genital contact (Bingham 1928; Goodall 1968), the frequency is very low, and the behavior disappears before maturity. By contrast, bonobos begin to show ventro-ventral genital contact before ventro-dorsal, and the ventro-ventral posture remains the most frequent type of genital contact throughout the immature period.

Another difference between species is seen in the reaction of immatures to the sexual behaviors of adults. Immature chimpanzees show great interest in copulation between adult males and females, and they frequently interrupt. Immature males interfere with copulation of their mothers more often than with copulation of other adult females, and immature females only disrupt copulation of their mothers. Such behavior seems to be related to the process of weaning, because it is frequently observed during the time for weaning. An immature individual does not interfere with copulation if its mother resumes estrous cycling well before the weaning period (Clark 1977; Tutin 1979). Immature bonobos also show great interest in sexual behavior between adults. However, they rarely disrupt such interactions and, instead, usually try to join them. They show more interest in the sexual behavior of adult females other than their mothers, and they sometimes engage in copulation-like behavior with adult females who have just engaged in sexual behavior with each other.

These differences may result because female bonobos usually resume estrous cycling long before they wean their offspring and most resulting copulations do not lead to conception (Furuichi 1992). Or, it may be that weaning proceeds more calmly in bonobos than in chimpanzees. Bonobo infants gradually become independent from their mothers, and bonobo mothers do not restrict demands by their offspring to suckle or to be carried as much as chimpanzees (Goodall 1986).

Although genital contact between immature bonobos is usually seen as a part of play, genital contacts between adult or adolescent individuals seem to serve an important role in facilitating the coexistence of individuals. Genital contact between males frequently occurs following display behavior by excited males or following agonistic interactions, and it seems to reduce tension. In chimpanzees, various social behaviors, such as pant-grunt, kiss, touch, and embrace, prevent or resolve conflict between males (de Waal 1989; Goodall 1986). Bonobos do not show such greeting behaviors; instead, they seem to substitute noncopulatory genital contact.

Genital contact between bonobo females also reduces tension—it is frequently performed during tense situations, such as group excitement involving all party members. However, female genital contact is rarely used to terminate display or agonistic behaviors as in males. The most important role of genital contact between females seems to be to reduce tension when one female approaches another and to facilitate maintenance of close proximity with one another. These functions seem related to the grouping pattern of bonobos, in which females usually remain with a mixed party and congregate near its core (Furuichi 1989; Kano 1982; Kitamura 1983; White 1988).

Genital contact in bonobos plays an important role in relations between groups. For example, during encounters with other groups, females (frequently) and males (occasionally) engage in genital contact with members of both groups. Once, when two groups met at a feeding site, the alpha males of both groups each displayed, then subsequently engaged in rump-rump contact. After this interaction, the members of both groups calmed down and then ranged and fed together.

Because of great individual variation, only a long-term study of several immature individuals could clarify the process by which sexual behaviors develop. However, the present data suggest some patterns in the development of noncopulatory sexual behaviors for each sex.

First, immature bonobo males show much more interest in sexual behavior than do immature females, and the frequency of the male's genital contacts increases with age. Although the males' genital contact abruptly decreases during early adolescence, it increases again when they enter adulthood because of frequent genital contact with other adult males. Moreover, during the sexually inactive period in adolescence, the role of genital contact changes from an element of play to a reducer of social tension.

Second, bonobo females show less interest than males in sexual behavior during immaturity, and the frequency of their genital contacts does not increase with age. The sexually inactive state continues almost until the end of the juvenile stage. Females leave their natal groups in early adolescence. Young females, when immigrating to a new group, tend to engage in genital contact more frequently than do older females (Furuichi 1989; Idani 1991). Genital contact between adult females seems to develop quickly as they transfer between groups, enabling them to cope with nonrelated females. Genital contact between adult females is much more frequent than between adult males.

Summary

Immature bonobos engage in genital contact with individuals of most age classes. Genital contact between immature individuals is most frequently performed in the ventro-ventral posture. It is usually observed during play sessions; males participate more frequently than females. Both immature males and females engage in genital contact with adult males, although not very frequently. On the other hand, only immature males frequently engage in genital contact with adult females, and its frequency increases with age until adolescence. Unlike chimpanzees, immature bonobos rarely disrupt with the sexual behavior of their mothers.

Genital-contact behaviors drastically change through adolescence. Males become sexually inactive during adolescence, but they resume frequent genital contact when entering adulthood. Females begin to show frequent genital contact after transfer to other groups. Genital contacts between adults follow fixed patterns functioning in terminating agonistic interactions, in reducing tensions, and in initiating or maintaining close proximity with one another.

Marked differences in the sexual behavior of chimpanzees and bonobos seem to appear during the course of maturation. Adult male bonobos seem to engage in genital contacts instead of nonsexual greetings, which, in contrast, are well developed in chimpanzees. The very high frequency of genital contacts between adult female bonobos seems related to their cohesive grouping patterns.

Acknowledgments

We thank Professor T. Kano, Dr. H. Ihobe, and other members of the Primate Research Institute, Kyoto University, for their valuable support and advice. We also thank the staff of the Research Center for the Natural Science of Zaire (C.R.S.N.Z.) and the research station at Wamba for their continuing cooperation. This study was financially supported by grants under the Monbusho International Scientific Research Program to T. Kano (#63041078) and T. Furuichi (#01790353).

References

Bingham, H.C. 1928. Sex development in apes. *Comp. Psychol. Monogr.* 5:1–165.

Clark, C.B. 1977. A preliminary report on weaning among chimpanzees of the Gombe National Park, Tanzania. In S. Chevallier-Skolnikoff and F.E. Poirier, eds., *Primate Bio-Social Development: Biological, Social and Ecological Determinants,* pp. 235–60. New York: Garland Publishing.

Furuichi, T. 1987. Sexual swelling, receptivity, and grouping of wild pygmy chimpanzee females at Wamba, Zaire. *Primates* 28:309–18.

———. 1989. Social interactions and the life history of female *Pan paniscus* in Wamba, Zaire. *Internat. J. Primatol.* 10:173–97.

———. 1992. The prolonged estrus of females and factors influencing mating in a wild group of bonobos *(Pan paniscus)* in Wamba, Zaire. In N. Itoigawa, Y. Sugiyama, G.P. Sackett, and R.K.R. Thompson, eds., *Topics in Primatology.* Vol. 2, *Behavior, Ecology, and Conservation,* pp. 179–90. Tokyo: Univ. of Tokyo Press.

Goodall, J. 1968. Behaviour of free-living chimpanzees of the Gombe Stream area. *Anim. Behav. Monogr.* 1:163–311.

———. 1986. *The Chimpanzees of Gombe: Patterns of Behavior.* Cambridge, Mass.: Belknap Press.

Hasegawa, T., and M. Hiraiwa-Hasegawa. 1983. Opportunistic and restrictive matings among wild chimpanzees in the Mahale Mountains, Tanzania. *J. Ethol.* 1:75–85.

Idani, G. 1991. Social relationships between immigrant and resident bonobo *(Pan paniscus)* females at Wamba. *Folia Primatol.* 57:83–95.

Kano, T. 1982. The social group of pygmy chimpanzees *(Pan paniscus)* of Wamba. *Primates* 23:171–88.

———. 1992. *The Last Ape: Pygmy Chimpanzee Behavior and Ecology.* Stanford, Calif.: Stanford Univ. Press.

Kitamura, K. 1983. Pygmy chimpanzee association patterns in ranging. *Primates* 24:1–12.

———. 1989. Genito-genital contacts in the pygmy chimpanzee *(Pan paniscus). African Study Monogr.* 10:49–67.

Kuroda, S. 1980. Social behavior of the pygmy chimpanzees. *Primates* 21:181–97.

Plooij, F.X. 1984. *The Behavioral Development of Free-Living Chimpanzee Babies and Infants.* Norwood, New Jersey: Ablex Publishing.

Thompson-Handler, N., R.K. Malenky, and N. Badrian. 1984. Sexual behavior of *Pan paniscus* under natural conditions in the Lomako Forest, Equateur, Zaire. In R.L. Susman, ed., *The Pygmy Chimpanzee,* pp. 347–68. New York: Plenum Press.

Tutin, C.E.G. 1979. Responses of chimpanzees to copulation, with special reference to interference by immature individuals. *Anim. Behav.* 27:845–54.

de Waal, F. 1987. Tension regulation and nonreproductive functions of sex in captive bonobos *(Pan paniscus). Nat. Geogr. Research* 3:318–35.

———. 1989. *Peacemaking Among Primates.* Cambridge, Mass.: Harvard Univ. Press.

White, F.J. 1988. Party composition and dynamics in *Pan paniscus. Internat. J. Primatol.* 9:179–93.

Grooming Relationships in Two Species of Chimpanzees

Yasuyuki Muroyama and Yukimaru Sugiyama

Introduction

Social grooming is the most common form of active affiliative behavior in nonhuman primates (Goosen 1987). This behavior is performed by one animal on another and consists of brushing and picking through the fur with the fingers and the toes. Social grooming appears in every aspect of social life in both chimpanzees *(Pan troglodytes)* (Goodall 1986) and bonobos *(P. paniscus)* (Kano 1986). Social grooming in chimpanzees and bonobos has been studied as it relates to social organization (Kano 1980; Nishida 1970; Sugiyama 1969; Sugiyama and Koman 1979); to social relationships between same- and opposite-sex members of a group (Furuichi 1989; Idani 1991; Ihobe 1992; Nishida and Hiraiwa-Hasegawa 1987; Simpson 1973; Takahata 1990a, 1990b); to social relations between neighboring groups (Idani 1990); to sexual behavior (McGinnis 1979); to the development of offspring (Nishida 1988; Pusey 1983); to reconciliation (de Waal and Roosmalen 1979); to reunion (Bauer 1979); and to reciprocity (Hemelrijk and Ek 1991). However, the nature and diversity of grooming across chimpanzee and bonobo populations and between the two species have rarely been studied or compared (Sugiyama 1988). There have been little data on social grooming and other interactions for bonobos comparable to data for chimpanzees.

In this study, we compare some measures of social grooming in several populations of chimpanzees and bonobos to elucidate the diversity of social grooming within and between the two species of *Pan.* We examine grooming patterns of males and females of each species within their groups. Also we examine how the grooming patterns for each species relate to the social relationships between group members.

Material and Methods

We used socioecological studies of chimpanzees that were carried out at Bossou (Sugiyama and Koman 1979), Gombe (Goodall 1965, 1983), Mahale (Nishida 1968, 1979), Budongo (Sugiyama 1969), Kibale (Ghiglieri 1984), as well as studies of bonobos at Wamba (Kuroda 1979), Yalosidi (Kano 1983), and Lomako (Badrian et al. 1981). Available published data on grooming frequencies between same- and opposite-sex pairs in a group, the number of grooming partners, reciprocity between grooming pairs, and grooming duration were analyzed and compared with each other. Data collected in different periods on the same populations were analyzed to examine consistency and trends in each population. Observed grooming frequencies between individuals were compared with expected frequencies, which were calculated from adult male-female sex ratios for each population. The measure used to compare grooming frequencies is the *standardized residual,* which is found by dividing the remainder of the observed value minus the expected value by the square root of the expected value (Sugiyama 1988). Study periods and adult male-female sex ratios of the populations used in this study are summarized in table 1. These study periods varied widely in sample method, analysis, and observation conditions.

Table 1
Summary of the subject populations of chimpanzees and bonobos.

Species	Study site	Study period	Male:Female	Male:Female	Sources
Pan troglodytes:					
	Bossou	1976–77	3:7	0.43	Sugiyama 1988
		1982–83	2:7	0.29	Sugiyama 1988
	Gombe	1960–62	15:16	0.94	Goodall 1965
		1969–70	15:13	1.15	Simpson 1973
		1978	8:20	0.40	Goodall 1983
	Mahale	1967–69	6:10	0.60	Nishida 1970
		1971–72	4:11	0.36	Tachibana 1991
		1973–74	5:15	0.33	Nishida 1979
		1981	12:49[a]	0.24	Takahata 1990a, 1990b
	Kibale	1976–78	13:17	0.76	Ghiglieri 1984
	Budongo	1966–67	18:14	1.29	Sugiyama 1969
Pan paniscus:					
	Wamba	1974–75	26:39	0.67	Kuroda 1979, 1980
		1975–77	12:17	0.71	Kano 1980
		1978–79	22:23[a]	0.96	Kano 1986
		1984–85	10:10[a]	1.00	Idani 1991
		1986–87(E1)	10:10[a]	1.00	Ihobe 1992
		1986–87(E2)	10:14[a]	0.71	Ihobe 1992
	Lomako	1980–82	11:18[b]	0.61	Badrian and Badrian 1984
	Yalosidi	1973–75	8:13[c]	0.63	Kano 1980, 1983

Note: Two populations (E1, E2) were observed at Wamba in 1986–87. a. These values include adolescents; b. The maximum number of adult males and females observed; c. Value calculated from the maximum group size and male-female ratio in the paper.

In the Bossou chimpanzee group, which has been studied since 1976 (Sugiyama and Koman 1979), adult female membership has remained almost constant, whereas the number of adult males has varied from one to four. For comparison with the other populations that have more than one adult male, data were analyzed from two study periods (1976–77 and 1982–83), when the Bossou group had two or three adult males and seven adult females (Sugiyama 1988).

Results

Grooming Frequencies Among Individuals

In all subject populations of chimpanzees, except for Bossou in 1976–77, grooming frequencies between males were higher than expected, particularly at Mahale (see table 2). By contrast, grooming frequencies between males and females were lower than expected in all populations. In addition, females groomed other females more frequently at Bossou than at Mahale and at Budongo, where females were much less likely to have any grooming interactions with either females or males.

Table 2
Frequency and expected frequency of grooming combinations by sex in chimpanzees and bonobos

Species	Study site	Period	Total	Grooming combinations		
				Male-Male	Male-Female	Female-Female
Pan troglodytes:						
	Bossou	1976	160	8(10.7)	60(74.7)	92(74.7)
		1982	181	13(5.0)	49(70.4)	119(105.6)
	Gombe	1960	98	43(22.1)	39(50.6)	16(25.3)
		1978	173	24(12.8)	71(73.2)	78(87.0)
	Mahale	1967	30	18(3.8)	12(15.0)	0(11.2)
		1971	167	97(9.5)	63(70.0)	7(87.5)
		1973	454	218(23.9)	187(251.0)	49(179.1)
		1981	635	389(63.3)	246(571.7)	—
	Kibale	1976	78	28(14.0)	24(39.6)	26(24.4)
	Budongo	1966	102	74(31.5)	26(51.8)	2(18.7)
Pan paniscus:						
	Wamba	1974	117	11(18.3)	70(57.0)	36(41.7)
		1975	155	19(25.2)	84(77.9)	52(51.9)
		1978	176	27(41.1)	103(90.0)	46(45.0)
		1984[a]	140	5(33.2)	98(73.7)	37(33.2)
		1986(E1)	107	16(33.2)	91(73.8)	—
		1986(E2)	119	77(28.9)	42(90.0)	—
	Lomako	1980	35	5(4.7)	18(17.1)	12(13.2)
	Yalosidi	1973	10	0(1.3)	7(5.0)	3(3.7)

Note: Cells show the actual frequencies of sessions, or bouts, of all adult grooming pairs observed. Figures in parentheses show expected frequencies calculated from sex ratios of adults in the populations. a. One-half of observation hours spent following females. E1 and E2 were two populations observed at Wamba.

Figure 1 represents grooming frequencies of same- and opposite-sex pairs. Chimpanzee populations differed greatly from one another in grooming frequencies. However, each population showed rather consistent grooming frequencies over different study periods. Also, across chimpanzee populations there was a significant tendency showing that when female-female grooming is high, male-male grooming is diminished, and where male-male grooming is high, female-female grooming is low (Kendall tau = -0.889, N = 9, p = .0003, two-tailed). However, the scores of the other combinations were not significantly correlated with each other.

Except for the E2 group of Wamba and the bonobo population of Lomako (see table 2), grooming frequencies between males were lower than expected; and frequencies between males and females were higher than expected (see table 2). Grooming interactions between females were as frequent as expected. In the E2 group of Wamba where social bonds between mothers and their adult male offspring were weak (Ihobe 1992), males had grooming interactions with other males much more often than with females. The Lomako and Yalosidi populations showed random distribution of grooming patterns relative to adult male-female sex ratios, although sample sizes were small.

Bonobos as compared to chimpanzees appeared to have less variation in grooming frequencies between populations or periods. As seen in figure 1, bonobo populations showed similar tendencies in comparison with chimpanzee populations. Unlike chimpanzees, the standardized scores of grooming between male bonobos had a negative correlation with the scores between males and females (Kendall tau = –0.467, N = 6, p = .056, two-tailed). The scores of the other sex combinations were not significantly correlated with each other.

Number of Grooming Partners

At Gombe, male chimpanzees appeared to have more female grooming partners than did females (10.7 male to female versus >6.7 female to male), whereas females had almost the same number of male partners as males did (6.0 female to male versus 5.7 male to male) (see table 3). At Bossou, however, both female and male chimpanzees had almost the same number of same- and opposite-sex partners. Bonobos had far fewer grooming partners than did chimpanzees, and bonobo males had fewer partners than did bonobo females.

Other Characteristics of Grooming

For chimpanzees at Gombe and Mahale, male-male one-way grooming pairs were more common than male-male two-way grooming pairs: male chimpanzee pairs at these sites were likely to groom each other during the

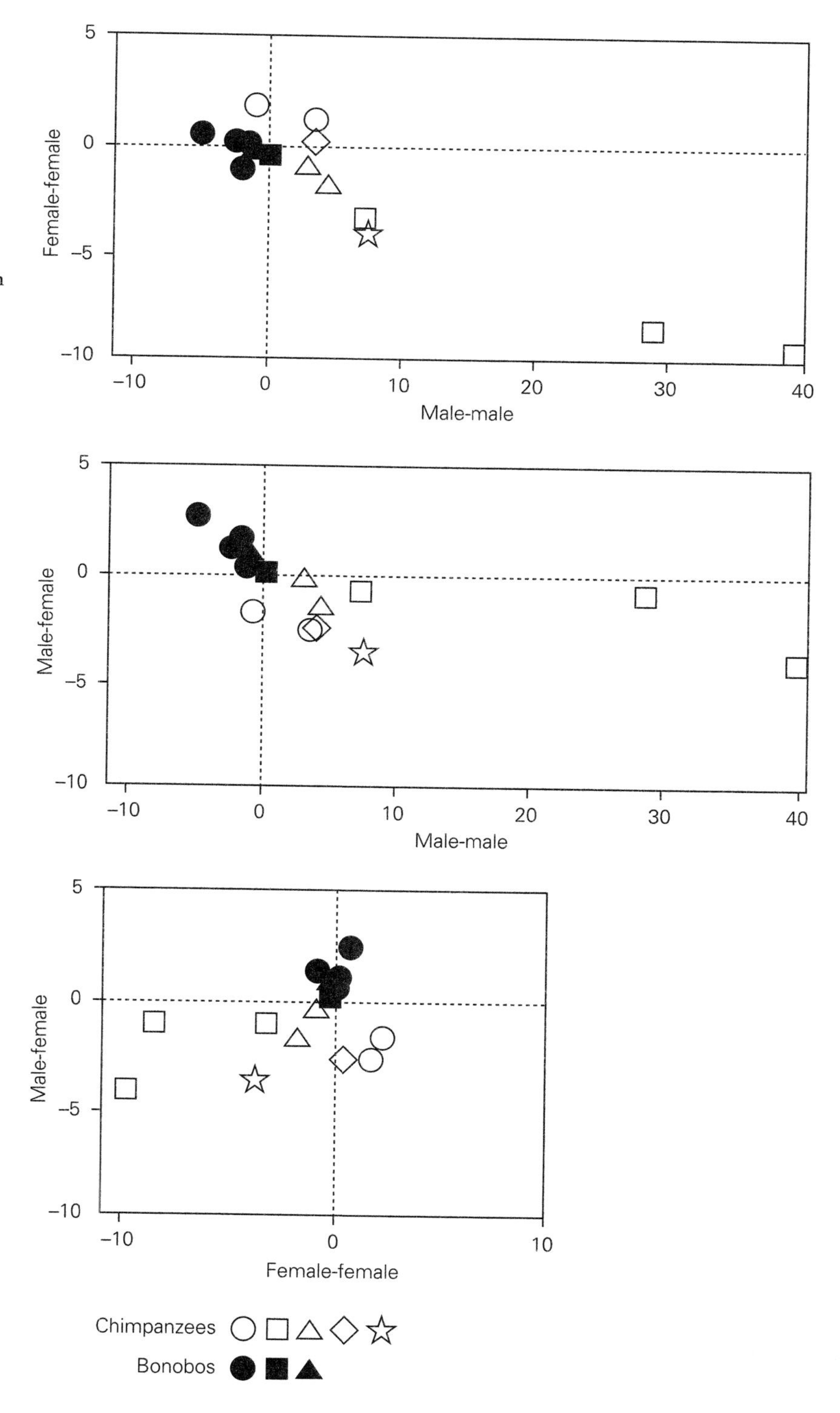

Figure 1
Grooming frequencies of same- and opposite-sex pairs. Same- and inter-sex grooming relations are represented by the scores of standardized residuals, observed-expected/ √expected, of grooming interactions. Expected values were calculated from adult sex compositions in the populations of chimpanzees and bonobos (see table 2). Data including all sex combination of grooming were used (N = 9 chimpanzees; N = 6 bonobos). Each symbol represents one study site of chimpanzees (○ : Bossou; □ : Mahale; △ : Gombe; ◇ : Kibale; ☆ : Budongo) or bonobos (● : Wamba; ■ : Lomako; ▲ : Yalosidi).

Table 3
Mean numbers of grooming partners in each grooming combination.

Species	Study site	Period	Mean number of grooming partners/potential partners			
			Male-to-Male	Male-to-Female	Female-to-Male	Female-to-Female
Pan troglodytes:						
	Bossou	1976	2.0/2	5.7/7	2.4/3	4.8/6
		1982	1.0/1	5.0/7	1.4/2	5.0/6
	Gombe	1969	8.5/10	—	—	—
		1983[a]	5.7/7	10.7/20	6.0/8	6.7/19
	Mahale	1981	6.6/9	—	—	—
Pan paniscus:						
	Wamba	1984	—	—	—	3.0/9
		1986(E1)	1.4/9	—	—	—
		1986(E2)	2.0/9	—	—	—

Notes: a. Goodall (1983) reported both the numbers of groomers and of groomees but not the numbers of grooming partners. Accordingly, the larger values of the two numbers were presented here. When the number of grooming partners is less than those being groomed, the values may be underestimated. E1 and E2 were two populations observed at Wamba.

Table 4
Percentage of bi- and uni-directional grooming dyads in each grooming combination.

Species	Study site	Period	Male-Male			Male-Female			Female-Female		
			N	bi-	uni-	N	bi-	uni-	N	bi-	uni-
Pan troglodytes:											
	Bossou	1976	3	33.3	66.7	21	61.9	19.0	21	52.4	28.6
		1982	1	0.0	100.0	14	28.6	42.9	21	57.1	28.6
	Gombe	1969	55	70.9	14.5	—	—	—	—	—	—
	Mahale	1981	45	48.9	24.4	—	—	—	—	—	—
Pan paniscus:											
	Wamba	1984	—	—	—	—	—	—	45	22.2	11.1
		1986(E1)	45	8.9	6.7	—	—	—	—	—	—
		1986(E2)	45	6.7	15.6	—	—	—	—	—	—

Note: E1 and E2 were two populations observed at Wamba.

study periods (see table 4). At Bossou, by contrast, male-male one-way grooming was more common than male-male two-way grooming. Both female- and male-female grooming relationships tended to be two-way. Data on female-female and male-female grooming at Gombe and Mahale were not available for comparison.

Durations of grooming sessions, or *bouts,* between male chimpanzees appeared to be the longest of all the sex combinations at Gombe, Mahale, and Kibale (see table 5). At Gombe, grooming durations between females

Table 5
Mean duration of grooming sessions, or bouts, in each grooming combination.

Species	Study site	Period	Male-Male		Male-Female		Female-Female	
			N	minutes	N	minutes	>N	minutes
Pan troglodytes:								
	Gombe	1983	24	25.9	61	13.6	36	6.3
	Mahale	1981[a]	389	3.0	110	1.9	—	—
	Kibale	1976	38	17.8	24	10.4	26	15.5
Pan paniscus:								
	Wamba	1974	11	14.5	70	21.3	36	15.5
		1975	19	25.5	84	22.8	52	16.1
		1978[b]	27	15.8	103	17.1	46	8.5
	Yalosidi	1973	0	—	17	33.4	7	30.3

Notes: a. Data from Mahale represent mean durations of grooming bouts, but data from other sites are length of grooming sessions; b. These values include sessions between adolescents.

were much shorter than between any other dyads, except for the duration between a mother and her adult daughter, which have been reported to be much longer (19.3 min, N = 42, Goodall 1986).

Unlike chimpanzees, two-way grooming relations between male bonobos were uncommon and less frequent than those between female bonobos (see table 4). The duration of grooming sessions between male and female bonobos was longer than any other combinations, although durations of grooming between males were about as long as those between males and females at Wamba in 1975 and 1978 (see table 5).

Discussion

For chimpanzees, many studies reported that higher grooming frequencies, larger numbers of grooming partners, and longer grooming durations were more likely between adult males than between males and females or between females (Nishida and Hiraiwa-Hasegawa 1986; Wrangham 1986). However, grooming frequencies between same- and opposite-sex partners, in particular those between males, varied greatly across different chimpanzee populations. In bonobos, by contrast, grooming interactions among males and females were more frequent and longer than those among males and among females. In general the variation of grooming among populations was much greater in chimpanzees than in bonobos.

Grooming Between Males

Social bonds between male chimpanzees, which are based on grooming and various other kinds of social interaction, are significant in two contexts (Wrangham 1986). The first context is cooperation during intragroup competition to achieve dominance. Dominance facilitates access to resources such as estrous females (Hasegawa and Hiraiwa-Hasegawa 1983; Nishida 1983; Sugiyama and Koman 1979; Tutin 1979). The second context is cooperation during antagonistic competition between neighboring groups (Goodall et al. 1979; Nishida 1979). When dominance relations between males are likely to be affected by alliance with other males or when relationships between neighboring groups are more hostile, cooperation among males within the group is more crucial. Under such circumstances male chimpanzees should have a larger number of grooming partners and should have longer grooming sessions to develop and maintain social bonds with those partners (Hutchins and Barash 1976; McKenna 1978) and to reduce tension (Terry 1970).

These hypothetical grooming patterns for male chimpanzees appear in populations at Mahale and Gombe. However, they do not appear in the Bossou group. The Bossou group had few male grooming partners within the group and few contacts with any neighboring group (Sugiyama 1988). The lack of competition at Bossou is likely to allow chimpanzee males to invest less in their grooming relationships with other males. Thus, diversity in male-male grooming patterns across chimpanzee populations may result when males vary grooming relationships to suit social circumstances. (In the present study, the effects of relationships between brothers on grooming relations in chimpanzees remain unknown.)

In bonobos, unlike chimpanzees, the prolonged receptivity of the female may reduce competition among males (Furuichi 1989). Male bonobos have far more opportunities to copulate with estrous females than do male chimpanzees (Furuichi 1987). Furthermore, as compared with chimpanzees, relations between the males of neighboring bonobo groups are less hostile (Idani 1991). Accordingly, as compared with chimpanzees, cooperation between bonobo males for dominance within a group or for intergroup interactions is less significant. In addition, the relations between bonobo brothers have no effect on their grooming (Ihobe 1992). This characteristic may result in less grooming between males and in less variation of grooming frequencies across bonobo populations.

Grooming Between Females

In chimpanzees, females spend much of their time alone with their immature offspring and, as compared with males, are less likely than males to obtain benefits from cooperation (Wrangham 1986). This suggests that grooming relationships between females, as compared with males, would be less diversified in accordance with social circumstances, which is consistent with the results of this study. However, kin relations of female chimpanzees may affect the frequencies and durations of female-female grooming. This is shown in populations at Gombe and Bossou where relationships between mothers and adult daughters appear to have higher grooming frequencies and longer durations (Goodall 1986; Sugiyama 1988).

In bonobos, by contrast, kin relations between females appear to have no effect on grooming frequencies. Bonobos leave their natal groups and transfer to another group when they are old juveniles or young adolescents (Furuichi 1989). There has been no evidence of cooperation between female bonobos for resources, although female bonobos, as compared with female chimpanzees, spend much more of their time with other females (Furuichi 1989). These factors may lead to little variation in female grooming patterns among bonobo populations, although the immigration of female bonobos may cause short-term variation within the groups (Idani 1991).

Grooming Between Males and Females

In chimpanzees, grooming relationships between males and females are significantly affected by the sexual status of females (Nishida 1979; Tutin 1979; Takahata 1990a). These relationships varied the least across populations. This may suggest that grooming between males and females should be less varied from one chimpanzee population to the next and less affected by social factors peculiar to each population.

Grooming frequencies between males and females varied as widely between bonobo populations as between chimpanzee populations. Female bonobos are likely to maintain strong bonds with their grown sons (Furuichi 1989), which increases male-female grooming frequencies. The intensity of social bonds between bonobo mothers and adult sons affects social relationships between males and between males and females in the group (Furuichi 1989; Ihobe 1992). By contrast, weakness of the mother-son bond owing to a mother's death, for example, may lead to establishment and maintenance of male-male grooming relationships as in the E2 group of Wamba in 1986–87 (Ihobe 1992). Thus, various kin relationships, such as those between bonobo mothers and sons, appear to diversify grooming between group members.

In conclusion, chimpanzees and bonobos differ strikingly in grooming relationships observed between both same- and opposite-sexes pairs. In chimpanzees, the variations in male-male grooming relationships and variations in matrilineal kin relations between females may result in the overall diversity of grooming behavior between chimpanzee populations. By contrast, in bonobos, who display far less diversity in grooming behavior between populations than do chimpanzees, the existence and intensity of kin relations between mothers and sons may diversify grooming relationships within groups.

Summary

The grooming behavior between same- and opposite-sex members of chimpanzee groups differed from grooming behavior in bonobo groups. In all of the chimpanzee groups except the one at Bossou, grooming frequencies between males were higher, numbers of male-male grooming partners were larger, and male-male grooming durations were longer than in any other sex combination. By contrast, in all studies of the bonobo groups except the one at Wamba, grooming interactions between males and females were more frequent and longer than in any other sex combination.

Diversity of grooming between different study populations and different periods was greater in chimpanzees than in bonobos. Grooming frequencies between chimpanzee males correlated negatively with those between females, whereas grooming frequencies between bonobo males correlated negatively with those between males and females. Differences in grooming patterns between chimpanzees and bonobos were discussed in relation to cooperation between males, female-female kin relationships, and kin relations between males and females.

References

Badrian, A. and N. Badrian. 1984. Social organization of *Pan paniscus* in the Lomako Forest, Zaire. In R. L. Susman, ed, *The Pygmy Chimpanzee: Evolutionary Biology and Behavior.* pp. 325–46. New York. Plenum Press.

Badrian, N., A. Badrian, and R.L. Susman. 1981. Preliminary observations on the feeding behavior of *Pan paniscus* in the Lomako Forest of Central Zaire. *Primates* 22:173–81.

Bauer, H.R. 1979. Agonistic and grooming behavior in the reunion context of Gombe Stream chimpanzees. In D.A. Hamburg and E.R. McCown, eds., *The Great Apes,* pp. 395–404. Menlo Park, Calif.: Benjamin/Cummings.

Furuichi, T. 1987. Sexual swelling, receptivity and grouping of wild pygmy chimpanzee females at Wamba, Zaire. *Primates* 28:309–18.

———. 1989. Social interactions and the life history of female *Pan paniscus* in Wamba, Zaire. *Internat. J. Primatol.* 10:173–97.

Ghiglieri, M.P. 1984. *The Chimpanzees of Kibale Forest.* New York: Columbia Univ. Press.

Goodall, J. 1965. Chimpanzees of the Gombe Stream Reserve. In I. DeVore, ed., *Primate Behavior: Field Studies of Monkeys and Apes,* pp. 425–73. New York: Holt, Rinehart and Winston.

———. 1983. Population dynamics during a 15-year period in one community of free-living chimpanzees in the Gombe National Park, Tanzania. *Z. Tierpsychol.* 61:1–60.

———. 1986. *The Chimpanzees of Gombe.* Cambridge, Mass.: Belknap Press.

Goodall, J., A. Bandora, E. Bergmann, C. Busse, H. Matama, E. Mpongo, A. Pierce, and D. Riss. 1979. Intercommunity interactions in the chimpanzee population of the Gombe National Park. In D.A. Hamburg and E.R. McCown, eds. *The Great Apes,* pp. 12–53. Menlo Park, Calif.: Benjamin/Cummings.

Goosen, C. 1987. Social grooming in primates. In G. Mitchell and J. Erwin, eds., *Comparative Primate Biology,* vol. 2B: *Behavior, Cognition, and Motivation,* pp. 107–32. New York: Alan R. Liss.

Hasegawa, T., and M. Hiraiwa-Hasegawa. 1983. Opportunistic and restrictive matings among wild chimpanzees in the Mahale Mountains, Tanzania. *J. Ethol.* 1:75–85.

Hemelrijk, C.K., and A. Ek. 1991. Reciprocity and interchange of grooming and "support" in captive chimpanzees. *Anim. Behav.* 41:923–35.

Hutchins, M., and D.P. Barash. 1976. Grooming in primates: Implications for its utilitarian function. *Primates* 17:145–50.

Idani, G. 1990. Relations between unit-groups of bonobos at Wamba, Zaire: Encounters and temporary fusions. *African Study Monogr.* 11:153–86.

———. 1991. Social relationships between immigrant and resident bonobo (*Pan paniscus*) females at Wamba. *Folia Primatol.* 57:83–95.

Ihobe, H. 1992. Male-male relationships among wild bonobos *(Pan paniscus)* at Wamba, Republic of Zaire. *Primates* 33:163–79.

Kano, T. 1980. Social behavior of wild pygmy chimpanzees *(Pan paniscus)* of Wamba: A preliminary report. *J. Human Evol.* 9:243–60.

———. 1983. An ecological study of the pygmy chimpanzees *(Pan paniscus)* of Yalosidi, Republic of Zaire. *Intern. J. Primatol.* 4:1–31.

———. 1986. *The Last Ape.* Trans. from Japanese into English by E.O. Vineberg. Stanford, Calif.: Stanford Univ. Press, 1992.

Kuroda, S. 1979. Grouping of the pygmy chimpanzees. *Primates* 20:161–83

McGinnis, P.R. 1979. Sexual behavior in free-living chimpanzees: Consort relationships. In D.A. Hamburg and E.R. McCown, eds., *The Great Apes,* pp. 429–40. Menlo Park, Calif.: Benjamin/Cummings.

McKenna, J.J. 1978. Biosocial functions of grooming behavior among the common Indian langur monkeys *(Presbytis entellus). Am. J. Phys. Anthro.* 48:503–10.

Nishida, T. 1968. The social group of wild chimpanzees in the Mahale Mountains. *Primates* 9:167–224.

———. 1970. Social behavior and relationship among wild chimpanzees of the Mahale Mountains. *Primates* 11:47–87.

———. 1979. The social structure of chimpanzees of the Mahale Mountains. In D.A. Hamburg and E.R. McCown, eds., *The Great Apes,* pp. 73–121. Menlo Park, Calif.: Benjamin/Cummings.

———. 1988. Development of social grooming between mother and offspring in wild chimpanzees. *Folia Primatol.* 50:109–23.

Nishida, T., and M. Hiraiwa-Hasegawa. 1987. Chimpanzees and bonobos: Cooperative relationships among males. In B.B. Smuts, D.L. Cheney, R.M. Seyfarth, R.W. Wrangham, and T.T. Struhsaker, eds., *Primate Societies,* pp. 165–77. Chicago: Univ. of Chicago Press.

Pusey, A.E. 1983. Mother-offspring relationships in chimpanzees after weaning. *Anim. Behav.* 31:363–77.

Simpson, M.J.A. 1973. The social grooming of male chimpanzees. In R.P. Michael and J.H. Crook, eds., *Comparative and Behavior of Primates,* pp. 411–505. London: Academic Press.

Sugiyama, Y. 1969. Social behavior of chimpanzees in the Budongo Forest, Uganda. *Primates* 10:197–225.
———. 1988. Grooming interactions among adult chimpanzees at Bossou, Guinea, with special reference to social structure. *Internat. J. Primatol.* 9:393–407.
Sugiyama, Y., and Koman, J. 1979. Social structure and dynamics of wild chimpanzees at Bossou, Guinea. *Primates* 20:323–39.
Tachibana, T. 1991. *The Frontiers of Primatology.* Tokyo: Heibonsha (Japanese).
Takahata, Y. 1990a. Adult males' social relations with adult females. In T. Nishida, ed., *The Chimpanzees of the Mahale Mountains: Sexual and Life History Strategies,* pp. 133–48. Tokyo: Univ. of Tokyo Press.
———. 1990b. Social relationships among adult males. In T. Nishida, ed., *The Chimpanzees of the Mahale Mountains: Sexual and Life History Strategies,* pp. 149–70. Tokyo: Univ. of Tokyo Press.
Terry, R.L. 1970. Primate grooming as a tension reduction mechanism. *J. Psychol.* 76:129–36.
Tutin, C.E., 1979. Mating patterns and reproductive strategies in a community of wild chimpanzees *(Pan troglodytes schweinfurthii). Behav. Ecol. Sociobiol.* 6:39–48.
de Waal, F.B.M., and A.V. Roosmalen. 1979. Reconciliation and consolation among chimpanzees. *Behav. Ecol. Sociobiol.* 5:55–66.
Wrangham, R.W. 1986. Ecology and social relationships in two species of chimpanzees. In R.I. Rubenstein and R.W. Wrangham, eds., *Ecological Aspects of Social Evolution: Birds and Mammals,* pp. 352–78. Princeton, N.J.: Princeton Univ. Press.

Reproductive Success Story

Variability Among Chimpanzees and Comparisons with Gorillas

Caroline E.G. Tutin

Introduction and Methods

The reproductive potential of females is defined by three parameters: age at birth of first offspring (first birth), interval between births, and life span. A study of reproduction must also include genetic, nutritional, and social factors. Genetic factors determine the limits of an individual's developmental rate, while nutritional and social factors influence the process of growth and maturation. Data from long-term field studies of chimpanzees have documented the low reproductive potential of females due to long interbirth intervals (Tutin 1980). Comparing the reproductive parameters of chimpanzees and gorillas, which despite their larger body size reproduce more quickly than chimpanzees (Harcourt et al. 1980), allows hypotheses to be developed about the impact of social and nutritional factors on reproduction.

Long-term field studies of the African chimpanzees and gorillas began in the 1960s. Research has been in progress for 26 to 34 years at three sites: at Gombe and Mahale, in Tanzania, with chimpanzees, *Pan troglodytes schweinfurthii,* and at Karisoke, Rwanda, with mountain gorillas, *Gorilla gorilla beringei* (Fossey 1983; Goodall 1986; Nishida 1990). Despite an enormous investment of time, effort, and money in these exceptional field studies, data on reproductive patterns remain limited. For example, at Mahale no female chimpanzee of known birth date has been observed at the time of her first birth, and at Gombe only eight females of known

age have been followed to sexual maturity, in 34 years of continuous observation. At Karisoke, only eight mountain gorilla females of known age have been followed to their first birth. The main reason for the paucity of data on reproductive parameters is that females of both species leave their natal group at adolescence (Harcourt et al. 1976; Pusey 1979), making it difficult to keep track of their lives. Few demographic data are available from other field studies of chimpanzees, as few have yet involved continuous, long-term observation of habituated individuals. But information from the small community of chimpanzees, *Pan t. verus,* at Bossou, Guinea, suggests that variability exists in reproductive parameters (Sugiyama 1989).

Many chimpanzees and gorillas in captivity are kept under conditions that preclude the expression of normal behavior, so data on reproduction are hard to acquire. For example, in considering age-at-first-birth, it is often impossible to determine whether a female was housed with a compatible, fertile male. Many captive chimpanzees and gorillas were raised in varying degrees of isolation, and this isolation can interfere with the development of normal reproductive behavior (Beck and Power 1988). Another difficulty arising when comparing the reproductions of wild and captive chimpanzees and gorillas is that the genotype of each species is different. All captive gorillas are of the western lowland subspecies, while mountain gorillas are the subjects of studies in the field. Similarly, few of the eastern subspecies of chimpanzee are represented in the captive population, while eastern chimpanzees are subjects of both of the long-term field studies in Tanzania.

Nevertheless, chimpanzees and gorillas are genetically close, displaying physical, ecological, and behavioral similarities as well as differences. For example, the life span of both species in the wild is thought to be about 40 to 45 years. Sexual dimorphism is marked in gorillas, with adult males weighing almost twice as much as adult females; chimpanzee males are bigger than females but to a lesser degree, with female adult weight about 75% of male adult weight (Jungers and Susman 1984).

The diets of chimpanzees and mountain gorillas differ greatly. The chimpanzees are mainly frugivorous (fruit-feeders), supplementing their diet with leaves, insects, and mammalian prey (Goodall 1986), while the mountain gorillas are folivorous (foliage-feeders), with 95% of their food being leaves, stems, and pith (Watts 1984). An important implication of this dietary difference is the extent to which food availability varies over time. The amount of food available to mountain gorillas changes little as, apart from bamboo shoots, their herbaceous food plants are abundant in all seasons. Chimpanzees, like all frugivores, are faced with both regular seasonal fluctuations in available fruits and unpredictable fluctuations throughout the year.

The social structure of chimpanzees differs from that of gorillas. Chimpanzee communities number from 20 to more than 100 members and include several adult males with as many, or more, adult females and their offspring. The members of a chimpanzee community are rarely all together but instead spend their time alone or in parties of variable size and composition. A party may change membership frequently during a single day (Goodall 1986). This social structure seems to occur throughout Africa except in the small remnant at Bossou, where the community of 20 chimpanzees is markedly more cohesive (Sugiyama 1981). Mountain gorillas, however, live in stable groups numbering between two and 35 individuals led by one to four adult males with several adult females and their offspring.

In both chimpanzees and mountain gorillas, individuals move between groups/communities at adolescence (Pusey 1979; Harcourt et al. 1976). Chimpanzee males remain in their natal community, while females often, but not always, migrate during adolescence. At Bossou, migration of female adolescents has not been seen (Sugiyama 1981), probably because the study community is isolated from other chimpanzee communities by human cultivation. The migratory movements of gorillas are more varied: most young adult males leave their natal group and become solitary for a time before attracting females from other groups and forming a new group. About half the adolescent female gorillas observed at Karisoke left their natal groups, and multiple transfers by females, where the females move from group to group until there is a compatible arrangement, were not uncommon (Stewart and Harcourt 1987).

Age-at-first-birth is a reproductive parameter that is similar in gorillas and chimpanzees. Age-at-first-birth is the sum of the stages of sexual development; that is, birth to menarche plus the period of adolescent sterility plus gestation length. Menstruation is rarely seen in field conditions, so menarche is equated with the first observed adult sexual behavior. Chimpanzee females have large, easily observed sexual swellings, but gorilla females have small sexual swellings, which are rarely observable in the field. Both species have a period of adolescent sterility when sexual cycles occur but conception does not, presumably because the cycles do not involve ovulation. Gestation length is longer in gorillas than in chimpanzees, the mean length being 256 days for gorillas and 228 days for chimpanzees.

Results

The age at first birth for wild chimpanzees at Gombe and at Mahale, for wild mountain gorillas at Karisoke, and for both species in captivity is shown on table 1. Most ages of wild females were estimated based on knowledge of developmental rates and behavior seen in others of known age. Many of the captive females were born in the wild so their ages were also estimated. To avoid a false impression of accuracy, all ages were rounded to the closest

Table 1
Age-at-first-birth of chimpanzees and gorillas.

	N	Mean years:months	Range years	Minimum years:months
Wild Chimpanzees:				
Gombe (Goodall 1986; J. Wallis pers. com.)	18	14:9	11–23	11:1
Mahale (Nishida et al. 1990)	22	14:6	12–20	12+
Wild Gorillas:				
Karisoke (Stewart et al. 1988; K. Stewart pers. com.)	8	10:2	9–13	8:9
Captive Chimpanzees:				
CIRMF (I. Orbell pers. com.)	19	11:2	8–17	7:6
Yerkes (Young and Yerkes 1943)	7	10:7	9–12	9:0
Holloman (Smith et al. 1975)	6	10:5	9–12	8:8
Captive Gorillas:				
International Studbook (Kirchshofer 1990; T. Strapac pers. com.)	44	9:5	6–19	6:8

year in the calculations of mean age. Minimum ages are in years and months when birth dates were known. In general, wild gorilla females give birth for the first time four years earlier than wild chimpanzee females. Four wild chimpanzee females (two at Gombe and two at Mahale) appear to be sterile, having not given birth by 23–40 years of age. No similar cases have been observed among mountain gorillas at Karisoke.

In captivity, chimpanzee females give birth three to four years earlier than wild chimpanzees. Average age-at-first-birth is similar for wild and captive gorillas, but the range of ages is much greater in the captive gorilla population. The higher age-at-first-birth of some captive female gorillas is likely to be the result of behavioral problems or lack of access to a fertile male.

Intervals between births, the length of time between successive live births when the first born was alive and with its mother at the time of the second birth, are compared in table 2. Wild chimpanzee females at Gombe and at Mahale had longer mean interbirth intervals than those of wild mountain gorillas and of both gorillas and chimpanzees in captivity.

Table 2
Interbirth interval of chimpanzees and gorillas.

	Intervals N	Mothers N	Mean Months	Range Months
Wild Chimpanzees:				
Gombe (Goodall 1986; J. Wallis pers. com.)	21	13	66	47–78
Mahale (Nishida et al. 1990)	19	16	72	53–88
Bossou (Sugiyama 1989)	8	6	51	36–84
Wild Gorillas:				
Karisoke (Stewart et al. 1988; K. Stewart pers. com.)	23	13	50	37–88
Captive Chimpanzees:				
CIMRF (I. Orbell pers. com.)	10	7	50	34–63
Captive Gorillas:				
North American Zoos (Sievert et al. 1991)	16	13	50	28–77
Jersey (R. Johnstone-Scott pers. com.)	4	2	53	47–61

Notes: Interbirth interval is defined as the length of time between successive live births when the first born was alive and with its mother at the time of the second birth.

At Bossou, the mean interbirth interval of the six mother chimpanzees was estimated at 4.25 years, or 51 months. The data from Bossou included two very short intervals of about three years. In both cases, the older infant died shortly after the birth of the second (Sugiyama 1981).

Discussion

Age-at-First-Birth

First, consider the differences between age-at-first-birth of wild versus captive females for each species keeping in mind the captive setting differs greatly from the wild setting and also the captive settings themselves may vary in terms of social life, space available, diet, and degree of stress. However, for chimpanzees the contrast between the age-at-first-birth of wild versus captive females is striking both in its scale and its consistency across different populations. Specifically, the sexual development of wild chimpanzee females, both at Gombe and Mahale, seems retarded with respect to the genetic potential of the species. In contrast, the similar age-at-first-birth of wild versus captive gorillas suggests that the reproductive development of gorillas is less affected by the conditions of captivity than is that of chimpanzees.

A relationship in humans between age at menarche, body weight, and body fat has been suggested by Frisch and McArthur (1974); a similar relationship may exist in chimpanzees. Studying chimpanzees and gorillas, Shea (1985) found that the growth spurt occurred earlier in female gorillas than female chimpanzees. While the relationship between the growth spurt and puberty is not yet clear (Watts and Gavan 1982), earlier sexual maturity in gorillas would explain why gorillas in captivity average at least one year younger than chimpanzees in captivity for their first births (see table 1). A relationship between puberty and either a critical body weight, a critical fat level, or the timing of the growth spurt may be a major factor delaying breeding in wild female chimpanzees, who weigh less than female chimpanzees in captivity.

Smith et al. (1975) reported a mean weight at menarche of 29 kilograms for six captive chimpanzees with a mean age of 8 years 9 months; an adolescent female at Gombe weighed only 21 kilograms at 9 years 6 months, while adult females had a mean weights of 30 kilograms at Gombe (Wrangham and Smuts 1980); and adult females had a mean weight of 35 kilograms at Mahale (Uehara and Nishida 1987). Considering the mean weight of four female chimpanzees *(Pan t. troglodytes)* from Zaire was 44 kilograms (Jungers and Susman 1984), the low weights of chimpanzees at Gombe and Mahale may not be typical for the species as a whole. However, little comparable data are available. In general, chimpanzees in tropical forest habitats appear larger than those at Gombe and Mahale, regardless of subspecies (R. Wrangham pers. com. for *Pan t. schweinfurthii* at Kibale, Uganda; C. Boesch pers. com. for *Pan t. verus* at Taï, Ivory Coast; and pers. observ. for *Pan t. troglodytes* at Lopé, Gabon).

Age-at-first-birth is a key variable in the lifetime reproductive potential of females but, in terms of reproductive success, to delay the first birth beyond the point when breeding becomes physiologically possible may be adaptive. Pregnancy entails a heavy maternal investment. Some data from captive chimpanzees and gorillas suggest that precocious breeding is risky: Young and Yerkes (1943) found prenatal and postnatal mortality was higher for offspring of chimpanzee females who became pregnant at age nine years or younger. However, this pattern did not occur in the CIRMF, Gabon, chimpanzee colony: only two of seven primiparous mothers aged nine years or less lost their infants, while four of eight primiparous mothers over nine years lost infants in the first week after birth (I. Orbell pers. com.).

Both diet and social factors vary between nature and captivity, so the effects of these two variables are difficult to separate. Chimpanzees at Gombe and Mahale grow more slowly than those in captivity, and this slow growth may delay menarche and/or prolong adolescent sterility. The similar reproductive development in wild and captive gorillas suggests the importance of regularity of nutritional intake, which characterizes both folivorous diets and food supply in captivity. As frugivores, chimpanzees face periodic food shortages because of great fluctuations in the availability of ripe fruit. Body weights of chimpanzees at Mahale decreased when fruit was scarce (Uehara and Nishida 1987).

In wild chimpanzees and gorillas, social factors may also contribute differences in age at first birth. While adolescent females of both species leave their natal groups, this social change may be harder for chimpanzees. The flexible association patterns of the chimpanzee community demand complex social relationships. Chimpanzees spend much time interacting with conspecifics, and social grooming is common, involving dyads of all age and gender classes (Goodall 1986). Immigrant female chimpanzees are readily accepted by males, but resident females usually react aggressively to new arrivals (Hasegawa 1989; Pusey 1979). Immigrant female gorillas, on the other hand, are not the targets of aggression from resident females (Stewart and Harcourt 1987). Stress resulting from social instability could delay the onset of fertile cycles or cause a high rate of failure in the first pregnancies that go undetected in field studies.

Interbirth Interval

A similar difference, as compared with age-at-first-birth, exists between captive chimpanzees and those in the wild (studied in Tanzania) in the length of the interval between successive live births: those in captivity having much shorter interbirth intervals. However, chimpanzees at Bossou have shorter interbirth intervals, comparable with those of captive chimpanzees (Sugiyama 1989). Variability represented by the range of interbirth interval exists within populations (see table 2). Preliminary data about Kibale Forest chimpanzees suggest long interbirth intervals of over seven years (R. Wrangham pers. com.), while for chimpanzees at Lopé, estimates of the interbirth interval range from four to over six years (Tutin unpubl. data). Interbirth interval is similar in wild and captive gorillas (Sievert et al. 1991).

The long interbirth intervals of some wild chimpanzees may result from the same nutritional factors that affect age-at-first-birth. But pertinent social factors may come into play as well. By the time a first-born is weaned, for example, the mother is integrated into the community. However, in the social structure of chimpanzees the mother is principally responsible for her offspring, who receive no direct and little indirect paternal protection. This lack of paternal protection is not so for gorilla offspring, as the adult male gorilla actively defends the members of his group (Fossey 1983). Data from Karisoke emphasize the importance of protection by male gorillas in preventing infanticide: "Unweaned infants who lose male protection are almost certain to be killed by other, unrelated males"(Watts 1990:39). The high risk of infanticide means that infant gorillas need protection, and the constant proximity of the adult male frees mothers from the constraints of vigilance.

In contrast, chimpanzee infants are much less vulnerable to infanticide by males, which accounted for only 5% of total infant mortality among chimpanzees at Gombe compared with 37% among gorillas at Karisoke (Goodall 1986; Watts 1991). Additional infant mortality at Gombe came at the hands of an adult female who killed at least three and possibly as many as eight young infants over a three-year period. Goodall (1986) reports that during that period, chimpanzee mothers with young infants associated more often than usual with adult males.

Raising an infant is more demanding for chimpanzee mothers than for gorilla mothers as, in addition to constant vigilance, chimpanzee mothers carry their infants for up to five years (Goodall 1986; Hiraiwa-Hasegawa 1990). In contrast, gorilla offspring develop independent locomotion earlier than chimpanzee offspring. Young gorillas change from ventral to dorsal travel by six months of age and, by the time they are two years old, are carried very little by their mothers (Fossey 1979). The greater dependency of chimpanzee infants on their mothers is illustrated both by the two cases at Bossou of three-year-old chimpanzees dying soon after the birth of a sibling (Sugiyama 1981) and by the contrasting fates of young, orphaned chimpanzees and mountain gorillas. Orphaned chimpanzees at Gombe or Mahale rarely survived if under five years of age at the time of the mother's death, whereas four orphaned gorillas, aged less than four years, were fostered by others in the group and survived (Fossey 1979, 1983; Goodall 1986; Nishida et al. 1990). Clearly, raising an infant in the wild is more demanding for a chimpanzee than for a gorilla mother, and these demands are likely to contribute to the longer interbirth intervals of chimpanzees. In the safety of captivity, female chimpanzees encourage independent locomotion in their infants from an early age (Nicolson 1977).

The strongest indication that social factors contribute to the observed differences in the interbirth intervals comes from Bossou, where these intervals are shorter than in other populations of wild chimpanzees (see table 2). The community at Bossou has been isolated by surrounding human settlement since at least 1967, so some degree of inbreeding has occurred. Social relationships at Bossou differ markedly from other populations of chimpanzees: namely, parties include a larger percentage of community members, parties are more permanent, and the dominant male actively maintains group cohesion (Sugiyama 1981, 1989). With females remaining in the natal group, some constraints on reproduction appear to be reduced. For gorillas, Watts (1990, 1991) suggested that social factors might affect reproduction, as birth rates decline with increasing group size, apparently due to competition between females for access to adult males.

Lifetime Reproductive Potential and Success

The consequences of the observed differences in terms of female reproductive potential for chimpanzees and gorillas in the wild are summarized in table 3. Over a lifetime, the differences are significant, with an average female gorilla able to produce 7.1 births compared to a female chimpanzee's 4.4 births. The best figures are based on the youngest recorded age-at-first-birth and shortest observed interbirth interval, giving theoretical maxima of 7.4 births for chimpanzees and 10.1 births for gorillas.

The difference in reproductive potential between wild chimpanzees and wild gorillas may not be a species-typical phenomenon, but instead an artifact of the few populations for which long-term data are available. Large differences in diet exist between mountain gorillas and western lowland gorillas *(Gorilla g. gorilla)* (Watts 1991), but no precise demographic data on western lowland gorillas are yet available. Similarly, chimpanzees live in a range of habitats in tropical Africa, from closed canopy forest to open savanna. Thus, for both species, significant differences in nutrition may exist between populations in the wild.

Compared with chimpanzees in captivity, the weights of adult chimpanzees at Gombe and Mahale are surprisingly low, so generalizations from these populations must be made with caution. Furthermore, chimpanzees in tropical forest habitats seem larger than those in woodland habitats at Gombe and Mahale. If this size difference results from poorer nutrition in the woodland habitat of western Tanzania, it could be due either to longer periods of fruit scarcity in these more seasonal habitats or to the relative rarity of the herbaceous plants that forest-dwelling chimpanzees exploit heavily when fruit is scarce. These herbs provide a year-round source

Table 3
Lifetime reproductive potentials of female chimpanzees and gorillas.

	Age-at-first-birth Years	Interbirth intervals Years	Births by 40 years of age N
Chimpanzees:			
Average	14.5	5.75	4.4
Best	11.1	3.9	7.4
Gorillas:			
Average	10.2	4.2	7.1
Best	8.75	3.1	10.1

of protein (Wrangham et al. 1991) and, at Lopé, gorillas eat these high-protein foods more regularly and in larger amounts than do sympatric chimpanzees (Tutin et al. 1991). However, nutritional factors do not explain why captive chimpanzees breed later than captive gorillas, nor why interbirth intervals are long among forest chimpanzees at Kibale. The later mean age-at-first-birth of chimpanzees as compared with gorillas is consistent and, thus, appears to be determined genetically.

Nutritional factors could act either directly, influencing the female's physiological ability to conceive or, as in the case of interbirth interval, indirectly via the rate of the first infant's development to a stage where its chances of survival are not impaired by the end of lactation. Suckling frequency appears to mediate the resumption of fertile cycles by affecting the hormonal status of the mother (Stewart 1988). Such a mechanism of control of interbirth interval allows input from both the infant (if suckling frequency decreases with increasing nutritional independence) and the mother (if she facilitates or blocks access to the breast) and it caters to individual variation in infants' developmental rates and temporal variation in maternal health. Maternal investment during both pregnancy and infant rearing is so high that strategies reducing infant and juvenile mortality would be selected for strongly.

Infant mortality is similar for both species in the wild, being 28% in the first year of life for chimpanzees at both Gombe and Mahale and 26% for mountain gorillas at Karisoke. Preweaning mortality is 40% at Gombe (Goodall 1986), 53% at Mahale (Nishida et al. 1990), and 34% at Karisoke (Watts 1991). Thus, the risk of losing an infant between one year of age and weaning is greater for chimpanzee than for gorilla mothers. If an infant of either species dies before weaning, its mother presumably will begin to cycle again, reducing the interbirth interval. However, the difference in lifetime reproductive success between chimpanzees and gorillas is likely to be even greater than the difference between the two species' theoretical reproductive potentials, especially if a chimpanzee female loses an infant during the late stages of lactation. In summary, it seems clear that female gorillas both produce and successfully raise more offspring than do female chimpanzees.

Social organization is generally similar within each species regardless of habitat, but the data from Bossou suggest that small changes in social relationships influence reproduction. Preweaning mortality is only 19% at Bossou (Sugiyama 1989) despite short interbirth intervals, indicating that this isolated group is well protected from hazards. Social effects on gorilla reproductive success are suggested by Watts' (1990) finding of negative correlations between the number of surviving births per female per year versus both group size and the number of females per group.

Thus, both nutritional and social factors are likely to affect reproduction in chimpanzees and gorillas. Nutritional factors affect developmental rates within genetically determined limits, and both regularity of intake and protein content of the diet probably contribute to the differences between chimpanzee populations. The data predict that age-at-first-birth will be lower in the heavier, forest-dwelling chimpanzee populations than in woodland-dwelling populations at Gombe and Mahale. Given adequate nutrition, social factors may explain some of the variation of interbirth intervals, the length of which is tailored to maximize reproductive success and, thus, is influenced by the risks of infant and juvenile mortality.

Summary

Data from long-term field studies of chimpanzees and mountain gorillas indicate that female gorillas have a higher lifetime reproductive potential than chimpanzees. On average, female gorillas have 7.1 births compared to 4.4 births for chimpanzees due to the gorilla's earlier first births and shorter interbirth intervals. Data suggest that wild chimpanzees, but not wild gorillas, reproduce below the species' genetic potential. Comparisons of diet and social organization suggest that social factors are at least partly responsible for the observed differences in reproductive performance.

Acknowledgments

This chapter could not have been written without the dedication of the leaders of the long-term field studies at Gombe National Park and Mahale National Park in Tanzania and at Karisoke in the Virungas National Park, Rwanda. I thank all those who contributed to data collection at these sites and particularly Jane Goodall and Janette Wallis at Gombe, Toshisada Nishida and Mariko Hiraiwa-Hasegawa at Mahale, Kelly Stewart and Sandy Harcourt at Karisoke, Richard Wrangham at Kibale and Christophe Boesch at Taï for generously providing unpublished data. I gratefully acknowledge information on reproduction in captive chimpanzees and gorillas from Richard Johnstone-Scott, Isabelle Orbell, Jorge Sabater Pi, and Sue Savage-Rumbaugh, and I particularly thank Judy Sievert and Tom Strapac for generously sharing their meticulous compilations of data on captive gorillas. Bill McGrew made constructive criticism, and I thank him for that and for much stimulating discussion.

References

Beck, B.B., and M.L. Power. 1988. Correlates of sexual and maternal competence in captive gorillas. *Zoo Biol.* 7:339–50.

Fossey, D. 1979. Development of the mountain gorilla *(Gorilla gorilla beringei):* The first thirty-six months. In D.A. Hamburg and E.R. McCown, eds., *The Great Apes,* pp. 139–86. Menlo Park, Calif.: Benjamin/Cummings.

———. 1983. *Gorillas in the Mist.* Boston, Mass.: Houghton Mifflin.

Frisch, R.E., and J.W. McArthur. 1974. Menstrual cycles: Fatness as a determinant of minimum weight for height necessary for their maintenance or onset. *Science* 185:949–51.

Goodall, J. 1986. *The Chimpanzees of Gombe.* Cambridge, Mass.: Belknap Press.

Harcourt, A.H., K.J. Stewart, and D. Fossey. 1976. Male emigration and female transfer in wild mountain gorillas. *Nature* 263:226–7.

Harcourt, A.H., D. Fossey, K.J. Stewart, and D.P. Watts. 1980. Reproduction in wild gorillas and some comparisons with chimpanzees. *J. Reprod. Fert. (Suppl.)* 28:59–70.

Hasegawa, T. 1989. Sexual behavior of immigrant and resident female chimpanzees at Mahale. In P.G. Heltne and L.A. Marquardt, eds., *Understanding Chimpanzees,* pp. 90–103. Cambridge, Mass.: Harvard Univ. Press.

Hiraiwa-Hasegawa, M. 1990. Maternal investment before weaning. In T. Nishida, ed., *The Chimpanzees of the Mahale Mountains: Sexual and Life History Strategies,* pp. 257–66. Tokyo: Univ. of Tokyo Press.

Jungers, W.L., and R.L. Susman. 1984. Body size and skeletal allometry in African apes. In R.L. Susman, ed., *The Pygmy Chimpanzee: Evolutionary Biology and Behavior,* pp. 131–77. New York: Plenum Press.

Kirchshofer, R. 1990. International Register and Studbook of the Gorilla (*Gorilla gorilla,* Savage and Wyman, 1847). Frankfurt: Frankfurt Zoological Garden.

Nicolson, N. 1977. A comparison of early behavioral development in wild and captive chimpanzees. In S. Chevallier-Skolnikoff and F.E. Poirier, eds., *Primate Biosocial Development,* pp. 529–60. New York: Garland.

Nishida, T. 1990. A quarter century of research in the Mahale Mountains: An overview. In T. Nishida, ed., *The Chimpanzees of the Mahale Mountains: Sexual and Life History Strategies,* pp. 3–36. Tokyo: Univ. of Tokyo Press.

Nishida, T., H. Takasaki, and Y. Takahata. 1990. Demography and reproductive profiles of the chimpanzees of the Mahale Mountains. In T. Nishida, ed., *The Chimpanzees of the Mahale Mountains: Sexual and Life History Strategies,* pp. 63–97. Tokyo: Univ. of Tokyo Press.

Pusey, A. 1979. Intercommunity transfer of chimpanzees in Gombe National Park. In D.A. Hamburg and E.R. McCown, eds., *The Great Apes,* pp. 465–79. Menlo Park, Calif.: Benjamin/Cummings.

Shea, B.T. 1985. The ontogeny of sexual dimorphism in the African apes. *Am. J. Primatol.* 8:183–8.

Sievert, J., W.B. Karesh, and V. Sunde. 1991. Reproductive intervals in captive female western lowland gorillas with a comparison to wild mountain gorillas. *Am. J. Primatol.* 24:227–34.

Smith, A.H., T.M. Butler, and N. Pace. 1975. Weight control of colony reared chimpanzees. *Folia Primatol.* 24:29–59.

Stewart, K.J. 1988. Suckling and lactational anoestrus in wild gorillas *(Gorilla gorilla). J. Reprod. Fert.* 83:627–34.

Stewart, K.J., and A.H. Harcourt. 1987. Gorillas: Variation in female relationships. In B. Smuts, D.L. Cheney, R.M. Seyfarth, R.W. Wrangham, and T.T. Struhsaker, eds., *Primate Societies,* pp. 155–64. Chicago: Univ. of Chicago Press.

Stewart, K.J., A.H. Harcourt, and D.P. Watts. 1988. Determinants of fertility in wild gorillas and other primates. In P. Diggory, M. Potts, and S. Teper, eds., *Natural Human Fertility: Social and Biological Determinants,* pp. 22–38. London: Macmillan Press.

Sugiyama, Y. 1981. Observations on the population dynamics and behavior of wild chimpanzees at Bossou, Guinea, in 1979–1980. *Primates* 22:435–4.

———. 1989. Population dynamics of chimpanzees at Bossou, Guinea. In P.G. Heltne and L.A. Marquardt, eds., *Understanding Chimpanzees,* pp. 134–45. Cambridge, Mass.: Harvard Univ. Press.

Tutin, C.E.G. 1980. Reproductive behaviour of wild chimpanzees in the Gombe National Park, Tanzania. *J. Reprod. Fert. (Suppl.)* 28:43–57.

Tutin, C.E.G., M. Fernandez, M.E. Rogers, E.A. Williamson, and W.C. McGrew. 1991. Foraging profiles of sympatric lowland gorillas and chimpanzees in the Lopé Reserve, Gabon. *Phil. Trans. R. Soc. London B* 334:179–86.

Uehara, S., and T. Nishida. 1987. Body weights of wild chimpanzees *(Pan troglodytes schweinfurthii)* of the Mahale Mountains National Park, Tanzania. *Am. J. Phys. Anthro.* 72:315–21.

Watts, E.S., and J.A. Gavan. 1982. Postnatal growth of nonhuman primates: The problem of the adolescent spurt. *Human Biol.* 54:53–70.

Watts, D.P. 1984. Composition and variability of mountain gorilla diets in the central Virungas. *Am. J. Primatol.* 7:323–56.

———. 1990. Ecology of gorillas and its relation to female transfer in mountain gorillas. *Internat. J. Primatol.* 11:21–45.

———. 1991. Mountain gorilla reproduction and sexual behavior. *Am. J. Primatol.* 24:211–25.

Wrangham, R.W., and B.B. Smuts. 1980. Sex differences in the behavioral ecology of chimpanzees in Gombe National Park, Tanzania. *J. Reprod. Fert. (Suppl.)* 28:13–31.

Wrangham, R.W., N.L. Conklin, C.A. Chapman, and K.D. Hunt. 1991. The significance of fibrous foods for chimpanzees. *Phil. Trans. R. Soc. London B* 334:171–78.

Young, W.C., and R.M. Yerkes. 1943. Factors influencing the reproductive cycle in the chimpanzee: The period of adolescent sterility and related problems. *Endocrinology* 33:121–54.

Chimpanzee children.
Photo by J. Goodall.

Ethological Studies of Chimpanzee Vocal Behavior

John C. Mitani

Introduction

Diversity is a fundamental feature of life and a theme that recurs throughout this volume. Recent field and laboratory studies of chimpanzees and bonobos have revealed behavioral and ecological diversity heretofore unsuspected among members of the genus *Pan*. Despite new data regarding virtually all aspects of the behavior and ecology of chimpanzees and bonobos, we currently lack detailed information regarding their vocal behavior in the wild. Pioneering field investigations of Peter Marler (Marler and Hobbett 1975; Marler and Tenaza 1977) over 25 years ago at the Gombe National Park have been followed by only a few systematic investigations of the vocalizations of chimpanzees (Clark and Wrangham 1993).

In an attempt to fill this gap in knowledge, I initiated a series of field investigations into the vocal behavior of the great apes. Here I summarize some results of my studies of wild chimpanzees by focusing on two traditional ethological questions; namely, the functional significance and the development of calls. As a first step in an ongoing and long-term project, I ask these questions with respect to the chimpanzee's species-typical, long distance call: the *pant-hoot.*

In keeping with another theme of this volume, my research relies on the comparative method to generate and to test hypotheses. By framing explicit questions and addressing them with comparative field data, the ethological approach adopted here serves to highlight behavioral diversity within and between individuals (Tinbergen 1963; Goodall 1986).

An Ethological Framework

I begin by assuming that an understanding of chimpanzee vocal behavior is impossible without detailed information regarding their social systems and behavior. An ethological framework that stresses the importance of interpreting behavior within the natural social and environmental settings of chimpanzees is, therefore, a fundamental first step in the analysis of the chimpanzee vocal communication system. Chimpanzees are unusual among diurnal anthropoid primates given that they do not live in stable groups. Instead, individuals form loosely organized unit-groups, or communities whose members associate in temporary parties that vary in size and composition (Goodall 1986; Nishida 1990). A tendency for males to aggregate contributes to the variable structure of chimpanzee parties. Mature males are more social than mature females; males are often found together and frequently engage in reciprocal grooming. In addition, males form alliances in which they direct aggression jointly toward conspecifics. Patterns of associations, grooming, and alliance formations are nonrandom, with specific individuals forming preferential relationships with others.

The Call

Within the unusually fluid, fission-fusion chimpanzee society, individuals emit a loud call known as the pant-hoot. Both males and females utter this call in a variety of situations, including in response to another calling individual, during travel, upon arrival at a particularly rich food source, and in response to strange conspecifics (Goodall 1986). Pant-hoots given by male chimpanzees can be divided acoustically into four distinct parts (see figure 4). Pant-hoots may begin with an introductory series of low-frequency tonal elements. These grade into a buildup phase, consisting of a series of shorter elements, delivered at faster rates and uttered both on inhalation and exhalation. Buildups are followed by a climax portion, including one or more high-frequency, high-amplitude elements whose acoustic properties resemble those of screams. Calls end with a brief letdown; letdown elements do not appear to differ significantly from those of the buildup.

Call Function

Previous fieldwork conducted at multiple study sites provides an important set of comparative observations from which to generate hypotheses regarding the functional significance of pant-hooting. The relative instability of chimpanzee parties has led some researchers to suggest that individuals use pant-hoots primarily to establish and maintain contact with spatially separated conspecifics (Reynolds and Reynolds 1965; Goodall 1968; Wrangham 1977; Ghiglieri 1984). Two lines of evidence, one social, the other ecological, have been cited to support this hypothesis. First, early field observations suggested that chimpanzees pant-hooted more often when they were in large parties and, presumably, spread out more than when they were in cohesive, smaller parties (Reynolds and Reynolds 1965). Second, fieldwork by Wrangham (1977) at Gombe and Ghiglieri (1984) in the Kibale Forest indicated that males who arrived at food trees and gave pant-hoots were joined more often by others than males who did not call. Recent fieldwork has failed to replicate either finding: observations of chimpanzees in the Mahale Mountains revealed no relationships between calling frequencies and party size variables (Mitani and Nishida 1993), and field research at the Kibale Forest indicated that males who called were no more likely to be joined than were noncallers (Clark and Wrangham 1993).

These conflicting observations led Toshisada Nishida and me to reopen an investigation into the functional significance of the male chimpanzee's pant-hooting behavior. As noted above, it is difficult to evaluate the function of nonhuman animal calls without information regarding the social behavior of individuals. Accordingly, I begin by examining social relationships between our male chimpanzee study subjects, and I follow this by presenting data regarding the behavioral contexts in which males pant-hoot. These analyses generate some novel, functional hypotheses regarding chimpanzee pant-hooting behavior. I conclude this section with some tests of those hypotheses derived from observational field data.

Nishida and I conducted observations of the M group chimpanzees at the Mahale Mountains National Park in western Tanzania. Mahale has been the site of long-term observations of chimpanzees during the past 27 years and, as a result of this fieldwork, the demographic and social histories of all study subjects were relatively well known (Nishida 1990). Seven of the ten adult males of M group were included in this study. Due to the fission-fusion nature of chimpanzee society, it was not possible to find and follow the three remaining males on a systematic and regular basis, and insufficient observations led to their exclusion from the following analyses. For this study, our field protocol included observing individuals during one-hour sampling periods in which we recorded, either notationally or on audio tape, all calling activity by focal males as well as details of social interactions that occurred.

Our behavioral observations accord with results from previous studies at Mahale by indicating that the seven focal males can be ranked in a linear hierarchy (Hayaki et al. 1989). In addition, further analyses revealed that males formed nonrandom patterns of alliance, association, and grooming (see figure 1). With these observations regarding male social relationships as a background, I proceed to explore the hypothesized spacing function of pant-hooting by examining the situations in which calls are given.

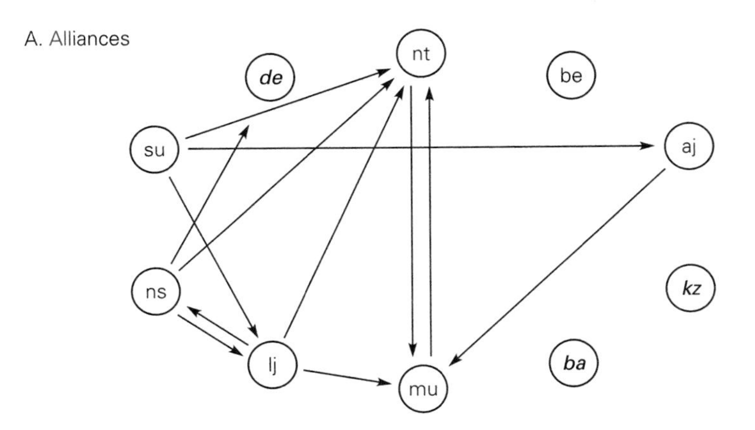

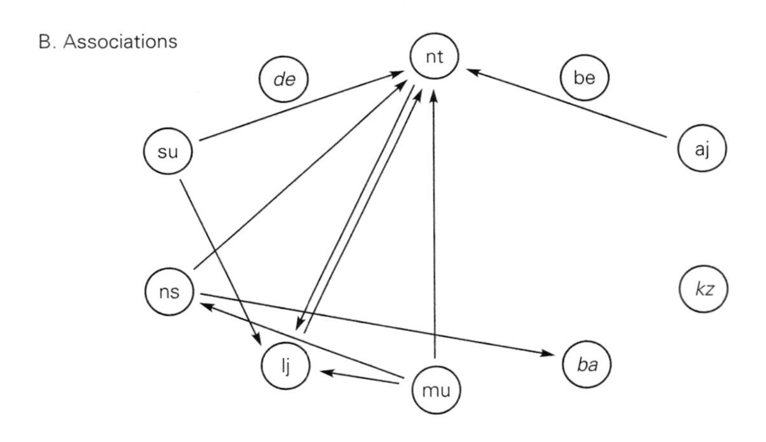

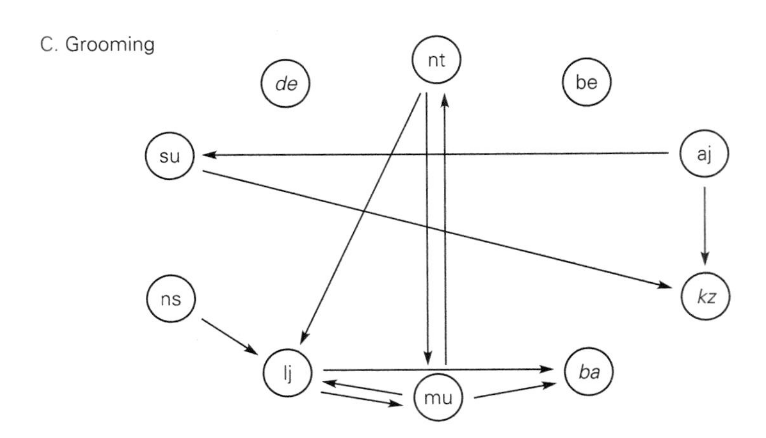

Figure 1
Alliances, associations, and grooming relationships among male chimpanzees.

Each circle represents one individual, with focal males depicted in plain letters, nonfocal subjects in italics. Males descend in rank counterclockwise, with the highest ranking individual, NT, indicated at the top.

A. Alliances. Arrows indicate those individuals with whom focal males formed alliances.

B. Associations. Arrows connect association partners; partnerships were defined between males whose time spent together exceeded one standard deviation above the population mean association value.

C. Grooming. Arrows indicate grooming partners, those males who groomed each other more than expected on the basis of chance.

Figure 2 shows the observed and expected behaviors of two high-ranking males immediately preceding and following calling. Expected behaviors were computed by assuming that males would call in proportion to the amount of time they spent in each activity. Both males traveled significantly more before and after calls than expected. Although sample

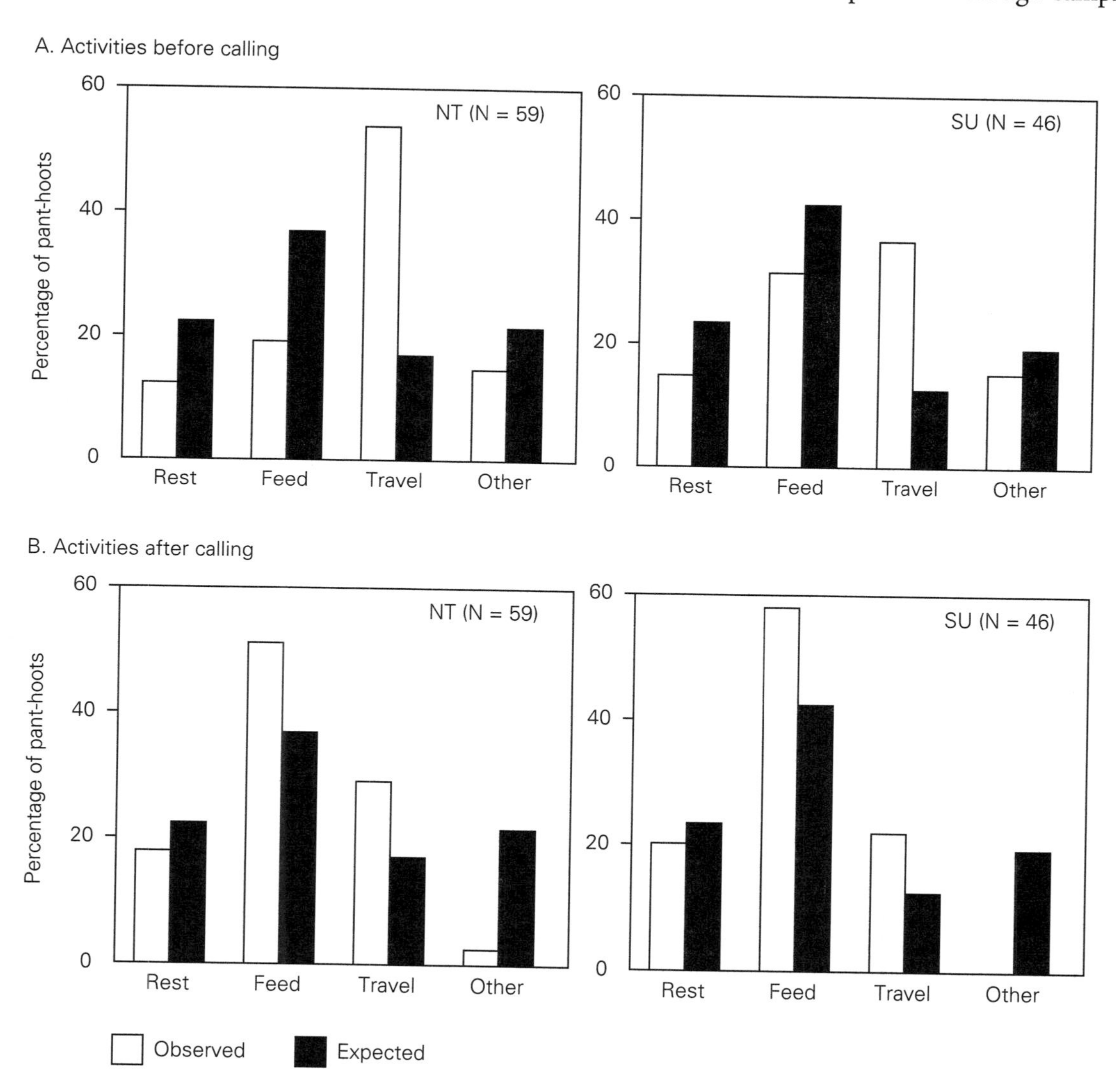

Figure 2

Contexts of pant-hoot production for two high-ranking males. A. Activities before calling. B. Activities after calling. Observed and expected percentages of pant-hooting are shown. Expected percentages were calculated by assuming that males called in proportion to the time they spent in each activity. Traveling preceded and followed calling more often than expected, and feeding followed pant-hooting more frequently than expected (χ^2 tests using call frequencies, df = 2, 2-tailed $p < .001$ for all comparisons). See text for further explanation.

sizes for the other five focal males did not permit statistical analysis, similar trends were evident for them as well. Both males for whom sufficient data are available (see figure 2) fed more often after calling than expected on the basis of chance. A similar tendency to feed after calling was not apparent for the other five focal subjects, however. These data regarding the behavioral activities associated with pant-hooting are consistent with the hypothesis that male chimpanzees call to maintain contact with conspecifics. By pant-hooting, males appear to signal a change in their locations to others (Boesch 1991).

If males call simply to maintain contact with all members in their group, one might expect all males to call equally often and to call more when spread out, as they are in large parties, compared with when they are moving in a more tightly clustered manner in smaller parties (Reynolds and Reynolds 1965). Additional observations did not support either prediction. First, males did not call equally often. We found a significant relationship between calling activity and rank: high-ranking males gave significantly more calls than low-ranking males (Mitani and Nishida 1993). Second, our attempts to correlate party size variables with calling activity did not reveal any relationships (Mitani and Nishida 1993).

This last result was particularly puzzling given prior suggestions, based largely on qualitative impressions, that calling frequency and party size are positively related. Our unexpected finding led us to entertain a new hypothesis, namely that instead of calling to indicate their spatial location to members in the unit-group, males pant-hoot only to maintain contact with particular individuals. Specifically, we propose that male chimpanzees may gain by maintaining contact with selected individuals—allies, frequent associates, grooming partners, and estrous females—from whom they receive fitness benefits.

To investigate this hypothesis, we examined the frequencies with which a male called in each of three situations: when all the focal male's preferred alliance, association, or grooming partner(s) were with him; when all the partner(s) or estrous female(s) were presumably nearby and within earshot; and when all the partner(s) or female(s) were presumed absent. For the first situation, we scored allies, frequent associates, and grooming partners as being with the focal male if they were within his sight during an entire one-hour sample period. Given the fission-fusion nature of chimpanzee society, this condition was rarely met, and a sufficient sample existed only for preferred associates (see figure 1). For the second situation, we assumed male partners or estrous females were nearby even if they were not within sight of the focal male during an entire sample period but were observed with the subject or heard during either the hour immediately preceding or the hours following the sample. For the third situation, we considered individuals absent if they were not observed during the day the sampling took place.

Results of these comparisons indicated that a male called significantly more often when his alliance partners were nearby compared with when they were absent (see figure 3a). Control comparisons involving randomly selected males who were not favored partners showed no differences in calling frequencies between the two conditions (see figure 3b). Similarly, a subject male called significantly more when his association partners were nearby compared with when they were either absent or with the focal individuals (see figure 3c). In contrast, controls involving males other than favored associates showed no differences among the three situations (see figure 3d). While these data indicate that males may call to attract potential allies or associates, grooming relationships did not appear to influence calling patterns: a male called equally often when his favored grooming partners were nearby compared with when they were absent (see figure 3e). Moreover, observations did not support the hypothesis that pant-hooting performs an intersexual function: calling by a male was not affected by the presence or absence of estrous females (see figure 3f).

I interpret these results within the ethological framework provided by our current understanding of the social lives of chimpanzees. Association and grooming patterns indicate the strong bonds that exist between male chimpanzees. These bonds are further strengthened by the cooperative relationships in which males engage while forming alliances. Males form selective coalitions that have significant reproductive consequences. Male rank is often determined by coalitionary behavior and, at the Mahale study site, rank is positively related to mating success (Nishida 1983). These observations lead one to consider how males maintain their cooperative relationships within a very fluid society where animals are often spatially separated. Given that communication processes mediate social interactions, the vocal behavior of chimpanzees may provide an effective means to maintain these important relationships. Field observations indicated that males may call to enlist the company and support of allies and frequent associates; males called more often when their preferred associates and allies were nearby compared with when they were absent. Note that a logical extension of the selective recruitment hypothesis is that males should show decreased levels of calling when they are with frequent associates and allies. Our data set did not permit a complete test of this prediction, but it did show that males decreased their calling rates when they were with associates compared with when those associates were nearby.

Call Development and Vocal Learning

Our current understanding of vocal learning processes in animals derives largely from ethological studies of male songbirds (Marler 1991). Laboratory and field evidence reveal that learning plays a major role in the

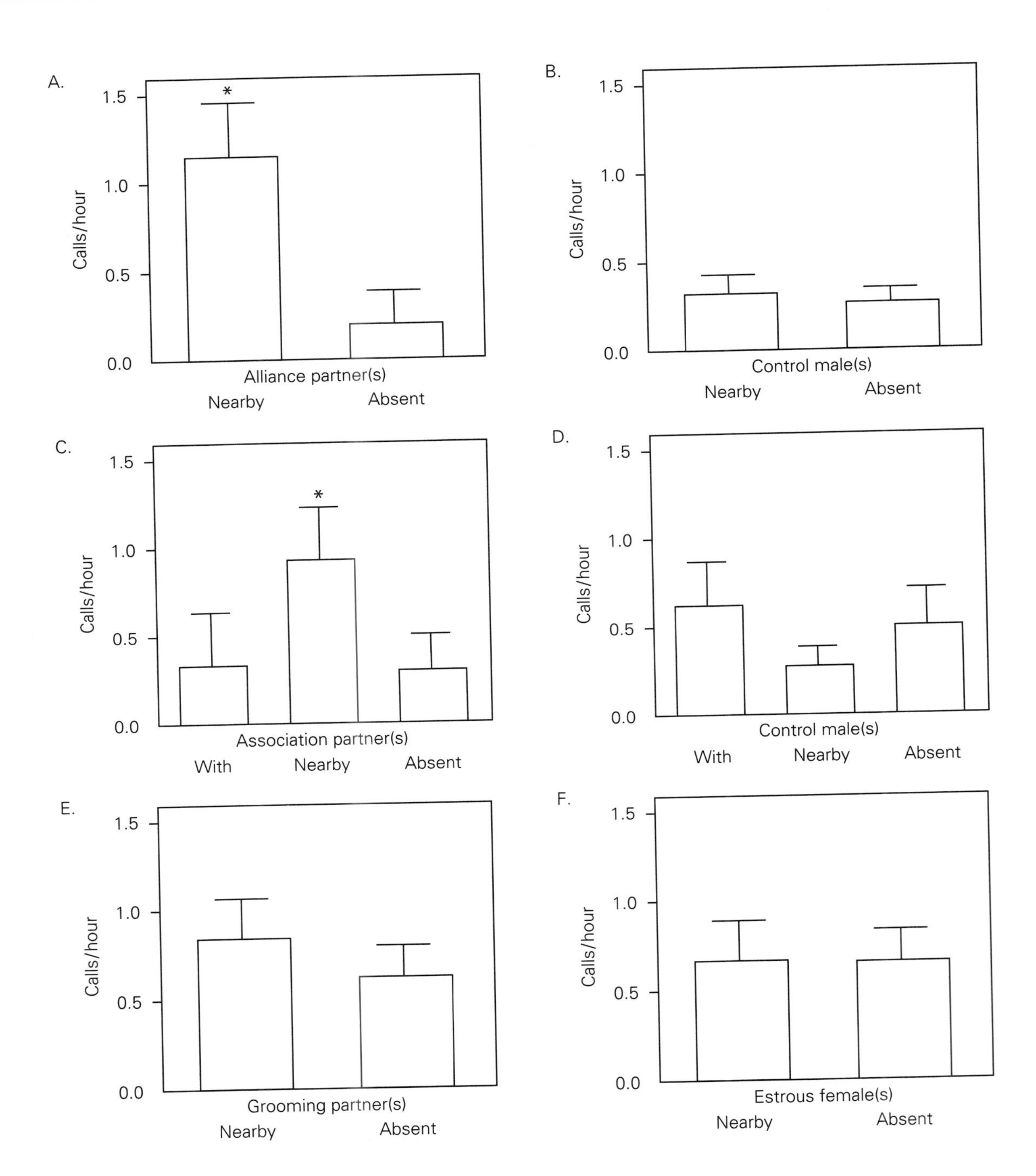
A.
1.5
1.0
0.5
0.0
Calls/hour
*
Alliance partner(s)
Nearby
Absent
B.
1.5
1.0
0.5
0.0
Calls/hour
Control male(s)
Nearby
Absent
C.
1.5
1.0
0.5
0.0
Calls/hour
*
Association partner(s)
With
Nearby
Absent
D.
1.5
1.0
0.5
0.0
Calls/hour
Control male(s)
With
Nearby
Absent
E.
1.5
1.0
0.5
0.0
Calls/hour
Grooming partner(s)
Nearby
Absent
F.
1.5
1.0
0.5
0.0
Calls/hour
Estrous female(s)
Nearby
Absent

◀ **Figure 3**
Effects of social contexts on pant-hoot production.

Alliance, association, and grooming partners were defined as in figure 1. Comparisons involving alliance and association partners were based on N = 6 focal males; grooming and estrous female comparisons included N = 7 focal males. * = 2-tailed $p<.05$, Wilcoxon matched-pairs signed-ranks tests between "nearby" and "absent" conditions and "nearby" and "with" conditions. See text for further explanation. (Figure adapted from Mitani and Nishida 1993.) The mean (±SE) number of pant-hoots produced by focal males when:

A. alliance partners were nearby and absent.

B. randomly selected control males were nearby and absent.

C. association partners were with them, nearby, and absent.

D. randomly selected control males were with them, nearby, and absent.

E. grooming partners were nearby and absent.

F. estrous females were nearby and absent.

development of the production of bird songs. In the laboratory, young birds raised in acoustic isolation from conspecific sounds typically develop abnormal song; in the field, neighboring male birds frequently share local song dialects.

Perhaps somewhat surprisingly, there is little evidence that the calls of nonhuman primates undergo significant acoustic modification during development (Snowdon and Elowson 1992). For example, in the case of nonhuman primates, both social isolates raised in the absence of any auditory feedback and cross-fostered infants grow up to produce species-typical sounds. In addition, conditioning has been shown to have only limited effects on altering the acoustic morphology of sounds produced by primates, and hybrid individuals give calls that do not resemble either of their parents.

The conclusion that the acoustic morphology of nonhuman primate calls is not subject to developmental modification is unexpected and puzzling given their long maturation periods. During these prolonged developmental periods, learning plays an important role in shaping the many behaviors that will contribute to growth, maintenance, and reproduction in later life. These considerations provided the impetus for a reexamination of the question of vocal learning by chimpanzees. One correlate of the process of vocal learning in humans and songbirds is the formation of local dialects and, as a first step in this investigation, I along with colleagues from Japan, Great Britain, and the United States have examined microgeographic variation in the calls of chimpanzees living in two neighboring populations (Mitani et al. 1992).

For this comparison we chose to analyze the pant-hoots of males from the two well-studied chimpanzee populations of the Mahale National Park and Gombe National Park. These populations are separated by approximately 150 kilometers along the eastern shore of Lake Tanganyika and belong to the same subspecies, *Pan troglodytes schweinfurthii*. As the result of the spread of human habitation, these populations are now isolated, although genetic continuity presumably existed between them until recently. Tape recordings of pant-hoots from Gombe were provided by Peter Marler, who in 1967 conducted a preliminary study of chimpanzee vocal behavior around the banana provisioning station (Marler and Hobbett 1975; Marler and Tenaza 1977). Calls from Mahale were tape recorded by a team of researchers associated with Nishida's long-term field research at the Kasoje Research Station (Richard Byrne 1984, Toshikazu Hasegawa 1988, and John Mitani 1989–90).

Our initial visual inspection of audio spectrograms and aural monitoring of pant-hoots indicated that calls of males from the two populations did not differ qualitatively. Closer examination of these calls, however, revealed differences in subtle features of the buildup and climax portions (see figure 4). Specifically, males at Mahale gave significantly shorter buildup elements at faster rates than males from Gombe (see figures 4a, b). In

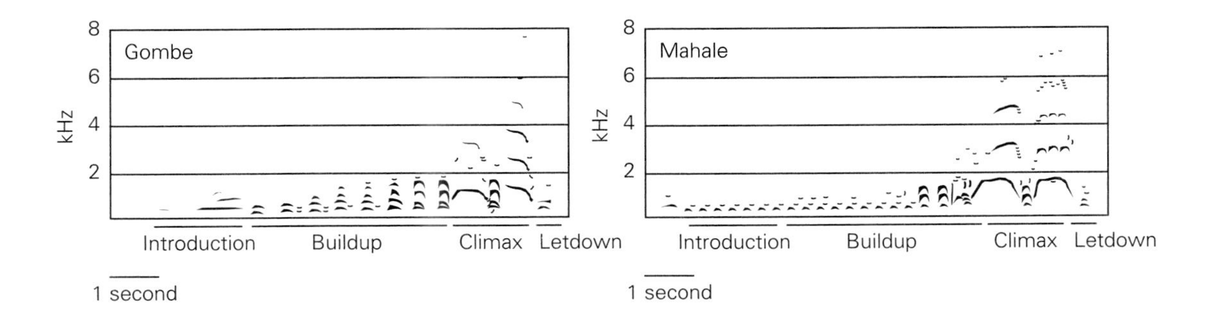

A. Durations of buildup elements.

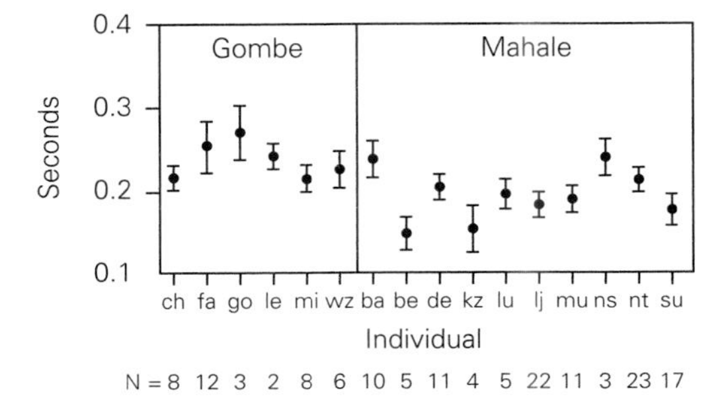

C. Average fundamental frequencies of climax elements.

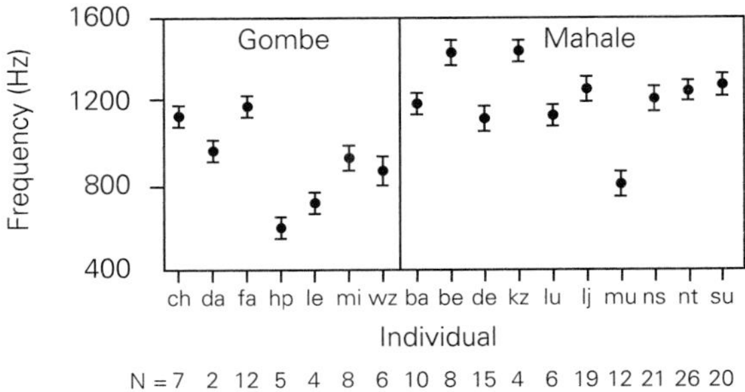

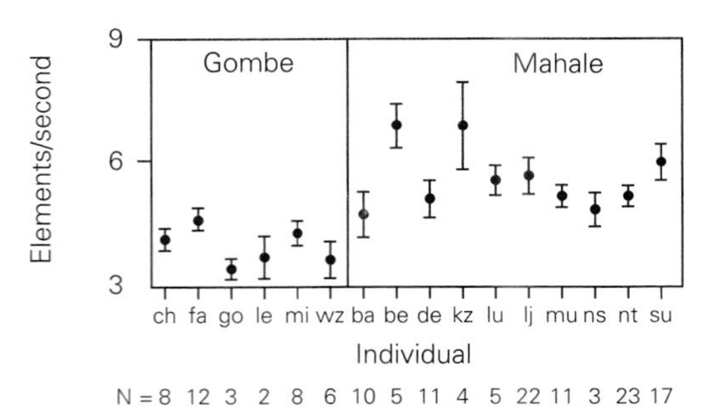

D. Frequency ranges spanned by climax elements.

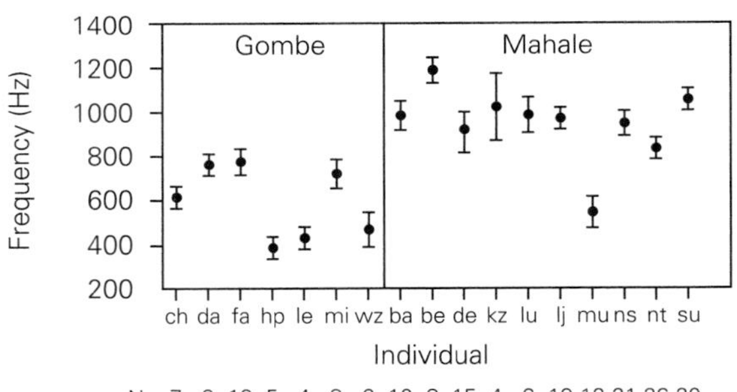

Figure 4

Audiospectrograms of representative pant-hoots from Gombe and Mahale, and comparisons of temporal and spectral features of calls from the two populations (A.-D.). Spectrograms in the top panels were produced on a MacIntosh Quadra 950 computer with MacRecorder sound analysis software. Analysis range=11 kHz. Frequency resolution=21 Hz. The spectrogram from Gombe illustrates the four typical stages of pant-hoots given by males. A. Buildup element durations. B. Rates of delivery of buildup elements. C. Average fundamental frequencies of climax elements. D. Frequency ranges spanned by climax elements. Means ±1 SE's are shown. Individuals and associated sample sizes for figures A. and C. are indicated in B. and D. respectively. Values from the Gombe and Mahale populations differed significantly in all features displayed in A.–D. Mann-Whitney U tests, 2-tailed $p<.05$ for all comparisons. (For analytical methods used in A.–D., see Mitani et al. 1992.)

addition, the Mahale males uttered higher-frequency climax elements that spanned greater frequency ranges than those produced by the Gombe males (see figures 4c, d). A subsequent analysis using recently recorded calls from Mahale confirmed that males from the two populations produce acoustically distinguishable calls (Mitani and Brandt in press).

While the results of our acoustic analyses suggest that the Mahale and Gombe chimpanzees utter their pant-hoots in subtly different ways, the causal factors underlying these differences are open to further empirical investigation. Two hypotheses subject to immediate test are that anatomical and habitat differences account for the vocal variability between the two populations. Chimpanzees at Mahale are significantly larger than those at Gombe (Uehara and Nishida 1987), and this variation in body size would lead one to predict that the larger Mahale chimpanzees would have deeper voices than the smaller Gombe males (Gouzoules and Gouzoules 1990). A similar prediction derives from consideration of the habitats occupied by the two study groups and the theory of signal detection. The Mahale study group inhabits a more densely forested area than the Gombe community (Collins and McGrew 1988) and, in these forested habitats, reflections and scattering off of multiple surfaces tend to degrade signals and hinder the efficient transfer of information (Wiley and Richards 1978). Under such circumstances, natural selection should favor individuals who produce low-frequency sounds, which degrade relatively less than higher frequency signals. Contrary to both expectations is the finding that the Mahale males utter climax elements with higher frequencies than those of the Gombe males (see figure 4c).

If anatomical and habitat differences do not underlie the vocal differences between the Mahale and Gombe males, three additional hypotheses may provide alternative explanations. First, vocal differences commonly exist between and within species of nonhuman primates (Snowdon 1986): and genetic variability may account for the differences in pant-hoots between the Mahale and Gombe males. The degree of genetic differentiation between the Mahale and Gombe populations remains an unanswered empirical question, whose resolution is currently being addressed using recently developed DNA amplification techniques (Takasaki and Takenaka 1991).

Second, one significant finding of recent research into the vocal behavior of nonhuman primates is that these animals vary the acoustic structure of their calls in subtle and semantically significant ways in different behavioral contexts (Cheney and Seyfarth 1990). In a similar fashion, Goodall (1986) has suggested that pant-hoots delivered in different behavioral situations vary acoustically. While our sample of calls from Gombe was recorded primarily around the provisioning station and the majority of pant-hoots from Mahale were taped in the chimpanzees' natural habitat, the observed vocal differences between Mahale and Gombe males may be due to variations in call usage. Our current, limited sample of tape recordings does not preclude this possibility, although preliminary analysis indicates that Mahale males do not alter the acoustic structure of their calls in different situations (Mitani unpubl. data). More recordings will be needed to test whether variations in call usage contribute to the vocal differences found here.

If anatomical, habitat, genetic, and contextual differences do not adequately account for the observed variability in pant-hoots between populations, then a third hypothesis is that these acoustic variations are dialectal. Dialectal differences are of special interest since they bear on the issue of vocal learning. Learning combined with limited dispersal after vocal acquisition has taken place commonly leads to dialects, and vocal learning in animals is frequently inferred from the existence of dialects (Kroodsma 1982). However, intraspecific population differences in vocalizations have been shown only rarely among primates (Green 1975), and the absence of dialects is consistent with the paucity of evidence for vocal learning in these animals.

Although the weight of evidence from both laboratory and field does not appear to be consistent with the suggestion of dialectal variation in the pant-hoots of chimpanzees, current theory ascribes special importance to the unusual nature of the acoustic differences between the Mahale and Gombe males. Unlike the discrete, large, and readily discernible acoustic differences that contribute to dialectal variations in songbirds (Marler and Tamura 1964), the vocal variations described here between two chimpanzee populations are apparent only on close inspection (see figure 4). The acoustic nature of bird song dialects and of variations in pant-hoots is consistent with the hypothesis that different learning processes may be involved in their acquisition (Marler 1991).

Many oscine birds memorize song patterns during a sensitive period early in life and, in these animals, song is recalled from memory at a later date and emerges gradually through vocal practice during a subsequent motor phase. Here I suggest that such memory based vocal learning may not occur in chimpanzees. Instead, I hypothesize that if the vocal differences between chimpanzee populations result from learning, they may reflect a process of selective reinforcement and conditioning over time. Young individuals may produce a range of vocal variants and then discard some to match a population-specific standard given their repeated exposure to the calls of others with whom they live (Nowicki 1989). This learning process would account for the statistical nature of the acoustic differences between the Gombe and Mahale populations, since the calls of all individuals would not be expected to converge in subtle features of pronunciation.

An ethological framework, one that views the behavior of animals within their natural environmental and social settings, may provide the means to interpret the results presented here. Viewed in such terms, the social system of chimpanzees may have created an appropriate selective milieu favoring the evolution of vocal learning. Male chimpanzees of neighboring groups are typically hostile toward one another; interactions between groups may lead to violent conflict during which animals are severely injured or killed (Goodall et al. 1979). Given these circumstances, it would be advantageous for chimpanzees to possess the ability to discriminate friend from foe at long distance. Acoustic differences between the calls of males from the different populations may provide a means for such discrimination, with vocal learning serving as the mechanism by which these variations arise.

Summary: Directions for Future Research

The studies reported here raise several issues that will provide fertile ground for future research. A strong test of the hypothesis that males call to maintain contact with selected individuals in their unit-groups requires additional observations regarding how chimpanzees respond to calls. Field playbacks using tape-recorded pant-hoots (Waser 1975) or systematic observations of more than one chimpanzee using multiple human observers (Uhlenbroek in progress) will provide the empirical means to investigate the responses to pant-hoots. Moreover, a much larger data set—to enable us to evaluate the effects of allies, associates, grooming partners, and estrous females in a multivariate model rather than independently—is needed for an empirically sufficient test of this hypothesis.

While comparative study of the calls of Mahale and Gombe males provides the first suggestive hint that learning influences vocal production in chimpanzees, this issue is far from resolved. Evaluating whether learning affects vocal development will also require information on how calls change during the maturation process. Systematic examination of the pant-hoots of juvenile and adolescent individuals promises to shed additional light on the issue of vocal learning in our closest living relatives. One hypothesis derived from the preceding analysis is that the calls of immature individuals will prove to be more variable than the calls of older individuals: cross-sectional and longitudinal comparisons involving young individuals will provide a means to test this prediction. Vocal plasticity may also lead to cultural drift in calling patterns. Our sample of pant-hoots from Gombe was recorded over 25 years ago and, if learning is involved in the vocal acquisition process, the calls of the Gombe males may have changed during the intervening period (Payne and Payne 1985). Acoustic analyses of more recent tape recordings from the Gombe population promise to resolve this question.

Finally, integrating the results of studies of call function and call development will form the basis for intriguing new lines of inquiry. For example, the suggestion that chimpanzees alter the acoustic morphology of calls during development raises the possibility that important social functions might underlie tendencies for individuals to copy the calls of others. Recently completed analyses suggests that acoustic similarities exist between the calls of individuals who preferentially associate with each other (Mitani and Brandt in press). Does similar sharing exist between allies and favored groomers? If so, what is the functional significance of this sharing? Questions such as these present a wealth of challenges to current ethologists and promise that additional sources of behavioral diversity within and between individuals will be uncovered by future field study of chimpanzees.

Acknowledgments

I thank J. Goodall, W. McGrew, T. Nishida, F. DeWaal, R. Wrangham, and the Chicago Academy of Sciences for invitations to participate in the "Understanding Chimpanzees: Behavioral Diversity" symposium and to contribute to this volume. Comments and editorial assistance by W. McGrew improved the manuscript. My fieldwork was sponsored by the Tanzanian Commission for Science and Technology and the Serengeti Wildlife Research Institute. I gratefully acknowledge the logistical support provided by A. Seki, E. Massawe, T. Nishida, M. Hamai, and the staffs of the Mahale Mountains Wildlife Research Centre and the Mahale National Park. Research reported here would not have been possible without the expert field assistance provided by M. Bunengwa, H. Bunengwa, M. Hawazi, and R. Hawazi. Fieldwork was supported by grants from the NSF BNS 8822764, BNS 8919726, and BNS 9021682, the Harry Frank Guggenheim Foundation, the National Geographic Society to John Mitani, and the Japanese Ministry of Education, Science, and Culture to Toshisada Nishida. Preparation of this manuscript was made possible, in part, by funds provided by a NSF Presidential Faculty Fellows Award, DBS-9253590.

References

Boesch, C. 1991. Symbolic communication in wild chimpanzees? *J. Human Evol.* 6:81–90.

Clark, A. 1993. Rank differences in the production of vocalizations by wild chimpanzees as a function of social context. *Am. J. Primatol.* 31:159–79.

Clark, A. and R. Wrangham. In Press. Chimpanzee arrival pant-hoots: Do they signify food or status? *Internat. J. Primatol.*

Cheney, D., and R. Seyfarth. 1990. *How Monkeys See the World.* Chicago, Ill.: Univ. of Chicago Press.

Collins, D.A., and W. McGrew. 1988. Habitats of three groups of chimpanzees (*Pan troglodytes*) in western Tanzania compared. *J. Human Evol.* 17:553–74.

Ghiglieri, M. 1984. *The Chimpanzees of the Kibale Forest.* New York: Columbia Univ. Press.

Goodall, J. 1968. The behaviour of free living chimpanzees in the Gombe Stream area. *Anim. Behav. Monogr.* 1:161–311.

———. 1986. *The Chimpanzees of Gombe.* Cambridge, Mass.: Belknap Press.

Goodall, J., A. Bandora, E. Bergmann, C. Busse, H. Matama, E. Mpongo, A. Pierce, and D. Riss. 1979. Intercommunity interactions in the chimpanzee population of the Gombe National Park. In D. Hamburg and E. McCown, eds., *The Great Apes,* pp. 13–54. Menlo Park, Calif.: Benjamin/Cummings.

Gouzoules, H., and S. Gouzoules. 1990. Body size effects on the acoustic structure of pigtail macaque *(Macaca nemestrina)* screams. *Ethology* 85:324–34.

Green, S. 1975. Dialects in Japanese monkeys: Vocal learning and cultural transmission of locale specific behavior? *Z. Tierpsychol.* 38:304–14.

Hayaki, H., M. Huffman, and T. Nishida. 1989. Dominance among male chimpanzees in the Mahale Mountains National Park, Tanzania. *Primates* 30:187–97.

Kroodsma, D. 1982. Learning and the ontogeny of sound signals in birds. In D. Kroodsma and E. Miller, eds., *Acoustic Communication in Birds,* vol. 1, pp. 1–23. New York: Academic Press.

Marler, P. 1991. Song learning behavior: The interface with neuroethology. *T. Neurosci.* 14:199–205.

Marler, P., and L. Hobbett. 1975. Individuality in a long range vocalization of wild chimpanzees. *Z. Tierpsychol.* 38:97–109.

Marler P., and M. Tamura. 1964. Culturally transmitted patterns of vocal behavior in sparrows. *Science* 146:1483–6.

Marler, P., and R. Tenaza. 1977. Signaling behavior of apes with special reference to vocalization. In T. Sebeok, ed., *How Animals Communicate,* pp. 965–1003. Bloomington: Indiana Univ. Press.

Mitani, J., and K. Brandt. In Press. Social factors influence the acoustic variability in the long distance calls of male chimpanzees. *Ethology.*

Mitani, J., T. Hasegawa, J. Gros Louis, P. Marler, and R. Byrne. 1992. Dialects in wild chimpanzees? *Am. J. Primatol.* 27:233–43.

Mitani, J., and T. Nishida. 1993. Contexts and social correlates of long distance calling by male chimpanzees. *Anim. Behav.* 45:735–46.

Nishida, T. 1983. Alpha status and agonistic alliance in chimpanzees. *Primates* 24:318–36.

———. 1990. *The Chimpanzees of the Mahale Mountains.* Tokyo: Univ. of Tokyo Press.

Nowicki, S. 1989. Vocal plasticity in captive black capped chick-a-dees: The acoustic basis and rate of call convergence. *Anim. Behav.* 37:64–73.

Payne, K., and R. Payne. 1985. Large scale changes over 19 years in songs of humpback whales in Bermuda. *Z. Tierpsychol.* 68:89–114.

Reynolds, V., and F. Reynolds. 1965. Chimpanzees of the Budongo forest. In I. DeVore, ed., *Primate Behavior,* pp. 368–424. New York: Holt, Rinehart, Winston.

Snowdon, C. 1986. Vocal communication. In G. Mitchell and J. Erwin, eds., *Comp. Primate Biol.,* pp. 495–530. New York: A. Liss.

Snowdon, C., and M. Elowson. 1992. Ontogeny of primate vocal communication. In T. Nishida, W. McGrew, P. Marler, M. Pickford, and F. deWaal, eds., *Topics in Primatology,* Vol. 1, *Human Origins,* pp. 279–92. Tokyo: Univ. of Tokyo Press.

Takasaki, H., and O. Takenaka. 1991. Paternity testing in chimpanzees with DNA amplification from hairs and buccal cells in wadges: A preliminary note. In A. Ehara, T. Kimura, O. Takenaka, and M. Iwamoto, eds., *Primatology Today,* pp. 613–6. New York: Elsevier.

Tinbergen, N. 1963. On the aims and methods of ethology. *Z. Tierpsychol.* 20:410–33.

Uehara, S., and T. Nishida. 1987. Body weights of wild chimpanzees *(Pan troglodytes)* of the Mahale Mountains National Park, Tanzania. *Am. J. Phys. Anthro.* 72:315–21.

Waser, P. 1977. Experimental playbacks show vocal mediation of intergroup avoidance in a forest monkey. *Nature* 255:56–8.

Wiley, R.H., and D. Richards. 1978. Physical constraints on acoustic communication in the atmosphere: Implications for the evolution of animal vocalizations. *Behav. Ecol. Sociobiol.* 3:69–94.

Wrangham, R. 1977. Feeding behaviour of chimpanzees in Gombe National Park, Tanzania. In T. Clutton Brock, ed., *Primate Ecology,* pp. 503–38. London: Academic Press.

Pacifying Interventions at Arnhem Zoo and Gombe

Christopher Boehm

An agonistic intervention takes place when one or more parties interfere in a dyadic conflict. In coalitions and alliances involving aggression (Harcourt and de Waal 1992) a third party may help an ally win a fight, and the results of the intervention are readily interpreted (de Waal 1982; Goodall 1986). It is more difficult to understand pacifying interventions, which do not favor the victory of an ally but merely dampen the level of aggression in a dyadic bout. Pacifying interventions may involve a *control role* for the intervenor (Bernstein and Sharpe 1966): a high-ranking individual manipulates other group members either when they are facing external dangers or when they are quarreling or fighting among themselves. Although the analytic control-role concept has met with problems when applied to some species (Erhardt and Bernstein 1992), I shall argue that this concept is highly explanatory when applied to interactions of chimpanzees. My argument is based on comparisons of the interactions of chimpanzees in the wild at Gombe and in captivity at Arnhem Zoo.

Pacifying-conflict intervention by dominant individuals is widely reported at the anecdotal level for chimpanzees and many other wild primates, but so far the only statistical data for free-ranging great apes comes from Gombe (Boehm 1992). Ultimately, this important behavior of chimpanzees in the wild merits the full statistical treatment that now should be within reach for the free-ranging inhabitants of Gombe, Mahale, and Taï Forest and for the captive chimpanzees at Arnhem Zoo. Each of these study sites has been established long enough to provide statistically adequate data for this rather rare behavior. The comparisons undertaken here using Gombe and Arnhem Zoo will examine similarities and differences in the ways that pacifying interventions are enacted in the two radically different environments. These comparisons should also be of use in clarifying the categories

of behavior useful to the analysis of pacifying intervention. This preliminary analysis is a balanced mixture of qualitative and quantitative methods, partly because de Waal's (1982) main presentation of his findings at Arnhem Zoo is not entirely quantified and partly because I believe that, given such a complex behavior, special emphasis on qualitative analysis is useful in refining categories for future analysis.

The interventions discussed below involve a strategy of pacification, a strategy that has the effect of dampening aggression. In these interventions, I shall refer to the animal initiating a dyadic conflict, usually the higher-ranking animal, as the *aggressor* and the other as the *victim*. An intervenor who is dominant to the aggressor can more easily manipulate a conflict. Then the intervenor's threat or attack behavior has the predictable effect of dampening the conflict quickly and of controlling attempts to continue it. If the intervenor is subordinate to the aggressor, intervention often takes the form of a diversion that either distracts the aggressor or balances power temporarily, making avoidance possible. Mothers often make such interventions to protect offspring and, as will be seen, they often succeed in stopping the conflict.

I shall break down pacifying interventions into two types: partial (de Waal 1992) and impartial (de Waal and van Hooff 1981, 1972; de Waal 1982; Boehm 1981, 1992). By partial, I mean that the pacifying intervention is directed at just one party, usually the aggressor. Partial intervenors can be either dominant or subordinate to the aggressor. By impartial interventions, I mean that the intervenor's aggression is directed at both protagonists or, perhaps, at the conflict or agonism itself: an individual target is not singled out and, usually, the intervenor is dominant to the aggressor.

Background for Comparison

At Gombe (Goodall 1986), three communities of wild chimpanzees currently coexist within a circumscribed area of some 30 square kilometers. A macrocoalition (Boehm 1992) of males patrols community boundaries. This coalition attacks individual male and anestrous female strangers when met either singly or in small groups but vocalizes hostility and retreats if it meets neighbors in force. All Gombe chimpanzees must cope with predators. Females often forage alone or in very small groups, while males more often gather in larger groups because of patrolling and promiscuous mating activities. The 50 members of the Kasakela community, one of the three at Gombe, share a single cultural and genetic history that is complicated by substantial female transfer. Kasakela will be the focus of this study.

At Arnhem Zoo (de Waal 1982), one group of 20 chimpanzees lives in an outdoor area of one hectare for the warmer half the year and in a barn for the inclement seasons. This group is mainly captive-born chimpanzees with a wide variety of zoo backgrounds. A few came to Arnhem Zoo in pairs. Thus, the Arnhem group's cultural and genetic history is diverse. The group has no predators or territorial competitors, and group members are usually together.

In number, Arnhem Zoo barely falls within the range of wild communities, but sex-age composition is similar to that of a wild community. Gombe's Kasakela community is often composed of more adult females than adult males. At both Arnhem and Gombe, there is a strict male hierarchy, which is traceable by directionality of submissive pant-grunting; the female hierarchy is less defined. Male rivalry for status is quite similar at the two sites and intensive. Fight-winning coalitions and alliances are prominent at Gombe, as are partial and impartial pacifying interventions. In such contexts, de Waal (1984) differentiates between three strategies of intervention: winner support, loser support, and control role. In this chapter, I shall focus on control-role strategies, which include not only all impartial interventions but also any partial interventions that support victims or that involve a fight-stopping rather than a fight-winning strategy.

At both sites, individuals solicit and join partners in aggression. Such coalition behavior is critical to gaining a high position, or status. At Gombe, adult kin tend to support one another, and brother coalitions have figured in determining alpha rank. In contrast, the Arnhem Zoo adults have no kin networks because of the group's recent provenance. But, in spite of radical differences in environment and recent history, the fundamentals of social organization in the two groups are similar in many basic respects.

Goodall (1986) describes some of the main contexts of attack behavior within the group at Gombe: to challenge for rank; to compete for some tangible resource; to coerce another, as with consortships; to retaliate for past behavior; and to redirect frustration. She also rates attacks within the group at three levels: level one includes hits, kicks, and pushes delivered in passing; level two includes pounding, dragging, and so on for under 30 seconds; and level three includes more violent attacks lasting from 30 seconds to five minutes with possible serious wounds (1986). The majority of attacks within the group take place within five minutes of a reunion involving more than four adults and involve either rivalry for status or redirection of frustration (1986). Other contexts for attacks are meat eating and prior agonism. At Gombe, in adult agonistic bouts that included everything from mere threats to level two and level three fights, males did most of the aggressing in contexts of reunion, social excitement, meat eating, and sexual behavior. In contrast, females did most of the aggressing in contexts of plant feeding and protecting the young. Overall, the adult males aggressed twice as often as adult females even though the numbers of males were significantly smaller than those of females. At both sites, fights involving the alpha male have led to mortal or near-mortal wounding (de Waal 1986).

De Waal and Hoekstra (1980) report that half of all conflicts at Arnhem Zoo (57% by males) follow other agonistic behavior and seem to involve social contagion or redirection of frustration. Nonagonistic contexts of aggression for males at Arnhem include sex, play, and competition over objects; for females, nonagonistic contexts of aggression include annoyance over minor provocations, protection of infants, competition over infants, and competition over objects. The Arnhem colony is fed in such a way as to minimize competition and human involvement. However, human management is intrusive: for example, a male who killed an infant was removed (de Waal 1978).

Some aspects of female aggression can be considered separately from males. Females at Gombe include individuals of high rank, one of whom may emerge as alpha female, and others who rank very low, with the majority of females falling in a middle-ranking category. In contrast, at Arnhem Zoo, the female hierarchy is more pronounced: the adult female Mama is the clearly identified alpha female with a female coalition behind her. Mama regularly intervenes in female conflicts over agonism between offspring. At Arnhem Zoo, unlike Gombe, female support is critical to the alpha male (de Waal 1982).

Pacifying Interventions at Gombe

The impartial pacifying interventions by Gombe chimpanzees have been outlined elsewhere in comparing them with humans (Boehm 1992). Here, I expand those data and supply data on partial pacifying interventions at Gombe. Over a two-year period (see table 1), 75 pacifying interventions, including both partial and impartial, occurred in conflicts involving at least one adolescent or adult. Of the 75 pacifying interventions, 80% were partial and 20% impartial. The success rate was substantially higher for the 15 impartial interventions because adult intervenors outranked aggressors in these interventions. These impartial interventions were mostly directed at fights between two adult males, between two adult females, or between an adult male and an adult female. Of the 60 partial interventions, 24 interventions protected victims who were nonkin, mostly in the same age-sex categories, while the other 36 partial interventions protected socially bonded kin, mostly juveniles or infants.

Impartial interventions occurring over an eight-year period at Gombe are shown on table 2. These 39 interventions served mainly to pacify fights between adult females or attacks on adult females by adult males and,

Table 1
Pacifying interventions (and attempts) for nine conflict dyads at Gombe, 1978–79.

Dyad	Partial-impartial interventions	H-V	H-V Kin	IMP	Stop	Cont	IAV	AIV	AVI
Male-male	4	1	—	3	4	—	4	—	—
Male-female	17	12	2	3	14	3	15	—	2
Female-female	10	4	—	6	10	—	10	—	—
Male-adolescent male	1	—	1	—	1	—	—	1	—
Female-adolescent male	2	1	—	1	2	—	—	1	1
Male-juvenile/infant	9	1	8	—	6	3	1	8	—
Female-juvenile/infant	18	2	15	1	11	7	18	—	—
Adolescent male-juvenile/infant	8	3	4	1	6	2	6	2	—
Adolescent female-juvenile/infant	6	—	6	—	3	3	5	1	—
Total	75	24	36	15	57	18	59	13	3

Notes: H-V = Help victim (nonkin) intervention; H-V Kin = Help victim (kin) intervention; IMP = Impartial intervention strategy; Stop = Agonism quickly damped; Cont = Agonism continues (counts as attempts to pacify); IAV = Intervenor dominant to both; AIV = Intervenor subordinate to aggressor; AVI = Intervenor subordinate to both.

sometimes, to pacify conflicts between two adult males. Of these 39 adult interventions, almost all succeeded in stopping the conflict. The two top males accounted for 23 interventions, with males intervening a total of 33 times, and females intervening a total of six times. Other high-ranking males intervened one to three times each, while four high-ranking females intervened one to two times each. Two protracted periods of instability in the male hierarchy (Goodall 1986) coincided with two years in which no impartial interventions were seen (Boehm 1992 and table 1). This suggests that the control-role behavior in impartial interventions depends on a stable hierarchy.

At Arnhem Zoo, a single male, not necessarily the alpha male, took the control role in the pacifying interventions. He prevented other males from intervening, but did not prevent the alpha female Mama from intervening with females. At Gombe, the control role was partly shared. However, this can be seen as environmentally determined: at Gombe, the alpha male often is not present when adult conflicts take place and, thus, others sometimes assume the control role. Videotape taken at Gombe shows that sometimes a high-ranking female may intervene in juvenile fights when the alpha male is present. At Gombe as at Arnhem Zoo (de Waal 1982), the adult females generally control the conflicts of offspring and other younger subadults. However, the division of labor is not total: at Gombe, the alpha male sometimes broke up fights between juveniles.

Table 2
Impartial pacifying interventions by individuals at Gombe, 1976–83.

	Total	Male	Female	Intervenors										
				FG	GB	EV	HM	SH	JJ	ST	FF	GG	MF	AT
Interventions	39	33	6	15	8	3	2	2	2	1	2	2	1	1
Dyads:														
Male-male	4	4	0	3	1	—	—	—	—	—	—	—	—	—
Male-female	10	7	3	3	2	1	—	—	1	—	1	1	—	1
Female-female	16	14	2	6	3	1	1	2	—	1	—	1	1	—
Male-juvenile/infant	2	2	—	1	—	—	1	—	—	—	—	—	—	—
Adolescent male-juvenile/infant	4	3	1	1	2	—	—	—	—	—	1	—	—	—
Adolescent male-adolescent male	3	3	—	1	—	1	—	—	1	—	—	—	—	—
Dominance triad:														
IAV	36	33	3	15	8	3	2	2	2	1	1	2	—	—
AIV	3	—	3	—	—	—	—	—	—	—	1	—	1	1
AVI	—	—	—	—	—	—	—	—	—	—	—	—	—	—
Fight level:														
1	13	10	3	6	3	—	1	—	—	—	1	1	—	1
2	20	17	3	7	3	3	1	1	2	—	1	1	1	—
3	6	6	—	2	2	—	—	1	—	1	—	—	—	—
Results:														
Stop	37	32	5	15	8	3	2	2	2	—	1	2	1	1
Continue	2	1	1	—	—	—	—	—	—	1	1	—	—	—

Notes: FG = Figan; GB = Goblin; EV= Evered; HM = Humphrey; SH = Sherry; JJ = Jomeo; ST = Satan; GG = Gigi; FF = Fifi; MF = Miff; AT = Athena; IAV = Intervenor dominant to both; AIV = Intervenor subordinate to aggressor; AVI = Intervenor subordinate to both.

In the conflicts shown at Gombe in table 2, the impartial intervenors were dominant to both protagonists in 36 out of 39 conflicts, with a top male or alpha male accounting for 23 interventions, all successful. In a few instances, multiple intervenors acted together in stopping a conflict, with up to four intervenors participating. Intervention tactics were varied (see table 3). In the case of charging and displays, Gombe Stream Research Centre records routinely indicate the targets of intervention displays when they are directed at individuals, but no targets were mentioned in these particular records. The most frequent tactic, used 22 times by males and three times by females, was to display at both protagonists without directing the display at either of them, causing them to remain focused on the intervenor's presence as he or she typically sat down. In a few cases the impartial intervenor displayed right between the protagonists, while in other cases the intervenor only needed to run or charge toward the conflict, which was the tactic of one male and three females. In table 3, I have recorded the tactics used in the 39 interventions. Staring, bristling, or approaching often occurred at the beginning of a successful sequence and preceded display, but these tactics as a single action were usually not sufficient to control the conflict. However, vocalizations such as coughs or pant-hoots or arm threats sometimes were successful without charging or displaying. These instances of control were effective from a distance and so cannot be classified as impartial interventions with complete assurance, but they definitely appear to have been pacifying.

Several interventions that involved more physical contact will help to demonstrate why some intervention strategies by a dominant individual can unambiguously be classified as impartial. Alpha male Goblin once made a series of attacks, which were videotaped, that separated adult females fighting in a tree; these attacks ceased immediately once the two females had been herded into different trees (Boehm 1992). I observed one unique intervention in which Satan charged at two adolescents who were in a level-two conflict grappling and fighting on the ground; his charge had no effect, so he forcibly pried them apart, which took four seconds (Boehm 1992). This unique intervention was paralleled by behavior at Arnhem Zoo (de Waal 1982).

Table 3
Tactics of impartial pacifying intervention at Gombe, 1976–83.

				Intervenors										
Tactic	M and F	Male	Female	FG	GB	EV	HM	SH	JJ	ST	FF	GG	MF	AT
Look	2	2	—	1	—	—	—	—	1	—	—	—	—	—
Erect hair	5	5	—	2	1	—	—	1	—	1	—	—	—	—
Approach	3	3	—	2	1	—	—	—	—	—	—	—	—	—
Run up	1	1	—	—	—	—	—	1	—	—	—	—	—	—
Charge	3	—	3	—	—	—	—	—	—	—	—	1	1	1
Display at	25	22	3	10	5	3	1	1	1	1	2	1	—	—
Display between	2	2	—	2	—	—	—	—	—	—	—	—	—	—
Fight	2	2	—	1	—	1	—	—	—	—	—	—	—	—
Arm threat	4	4	—	1	2	—	1	—	—	—	—	—	—	—
Cough	1	1	—	1	—	—	—	—	—	—	—	—	—	—
Pant-hoot	1	—	1	—	—	—	—	—	—	—	1	—	—	—
Total	49	42	7	20	9	4	2	3	2	2	3	2	1	1

Notes: FG = Figan; GB = Goblin; EV = Evered; HM = Humphrey; SH = Sherry; JJ = Jomeo; ST = Satan; GG = Gigi; FF = Fifi; MF = Miff; AT = Athena.

With impartial interventions, the success rate in stopping conflicts was 37 out of 39 (see table 2). This success rate is not remarkable considering that the intervenors were dominant to the aggressors. Partial interventions observed in 1978–79 were directed toward protecting kin, primarily mothers protecting juveniles and infants, 60% of the time. By contrast, impartial interventions showed no tendency to be directed at conflicts that involved either bonded kin or coalition allies.

Pacifying Interventions at Arnhem Zoo

De Waal's published studies of intervention behavior at Arnhem Zoo are based on large numbers of observations with a second observer and videotaped records as experimental control on accuracy. Virtually all interventions were recorded. Although the data for Arnhem are not broken down into the partial and impartial pacifying categories used for Gombe, de Waal (1982) nevertheless provides good qualitative descriptions of these frequent behaviors. For example, he considers interventions involving a control role that is played by one of the high-ranking individuals to be one of several types of intervention: "This individual breaks up fights by pulling or beating the two combatants apart, after which he stands between them to prevent further aggression. If he intervenes partially, it is usually on behalf of the individual under attack" (de Waal 1992). De Waal goes on to say that intervening individuals "seem to place themselves *above* the conflicting parties in the sense that their interventions are not guided by affiliative preferences." This behavior observed at Arnhem Zoo is similar to that observed at Gombe. Unlike Gombe males, however, the Arnhem males seemed to compete decisively over the control role, insofar as the right to intervene is treated possessively and defended aggressively. De Waal (1992) suggests that fight-stopping behavior may well create a political constituency among lower-ranking individuals who, as likely victims, support their protector.

De Waal's accounts (1982) of control-role behavior include one episode in which the new alpha male Luit breaks up a level-three fight between two females, whose allies had entered the conflict:

> Luit leapt in and literally beat them apart. He did not choose sides in the conflict, like the others; instead, anyone who continued to fight received a blow from him . . . On other occasions he put a stop to serious conflicts less heavy handedly. When Mama and Puist were locked in a fight he put his hands between them and simply forced the two large females apart.

He then stood between them until they had stopped screaming . . . Besides such impartial interventions Luit also intervened on behalf of one or other party . . . Instead of a *winner-supporter,* he became a *loser-supporter.*

Luit's impartial prying-apart tactic sounds very much like the unique strategy employed by Satan at Gombe, while Goblin's separation of the two females at Gombe corresponds tactically to beating apart at Arnhem Zoo. Such brief, serial attacks have been reported several times at Gombe.

Interventions at Arnhem can be illustrated further. De Waal and Hoekstra provide data on mothers coming into conflict because their infants scream, a situation also described by Goodall (1986). De Waal and Hoekstra (1980:933) write that:

Jonas and Wouter (aged 2.5 and 3 years, respectively) play-wrestled in front of their mothers, Jimmie and Tepel. They sat together with Mama, the much-respected alpha-female, who was sleeping. Suddenly the wrestling between the two infants turned into fighting and screaming. Both mothers alternately looked at each other and at the ongoing fight. Jimmie gave a soft grunt and Tepel restlessly shifted her position. Eventually, Tepel woke up Mama by poking several times in her side. She pointed in the direction of the infants with her hand. Mama got up slowly, seemed to take in the situation at a glance and stopped the fight by an upsway with her arm and a loud grunt-bark. The two infants separated for a while and Mama soon continued her siesta. Thus Tepel's behavior seemed to be aimed at the regulation of a conflict involving her child without risking a conflict with Jimmie.

This situation nicely illustrates group awareness of the control role; indeed, de Waal (1984) has suggested that the control-role concept in humans, with its concomitant notions of expectations and norms, may apply directly to control-role behavior of chimpanzees. The fact that adult females at Arnhem Zoo were responsible for 80% of all interventions in juvenile or infant conflicts shows a marked division of labor by sex in the control-role, with females adjudicating squabbles and fights of younger individuals. While a similar pattern would appear to exist at Gombe, data are still being compiled.

In an interesting and unique pattern occurring at Arnhem, a third party takes away an object to be used in a display or attack from an adult aggressor as a means of dampening conflict (de Waal and van Hooff 1981; de Waal 1982). Although Gombe chimpanzees predictably scoop large stones and hurl sticks or stones in the course of displays, and objects for display are everywhere, they rarely engage in confiscation as do the Arnhem chimpanzees. Direct physical intervention by females in conflicts of adult males is quite rare at Gombe but, at Arnhem Zoo, females frequently confiscate objects from males who are on the verge of attacking someone.

Comparison

Similarities in the behavior of chimpanzees between Gombe and Arnhem Zoo are notable given the differences in the environment, which at Arnhem prevents fissions within the group and, thereby, keeps males and especially females together more than at Gombe. Avoidance is difficult at Arnhem because the living area lacks a canopy or peripheral areas. On this basis, Arnhem's high rates of pacifying interventions may result from a combination of inability to avoid conflict and more time for "politicking" (de Waal 1992; Goodall 1986).

Published data do not permit a full statistical comparison of pacifying intervention rates at the two sties, but the Gombe rates can be assessed as follows. Over an eight-year period at Gombe there were about five impartial intervention attempts each year (see table 2), almost always successful. In addition, as seen in table 1, for a two-year period partial interventions average 30 each year, with a 70% success rate. Combining these figures, the observed yearly average at Gombe is 35 attempts of pacifying interventions with an overall success rate of about 75%. At Arnhem Zoo, while the intervention attempts would appear to be much more frequent for a group less than half the size of the Gombe Kasakela study group, a great proportion of the major agonistic interactions are recorded because of continuous observer presence and an unimpeded view. Thus, observed frequencies at Arnhem Zoo are close to true frequencies. By contrast, I estimate that at least two-thirds of the adult fighting behavior at Gombe is not observed because of the way that the observations are made (Goodall 1986). Furthermore, observers frequently hear fights but cannot see them.

Focusing on just those interventions that actually stop conflict, the true annual frequencies at Gombe can be estimated at 20 impartial and 84 partial pacifications, for a total of 104 pacifying interventions in conflicts involving at least one adolescent but very often involving a pair of adults. The amount of effective intervention and its impact upon the social organization at Gombe must be deemed substantial. Classical control-role interventions of the impartial type take place perhaps twice a month at Gombe, while some type of successful intervention takes place about twice a week. It must be kept in mind that, at Gombe, many dyads prone to fight are not in proximity much of the time because the chimpanzees often are in small subgroups.

Two patterns emerge at both sites: intervention is used both to ally and to pacify; and one high-ranking male does most of the impartial pacification. What is different between the sites? At Arnhem Zoo, a female coalition supports the male who plays the control-role, and this male is not necessarily the alpha male. This situation may be a consequence of Arnhem's special history, in which the adult female Mama and other females established a dominance hierarchy that continued to be strong after males were introduced.

While partial and impartial intervention strategies are used at both sites, tactics differ somewhat. Confiscation, prying apart, and beating apart were frequent tactics at Arnhem Zoo and rare at Gombe. At Gombe, less physical contact was involved, with displays being the chief intervention tactic.

Explaining Continuity and Diversity

I began a with a discussion of continuity and the presumption of genetic dispositions for intervention behavior. In spite of major differences in environment, in group provenence, and in recent social history, the behavior patterns are basically quite similar at Gombe and Arnhem Zoo. Competition involving conflict expresses itself very similarly at both sites, and this similar behavior appears to be species specific. However, additional comparisons are needed for falsification. For example, it will be of particular interest to see Christophe Boesch's results for the long-term study at Taï and consolidated results from researchers at Mahale. My prediction is that partial pacifying interventions will be found universally among chimpanzees because of the innate protective tendencies of mothers and others, while impartial pacifying interventions will be found wherever high-ranking chimpanzee males are in a strong political position so that they can act as arbitrators.

Aside from fear, avoidance, submissive signals, and reconciliation (de Waal 1989), at least three other factors limit wounding and loss of life in conflicts occurring within chimpanzee groups. One factor is an inhibition that may be species specific that directs most biting behavior at fingers or toes. This inhibition of biting protects the biter because fights are not escalated to dangerous levels. The other two factors are partial and impartial pacifying interventions. Maternal protection of offspring obviously helps inclusive fitness; protection of male subadults by other males and females could result in weak yet significant group-selection advantages by augmenting community size. On the other hand, dominant interventions in a control role are less readily explained (Boehm 1981, 1992). I have suggested scenarios that might account for such a control-role behavior with its appearance of genetic *altruism.* Impartial intervention could be a special extension of the protective tendency toward subadults, or an adverse reaction to the noise and stress that accompany agonism, or both. In any event, if control-role behavior is found at all study sites in the wild, the general case for control-role behavior being species-specific will be strengthened.

Given the overall behavioral flexibility for which chimpanzees are well known, the similarities at Arnhem and Gombe are remarkable. Diversity has been noted as well, and a plausible explanation is adaptive modification (Kummer 1971), which stems from environmental factors that influence group traditions far more immediately than genes. For Arnhem Zoo, I have mentioned the absence of a canopy, of peripheral refuge areas, and of a need to forage as factors that may help to explain the higher rate of conflict, the pattern of confiscation, and the greater tendency for interventions to become physical (Goodall 1986).

The recent cultural history of Arnhem Zoo also must be considered. For 18 months, the adult female Mama ruled the Arnhem group, and the subsequently introduced males remained subordinate until she and her ally Gorilla were temporarily removed. While group tradition was changed radically by human intervention to favor a Gombe-type male-dominated social hierarchy, certain aspects of the initial structure showed staying power. The female coalition at Arnhem continues to influence the male hierarchy far more than anything seen at Gombe. The special history of Arnhem Zoo, combined with environmental factors that include females in political life, made this possible.

Cultural diffusion at Arnhem Zoo is also a possibility. The Arnhem group's contact with humans provides the opportunity to observe human styles of conflict intervention. Arnhem chimpanzees may have observed confiscation behavior in humans and adopted it. However, such behavior has been observed once among Gombe chimpanzees (Goodall pers. comm.).

In revisiting Bernstein and Sharpe's (1966) concept of the control-role, if one leaves aside protection of group members from external predators, this characterization fits the behavior of both wild and captive chimpanzees. Yet, an obvious question remains: Are humans projecting their own leadership strategies onto other primates? I argue that with chimpanzees—the primates with whom we have the best opportunity to explain similarities to ourselves by homology rather than by analogy—one can, indeed, distinguish objectively between pacifying interventions, which involve strategies likely to dampen conflicts, and fight-winning interventions, which involve strategies likely to continue or exacerbate conflicts. Furthermore, at both Gombe and Arnhem Zoo, one individual (a male) tends to play the control role most of the time, while at both sites females act in this role only in certain contexts. These are important similarities.

At both Arnhem and Gombe, chimpanzees employ a wide variety of similar tactics in pacification. This flexible approach to problem solving suggests that the control of conflicts involves fairly sophisticated cognitive calculations, which amount to a general strategy of pacification that is both behavioral and intentional (Boehm 1978). The motivation may be complex: protection would be a likely motive where both partial and impartial interventions terminate fights quickly. Elimination of disturbance-related stress in the control individual is another possible motivation; enhancement of dominance standing in the hierarchy and delayed reciprocation may also provide motivation.

Given the findings at these two sites, the contradictions and complications that have beset application of a control-role concept to other species (Erhardt and Bernstein 1992) seem to be obviated for chimpanzees. Were the control role established only for captive chimpanzees, it could be argued that crowding and other effects are bringing out a behavior that is not naturally adaptive (McGrew and Tutin 1978). However, the data for Gombe parallel the Arnhem findings quite closely, even though quantitatively the true rate of intervention would seem to be considerably higher at Arnhem where the group is smaller and even though some tactics are used very rarely at Gombe but frequently at Arnhem. Based on the parallels, it can be argued that the ability to readily learn intervention behavior is genotypic. But intervention behavior is flexible enough to be subject to rapid adaptive modification. Such modifications can be attributed to differences in physical environment, to absence of territorial competitors, to differences in recent social history and, possibly, to influence from a human cultural environment.

At both Gombe and Arnhem, it is clear that the concept of the control role in dampening conflict within the group is worth revisiting. The genetic provenience of control-role behavior is not readily explained, but its function is to protect community members with little risk to a dominant intervenor. Humans, too, seem to intervene in conflicts everywhere, even though with nomadic foragers such efforts are weak because protagonists in the conflict can readily change groups (von Furer-Haimendorf 1967). Furthermore, where humans remain strictly confined to one place for ecological or political reasons, as with chimpanzees at Arnhem Zoo, their means of conflict resolution become more inventive; they try much harder to manage conflicts preemptively or definitively because they can no longer rely upon avoidance. This, however, is a topic for future research.

This comparison of chimpanzee research at two well-studied sites is not an end but a beginning. First, the control-role concept has proved to be useful analytically. What can be learned from applying it successfully to understanding chimpanzees may clarify its potential applicability to other species. Second, substantial further comparison of chimpanzees would be desirable. Standardization of behavior categories and statistical comparison of Gombe and Arnhem Zoo could be readily accomplished. It would be useful also to compare Gombe and Mahale, to see whether behavioral diversities exist between two groups whose environments and histories of provisioning are quite similar. Eventually, it would be useful also to compare these sites with the unhabituated sites in East and West Africa to factor out possible effects of provisioning. The chimpanzee community in Taï Forest, a group distant from Gombe and Mahale, is of particular interest: the long-term study there makes possible a full evaluation of both fight-winning and pacifying intervention in the absence of provisioning.

I hope that this preliminary comparison, along with the refinements of methods and concepts I have attempted, will make possible further comparisons of a control-role behavior, which is both functionally important for chimpanzees and intrinsically fascinating because of its parallels to the management of human conflict.

Summary

The control-role concept has been explored at Gombe and Arnhem Zoo to see if diversity in environmental circumstances is accompanied by diversity of fight-stopping behavior. Similarities of social structure and context of agonism have served as an experimental control for the comparison. The basic pattern of intervention was found to be similar, while differences in the tactics of intervention are discussed in terms of major differences in physical environments. At Arnhem Zoo, interventions appear to be more frequent, to involve more physical contact, and to involve a female coalition that supports the control male who is not necessarily the alpha male. At Gombe, the alpha male does most of the conflict resolution, or pacifying intervention, without any female coalition to support him. At both sites, females intervene in subadult conflicts and, sometimes, in adult conflicts. However, at Arnhem Zoo, females confiscate objects from adult males before these objects can be used in displays, while males frequently pry apart or beat apart combatants. Both behaviors are extremely rare at Gombe, where displays are the main tactic.

Cultural similarity and diversity have been discussed in terms of genetic disposition, environmental difference, and differences in recent history. The conclusion is that the control-role concept applies well to the pacifying interventions of chimpanzees, be they partial or impartial, and that such behavior would appear to be species specific because it has remained relatively uniform in two radically different environments. This hypothesis merits further testing.

Acknowledgments

I am grateful to Frans de Waal for generously helping to orient this study, to Craig Stanford for critical comments, to Bill McGrew for substantial editorial assistance, to Jane Goodall for making available data and facilities at the Gombe Stream Research Centre and for critical comments on an earlier version of this paper, and to Tanzanian National Parks and other Tanzanian governmental agencies. I express my gratitude also to Yahaya Almasi, Hilali Matata, Hamisi Mkono, and Eslom Mpongo for their cooperation and hard work in securing videotapes of conflict intervention; to the L.S.B. Leakey Foundation, the Harry Frank Guggenheim Foundation, and the Jane Goodall Institute for financial support of the long-term videotape project at Gombe; and to the Harry Frank Guggenheim Foundation for supporting my field research on conflict intervention.

References

Bernstein, I.S., and L. Sharpe. 1966. Social roles in a rhesus macaque group. *Behav.* 26:91–103.

Boehm, C. 1978. Rational preselection from hamadryas to *Homo sapiens:* The place of decisions in adaptive process. *Am. Anthro.* 80:265–96.

———. 1981. Parasitic selection and group selection: A study of conflict interference in rhesus and Japanese macaques. In A.B. Chiarelli and R.S. Corruccini, eds., *Primate Behavior and Sociobiology.* pp. 161–82. New York: Springer-Verlag.

———. 1992. Segmentary "warfare" and the management of conflict: Comparison of East African chimpanzees and patrilineal-patrilocal humans. In A. Harcourt and F.B.M. de Waal, eds., *Us Against Them: Coalitions and Alliances in Humans and Other Animals.* pp. 137–73. Oxford, England: Oxford Univ. Press.

Erhardt, C.L., and I.S. Bernstein. 1992. Intervention behavior by adult male macaques: Structural and functional aspects. In A. Harcourt and F.B.M. de Waal, eds., *Us Against Them: Coalitions and Alliances in Humans and Other Animals.* pp. 83–111. Oxford, England: Oxford Univ. Press.

von Furer-Haimendorf, C. 1967. *Morals and Merit.* Chicago: Univ. of Chicago Press.

Goodall, J. 1986. *The Chimpanzees of Gombe.* Cambridge, Mass.: Belknap Press.

Harcourt, A.S., and F.B.M. de Waal, eds., 1992. *Us Against Them: Coalitions and Alliances in Human and Other Animals.* Oxford, England: Oxford Univ. Press.

Kummer, H. 1971. *Primate Societies: Group Techniques of Ecological Adaptation.* Arlington Heights, Ill.: Harlan Davidson.

McGrew, W.C., and C.E.G. Tutin. 1978. Evidence for a social custom in wild chimpanzees? *Man* 13:234–51.

de Waal, F.B.M. 1978. Exploitative and familiarity-dependent support strategies in a colony of semi-free living chimpanzees. *Behav.* 66:268–312.

———. 1982. *Chimpanzee Politics: Power and Sex among Apes.* New York: Harper and Row.

———. 1984. Sex differences in the formation of coalitions among chimpanzees. *Ethol. and Sociobiol.* 5:239–55.

———. 1986. The brutal elimination of a rival among captive male chimpanzees. *Ethol. and Sociobiol.* 7:237–51.

———. 1989. *Peacemaking Among Primates.* Cambridge, Mass.: Harvard Univ. Press.

———. 1992. Coalitions as part of reciprocal relations in the Arnhem chimpanzee colony. In A.H. Harcourt and F.B.M. de Waal, eds., *Us Against Them: Coalitions and Alliances in Humans and Other Animals* pp. 233–58. Oxford, England: Oxford Univ. Press.

de Waal, F.B.M., and J.A. Hoekstra. 1980. Contexts and predictability of aggression in chimpanzees. *Anim. Behav.* 28:929–37.

de Waal, F.B.M., and J.A.R.A.M. van Hooff. 1981. Side-directed communication and agonistic interactions in chimpanzees. *Behav.* 77:164–98.

Social Relationships of Female Chimpanzees

Diversity Between Captive Social Groups

Kate C. Baker and Barbara B. Smuts

Introduction

Research on chimpanzees, both in the wild and in captivity, consistently indicates two salient principles underlying social interactions between males. First, males employ complex social strategies to compete intensively for dominance status. Second, males exhibit a variety of behaviors that function to resolve the social tensions caused by this intense competition. These conciliatory behaviors allow males to maintain strong, cooperative relationships in the face of chronic competition. With these two basic principles in mind, the otherwise bewildering variety and complexity of male-male interactions and relationships have become understandable and, to some extent, predictable (de Waal 1982, 1989; Nishida 1983; Goodall 1986).

In contrast, our understanding of social relationships between female chimpanzees lags far behind our understanding of social relationships between male chimpanzees. There are two reasons for this deficit. First, female chimpanzees in the wild typically forage alone or in small parties with close kin and, therefore, spend much less time together socializing than do male chimpanzees (Wrangham and Smuts 1980). For this reason, observations of social interactions between females are relatively rare. Second, even in captive groups where females are forced into close proximity, dominance interactions between females tend to be far less common and less dramatic than those of males, and therefore, female-female relationships have received less systematic attention (de Waal 1982).

The little information we do have about female-female relationships is intriguingly inconsistent. In particular, it is unclear under what conditions, or even whether, female chimpanzees strive to dominate each other. The view that female chimpanzees show little concern for achieving dominance has been most clearly expressed by de Waal (1982, 1984, 1993), who reported that stable relationships between captive females involved little display and intimidation, but rather an acceptance of rank based upon respect, perhaps elicited by the individual's personality or age. De Waal's conclusions are concordant with Bygott's (1974) analysis of female-female relationships in the wild at Gombe. On the other hand, dominance striving between females in the wild has been reported by Nishida (1989) and Goodall (1986), who described several examples of repeated, antagonistic interactions between certain females that appeared to result in reversals of status. In addition, Wrangham et al. (1992) found no differences between the sexes in rates of dominance interactions in the Kibale Forest, after correcting for the fact that females had fewer opportunities to socialize than did males. These inconsistent results indicate considerable, but unexplained, diversity in the social relationships of female chimpanzees.

This chapter will attempt to provide an explanation for the apparent inconsistencies in the behavior of female chimpanzees. The purpose of this chapter is threefold. First, we provide further evidence of behavioral diversity in female-female interactions by comparing female behavior in two captive groups of chimpanzees. One is the Arnhem Zoo colony in the Netherlands, which is studied by de Waal and colleagues; the other is the captive colony recently established at the Detroit Zoo.

Second, we use these comparisons to identify the importance of social stability, a factor that may help to explain why female chimpanzees exhibit so much behavioral diversity in their interactions with one another. In particular, we hypothesize that when social stability is unstable, females tend to compete intensively for dominance, and they do so through use of complex social strategies reminiscent of strategies shown routinely by males; in contrast, when social stability is stable, females are much less likely to compete with each other for dominance. For current purposes, social stability is high when all adult group members have known each other for several years as in the Arnhem Zoo colony; whereas social stability is low when the group includes many adults that have met only recently as in the Detroit Zoo colony.

Third, we conclude by considering how behavioral diversity of female chimpanzees in captivity at Arnhem Zoo and Detroit Zoo can help to illuminate the functional significance of female-female social relationships in the wild.

Methods: Comparison of Two Captive Colonies

Our analyses of behavioral diversity involve three aspects of behavior: *aggression, reconciliation,* and *coalition.* For each aspect, we compare results from Detroit Zoo and Arnhem Zoo. We also include data from field studies when available. For some analyses, we compare both female-female and male-male interactions to determine whether females exhibit social strategies similar to those employed by males competing for rank.

Research sites

Detroit Zoo. In 1989, the Detroit Zoo in Detroit, Michigan, established a large chimpanzee colony modeled on the Arnhem Zoo colony in the Netherlands. The Detroit Zoo's chimpanzee facility consists of a large (greater than one hectare) outdoor exhibit area and an indoor facility with two large identical day rooms and 12 night cages. Eleven chimpanzees including two adult males, six adult females, one juvenile female, and two orphaned infant females formed the colony. Six of the eight adults were wild-born and two were captive-born. All adults had been housed long-term with a same-sex partner prior to group formation. The two adult males had been residents at the Detroit Zoo. The six adult females came as three female-female pairs from other zoos.

Arnhem Zoo. The Arnhem colony, established in 1971, consists of approximately 20 to 25 individuals in facilities similar to those in Detroit Zoo. Colony formation and colony characteristics are described by van Hooff (1974), de Waal (1982, this volume), and Adang et al. (1987).

Data Collection and Analysis in Detroit Zoo

Data collection occurred during a 14-month period: an eight-month introduction period and a six-month period following group formation (Baker 1992; McDonald Black 1992).

Three trained observers collected 580 hours of focal observations (Altmann 1974) from July 1989 to August 1990. One chimpanzee, chosen at random, was observed. Social interactions, time of occurrence, responsibility for initiation, and terminations were recorded for the focal chimpanzee and others involved. Fifteen-minute focal samples also included on-the-minute scans of the focal individual's behavior and the proximity of other group members, plus continuous records of approaches and leaves, and a variety of behaviors defined in the *ethogram,* a modification and compilation of behavioral descriptions (Goodall 1968; Bygott 1979; van

Hooff 1974). Agreement between observers was determined using *Cohen's kappa,* an agreement statistic that corrects for chance (Bakeman and Gottman 1986). All values of Cohen's kappa exceeded .70 (Baker 1992).

Information about coalitions (de Waal 1978, 1984) was drawn from videotaped records of agonistic interactions in the indoor facility. A reconciliation protocol lasting 30 minutes was conducted after each incident of dyadic aggression involving contact, following the methodology of de Waal and van Roosmalen (1979). As close as possible to 24 hours after the reconciliation protocol, observations were conducted on the same individual using the same methods to serve as a control. Control observations were conducted in order to correct for baseline differences in sociability (de Waal and Yoshihara 1983).

Female dominance rank was defined in terms of *agonistic dominance* (Nöe et al. 1980), except that we excluded all facial and vocal elements except pant-grunting, and we added the submissive behaviors bowing, bobbing head and trunk. Within a dyad, an individual was considered dominant if she received all the bows or all the pant-grunts (always unidirectional), or if she was responsible for more than 51% of attacks (Strum 1982).

Results

Contexts of Aggression

Analysis of agonistic dominance of females at Detroit Zoo indicated that during the first six months after colony formation, female dominance relationships were unstable (Baker 1992). In contrast, at Arnhem Zoo female dominance relationships were stable (de Waal 1982). We describe here differences in aggressive behavior of females that appear to be related to these differences in hierarchy stability.

Among male chimpanzees, aggression frequently occurs during reunions or *in no apparent context,* that is, when no resource is at stake. Particularly during reunions, males are prone to display or to attack those males over whom they are already dominant (Goodall 1986). Such aggression is thought to be related to competition for status: it is used to reemphasize and maintain rank, even in the absence of immediate need to contest resources (de Waal and Hoekstra 1980; Goodall 1986; Nishida 1989).

In previous studies at Gombe and Arnhem Zoo, aggression in no apparent context accounted for a much smaller proportion of female-initiated aggression compared with male-initiated aggression (see table 1), suggesting that in these two groups, females rarely competed for status. In

Table 1
Percentage of aggression occurring in no apparent context: Gombe, Arnhem, and Detroit.

Initiator	Gombe (%)	N	Arnhem (%)	N	Detroit Zoo (%)	N
Females	13	159	38	112	65	104
Males	58	319	80	216	62	21

Notes: Results include all aggression initiated by adult females or by adult males, regardless of age/sex of recipient. Amount of female-initiated aggression occurring in no apparent context: Detroit vs. Gombe χ^2=78.8, p < .001; Detroit vs. Arnhem, χ^2=16.8, p < .001 (two-tailed). (Goodall 1986:342; de Waal and Hoekstra 1980:932; Baker 1992.)

contrast, at Detroit Zoo females initiated aggression in no apparent context about as often as did males, and significantly more often than at Gombe and Arnhem Zoo (see table 1).

Contexts of aggression can be further evaluated by considering the sexes of both participants (Nishida 1989). At Mahale, significantly less (32%) aggression occurred between adult females in no apparent context compared with aggression at Detroit Zoo (92%) (χ^2=41.5, N=105, p < .001). The percentage of aggression in no apparent context at Detroit Zoo was lower for female-male fights (67%) as well as for fights involving the juvenile female and adult females (37%), indicating that the high proportion of adult female-female aggression occurring in no apparent context at Detroit Zoo was not merely an artifact of the ethogram, observer bias, or some colony-wide characteristic.

We must consider carefully, then, the differences other than those related to social stability between Arnhem Zoo and Detroit Zoo that might account for the differences in female-female aggression reported here. For example, fewer fights can be expected in the context of infant protection when natural mother-infant bonds are lacking, as was the case at the Detroit Zoo. This lack of mother-infant bonds may explain some of the differences between female aggression at Arnhem Zoo and at Detroit Zoo. However, opportunities for protectiveness were not absent at the Detroit Zoo. First, the group included two orphaned infants with whom all the females were bonded to some degree and with whom relationships bearing hallmarks of natural mother-infant relationships were formed by adoptive mothers (McDonald Black et al. in prep). Second, the female/infant ratio was the same at Arnhem Zoo and Detroit Zoo. Finally, aggression arising to protect familiar partners may have been even more important in the unsettled Detroit Zoo group, with its three pairs of long-term familiar partners and relatively high rates of aggression between females (Baker 1992). Although the relative strengths of these factors cannot be assessed directly with the present data, it is reasonable to hypothesize that the greatly elevated percentage of female aggression in no apparent context in the Detroit Zoo colony may be attributed, at least in part, to female use of aggression as a tool in dominance competition during the formation of relationships.

Reconciliations

De Waal and van Roosmalen (1979) and de Waal (1993) use patterns of post-conflict behaviors between former opponents to suggest that the female-female sphere in chimpanzees is relatively inactive in terms of dominance and aggression. They report that *post-conflict contact,* which involves kissing, touching, stretching hands toward one another, and submissive vocalizations, occurs more often between recent former opponents than would be predicted by chance alone. These reconciliations calm the opponents, appear to repair social bonds, and to create mutual trust (de Waal 1989). At Arnhem Zoo, opponents reconciled most often following male-male fights, less often following male-female fights, and least often following female-female fights (de Waal and van Roosmalen 1979). De Waal argues that males reconcile with males much more often than females reconcile with females because males as compared to females are more competitive and more highly motivated to maintain cohesive relationships threatened by competition (de Waal 1986).

Based on de Waal's argument, reconciliation patterns can be used as a yardstick to measure activity in the domains of dominance and aggression. Table 2 compares the percentage of fights reconciled at Detroit Zoo and Arnhem Zoo. The sole pair of Detroit Zoo males rarely reconciled. This result, which presumably reflects the males' unresolved dominance relationship, unfortunately precludes useful comparisons between male-male and female-female reconciliations at Detroit Zoo. However, table 2 reveals two interesting results. First, the percentage of fights followed by a reconciliation was four times greater among Detroit Zoo females than among Arnhem Zoo females (χ^2=16.2, n=51, $p < .001$). Second, at Detroit Zoo, there were no significant differences in the proportion of fights reconciled between female-female opponents as compared to female-male opponents; whereas at Arnhem Zoo, female-female opponents reconciled significantly less than female-male opponents (de Waal 1986). Because reconciliation is thought to reflect a need for social stability and tension reduction in the face of competition and aggression that jeopardizes relationships, these results suggest that the female-female interactions at the newly formed Detroit Zoo colony were considerably more active in terms of aggression, hierarchical relationships, and tension reduction than expected based on previous studies of the well-established Arnhem Zoo colony.

Table 2
Reconciliation behavior at Detroit and Arnhem: protocols containing affiliative contact.

	Detroit[a]		Arnhem[b]
	Aggression (%)	Control (%)	Aggression (%)
Female-female	74	35	18
Female-male	64	27	39
Male-male	11	44	47

Notes: a. Sample sizes: Female-female fights 23; female-male fights 11; male-male fights 9. b. No control follows were conducted at Arnhem. Arnhem data based on de Waal 1986:470.

Coalitions

In the wild and in captivity, both males and females form *coalitions,* temporary alliances, often providing agonistic aid toward one participant in a dispute. At Arnhem Zoo, patterns of coalition formation varied depending on the sex and the social conditions. When male dominance relationships were in flux, a male tended to form reciprocal coalitions with those whose support could benefit his struggle to rise in rank. These coalitions required the suspension of social preferences to recruit the most effective allies (de Waal 1978, 1984). Nishida (1983) reported similar opportunistic, flexible coalition formation between males in the wild. However, when the male hierarchy was stable males tended to form coalitions that protected familiar partners. This pattern of forming coalitions also characterized female-female coalitions at Arnhem (de Waal 1978, 1984). As noted earlier, female dominance relationships were consistently stable at Arnhem Zoo. We use data from Detroit Zoo to test the hypothesis that females, like males, use opportunistic coalitions when dominance relationships are unstable.

Each coalition involves three individuals: A, B and C. If B and C are fighting and A attacks C, then A is considered to have formed a coalition with B against C. De Waal (1984) compares the percentage of time A spent in proximity to B (the individual supported) versus C (the individual opposed) and, for each individual, calculates the mean proportion of support choices that favor the individual with a higher proximity score. The higher this proportion, the more the individual formed coalitions based on *familiarity,* that is, preexisting social bonds.

We repeated de Waal's methods to analyze coalitions of Detroit Zoo females. In our analysis, preexisting social bonds were measured by *C scores,* which are weighted measures of time spent in proximity (Smuts 1985; Baker

Table 3
Coalitionary behavior: support choices among Detroit chimpanzees.

	Proximity score (%) (greater for animal supported)	Coalitionary acts
Sarah (adult female)	67	55
Peggy (adult female)	62	29
Tanya (adult female)	97	31
Trixi (adult female)	88	16
Bubbles (adult female)	70	20
Beauty (adult female)	57	79
Abby (juvenile female)	45	33
Joe (adult male)	57	7
Chuck (adult male)	74	38

Notes: Adult females listed from highest to lowest rank. Proximity score: mean proportion of support choices in favor of the most familiar individual.

1992). At Arnhem Zoo, all females supported the more familiar partner in at least 70% of their coalitions; all male scores fell below this threshold (de Waal 1984). In contrast, at Detroit Zoo, three of the six adult females choices for support fell below the 70% threshold established at Arnhem Zoo, and one of the male support choices exceeded the threshold (see table 3). These results indicate less differentiation between female and male support patterns than was found at Arnhem Zoo and a weaker link between support choices and social bonds for several Detroit Zoo females than for Arnhem Zoo females. Thus, females, like males, appear to form opportunistic, flexible coalitions to a greater degree when rank relationships are unstable as at Detroit Zoo than when they are stable as at Arnhem Zoo.

Discussion

Our analysis of aggression, reconciliation, and coalition demonstrates considerable diversity in the social behavior of captive female chimpanzees. In particular, our findings suggest that the competitive strategies adopted by female chimpanzees depend on the degree of familiarity and stability of their relationships. The Detroit Zoo data show that, when developing relationships with other female adults, females compete intensively and sometimes employ strategies similar to those seen in dominance struggles between adult males (Yerkes 1943). The Arnhem Zoo data, in contrast, show that when females become familiar with one another, their dominance relationships stabilize and overt female-female competition is rare. In contrast, overt competition and striving for dominance between captive males persist indefinitely (de Waal 1982). To understand differences in the social behavior of male and female chimpanzees in captivity, we need to reconsider how and why females and males compete in the wild.

Observations at Gombe (Pusey 1980; Goodall 1986) and at Mahale (Nishida 1989) suggest that female-female competition is most intense in the context of establishing and protecting *core areas.* Pusey (1980) reported that at Gombe seven out of eight immigrant females were seen being attacked or chased by resident females. Extensive observations of one immigrant female documented several attacks by coalitions of resident females in the first few days after her transfer. At Mahale, systematic observations of adult females showed that most female-female aggression occurred between residents and recently immigrated females (Nishida 1989). Long-term residents consistently dominated immigrants and formed coalitions against them. Immigrants attacked other immigrants, but aggression between resident females was very rare. Nishida (1989) concludes that young females competed with one another and with older residents to establish core areas; whereas older, resident females with established core areas tried to prevent immigrants from taking portions of their core areas to form new ones. These findings suggest that immigrant females derive important benefits from establishing core areas and that such efforts by immigrant females inflict a cost on resident females.

Variation in the location and quality of core areas might influence female reproductive success in several ways. First, data from Mahale indicate that male infants of recently immigrated females living near the periphery of the range of the community are at risk of infanticide by community males (Nishida et al. 1990). Second, females with core areas at the center of the community range may be less vulnerable to severe attacks by males from neighboring communities (Goodall 1986; Nishida 1989). Third, core areas may differ in foraging quality (Pusey 1978); however, no data are currently available to evaluate this suggestion.

Once a female becomes a resident and succeeds in establishing a core area, her rank appears to become relatively fixed (Nishida 1989), and her dominance relationships between other resident females are generally quite stable, especially when compared with relationships between males. Nishida suggests that once a core area is established, females "have no pressing need to strive for higher rank" (Nishida 1989).

Several factors might operate, either individually or in concert, to reduce the *payoffs,* the benefit minus cost, to females for pursuing female-female competition beyond the minimum required for establishing an initial core area. The benefits associated with improving the quality of a core area once that area has been established may be low compared with the benefits associated with acquiring a core area. Also, once a female immigrant has settled into a core area, further competition with long-term residents might be costly as a result of reduced agonistic support from males once the female conceives and loses her sexual appeal; of risks of injury to an infant once the female gives birth; and of increased resistance by established resident females if, and when, younger immigrants try to encroach further on the established core areas.

These considerations suggest that the payoff curve associated with competition between female chimpanzees rises steeply until a female develops a sufficiently high rank to establish a functional core area. Then, the payoffs associated with further, incremental rises in rank are probably rather small, and the curve begins to flatten out, that is, benefit minus cost approaches zero (see figure 1). Similarly, the costs associated with a loss of rank (movement from right to left in figure 1) for dominant females may be negligible until and unless the female's status drops to the point where her core area is compromised. Accordingly, in situations associated with potential movement into or out of the steep portion of the payoff curve, that is, opportunities to establish core areas or to protect already established core areas, we expect females to engage in intense competition resulting in changes in dominance rank. In contrast, in situations associated with the flatter portion of the payoff curve, such as opportunities to improve a core area once established, we expect to see greatly reduced competition, resulting in stable dominance relationships.

Among male chimpanzees the payoff curve associated with competition appears to rise quite slowly until the male achieves high rank. At that point, incremental changes in rank tend to be associated with steep improvements in reproductive success (de Waal 1982) (see figure 1). Indeed, at Mahale, records of copulations with fertile females indicate that only the alpha male derived significant reproductive advantages from status competition (Hasegawa and Hiraiwa-Hasegawa 1990). This situation indicates the existence, at least under some conditions, of the most extreme version of a payoff curve. A concave payoff curve that rises very steeply toward the right is associated with intense competition, a chronic concern with status, and periodically unstable hierarchies where many males will compete for alpha rank. In addition, even *postprime males,* who have little chance of achieving alpha status, sometimes gain reproductive advantages by manipulating hierarchical relationships (de Waal 1982, 1993; Nishida 1983). This situation should further decrease the stability of the hierarchy.

These proposed differences in payoff curves for males and females may influence social relationships in ways that reinforce stability in female-female relationships and instability in male-male relationships. If the best strategy for a female chimpanzee is to maintain a core area of adequate quality as opposed to continually striving for a better core area, then older, long-term resident females can perhaps best achieve this goal by largely suspending competition between themselves and by forming stable, reliable alliances to resist competition from younger, more recent immigrant females. If such alliances guarantee each female what she most needs—maintenance of her core area—then resident females would not be tempted, as males are, to

Figure 1
Payoff curves associated with competition.

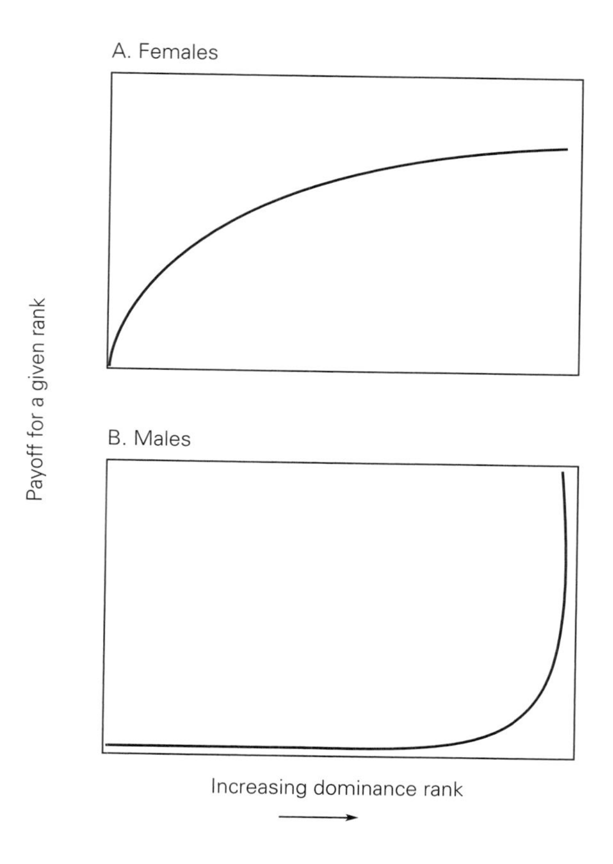

opportunistically change alliance partners to improve their competitive standing. Observations from both Gombe (Pusey 1980; Goodall 1986) and Mahale (Nishida 1989) indicate a consistent tendency for dominant, resident females to form alliances against subordinate immigrant females. Equally significant, during extensive observations of adult females, Nishida (1989) never saw a resident support an immigrant against another resident. We know from studies of macaques, in which high-ranking females consistently support one another against subordinates, that reliable alliances of this sort lead to extremely stable dominance relationships, because ambitious subordinates are unable to rise in rank by playing dominant animals off against one another (Chapais 1992). Similar support choices by chimpanzee females could explain why their dominance relationships are usually so stable (de Waal 1984).

In contrast, among male chimpanzees, all alliances are vulnerable to defection for strategic purposes (de Waal 1982; Nishida 1983; Goodall 1986). Such opportunism is expected when, by switching partners, a *prime male* has a greater chance of achieving or maintaining alpha rank; or when, by becoming a critical ally of a dominant younger male, a postprime male gains significant reproductive privileges. Moreover, strategic defection makes for extremely unstable hierarchies (de Waal 1984), as shown by the dramatic changes in rank relationships between female macaques that occur after the rare breakdowns in their reliable support patterns (Chapais 1992).

De Waal's analysis of differences between male and female chimpanzees focuses on two aspects of dominance and competition (de Waal 1982, 1984, 1993). The first aspect concerns the much greater importance that males, as compared with females, place on formal rank. This importance is reflected in the frequent, ritualized encounters that confirm an asymmetrical power relationship between males. Such acknowledgments of formal rank are associated with male strategies directed toward continual improvement of status (de Waal 1986, this volume). Our research does not speak to this first aspect; it remains unclear whether circumstances exist under which female chimpanzees show the concern for formal rank that males show.

The second aspect of differences between males and females concerns the degree of motivation to compete and the use of competitive strategies. (This aspect is distinct from formal rank because, although the outcomes of competitive encounters are frequently used by researchers to define dominance rank, differences in agonistic power are not necessarily accompanied by formal recognition of status by conspecifics.) De Waal highlights three sex differences in the competitive domain. First, in contrast to males, female chimpanzees lack a strong motivation to compete with other females, and dominance standing is not important to their reproductive success. De Waal attributes these differences to the tendency of wild females to disperse in the face of feeding competition (de Waal 1978, 1993). Second, females fail to show the elaborate, competitive strategies characteristic of males, including opportunistic formation of coalitions and "the set of dominance-related mechanisms required to maintain social bonds in the face of serious competition" (de Waal 1986). Third, female agonistic interventions function to protect the well-being of friends and relatives and to promote social stability (de Waal 1993), rather than to improve competitive ability, as among males. De Waal argues that all these differences reflect sex differences in "genetically implanted value systems," such that males value high rank and cohesive relations with other males, whereas females value supportive, stable social relationships (de Waal 1993).

We use our payoff model, combined with evidence of behavioral diversity among captive females and with recent field data, to suggest an alternative perspective on the contrasting behaviors of chimpanzee males and females. Although females in the wild sometimes benefit from foraging independently (Wrangham and Smuts 1980), this does not imply that females consistently avoid competitive interactions with other females or that females lack concern for dominance standing. Wrangham et al. (1992)

reported that when females in the wild have opportunities to interact, they show rates of dominance interactions comparable to those among males. Furthermore, the evidence reviewed suggests that successful competition against other females may be critical to establishing core areas and, therefore, to reproductive success. The second aspect of sex difference described by de Waal is inconsistent with evidence from the Detroit Zoo and from the wild, which indicates that, in situations characterized by potentially great payoffs for winning competitive encounters such as situations involving formation of new relationships with competitors, females adopt the values typically shown by males and compete intensively, sometimes employing the male-like strategies of opportunistic coalitions and frequent reconciliations with rivals. De Waal (1984) emphasized that his finding that females tend to support long-term familiar partners in agonistic disputes differs dramatically from patterns of agonistic support observed between males striving to improve their status. He interpreted this incongruity to mean that females were relatively uninterested in competitive goals. However, the fact that female patterns of support during conflict differ from those of males does not necessarily mean that females, are not competitive. We have suggested that consistent and reliable support between familiar females such as long-term resident females in the wild functions to maintain the high competitive ability of these resident females, thereby protecting the integrity of their core areas against potential incursions by recent immigrants. Thus, far from constituting evidence for the absence of female-female competition, reliable coalitions between familiar female chimpanzees can be viewed as manifestations of female competitive strategies designed to maintain access to important, contested resources. Our model suggests that natural selection has favored the ability of both sexes to modify their own competitive strategies depending on the nature of the payoff curves that characterize a given situation.

Although our model is consistent with the available data, it requires further evaluation. Additional observations of newly forming relationships and of introductions of strange females into established groups in captivity should prove useful in this regard. However, a deeper understanding of female-female social relationships depends primarily on further information on chimpanzees in the wild. Unfortunately, apart from the studies in Mahale, much of the available information about the social behavior of females in the wild has been gathered opportunistically, that is, when observers were focusing on some other topic.

We lack information of the most basic kind, including the frequency with which unprovisioned females encounter one another, when such encounters occur, and the frequency and nature of competitive interactions. We need to know about the core areas: whether and how these areas change over time, whether and how variation in location and quality of core areas reflects differences in dominance status and how that variation influences female reproductive success. We need to know whether female chimpanzees reap important benefits from competition other than benefits associated with establishment of core areas and whether familiar females sometimes compete with one another, as suggested by observations of occasional aggression between resident females at Gombe (Goodall 1986). As we learn more about female chimpanzees in the wild, increasing our awareness of their remarkable capacities for behavioral flexibility, we will likely discover that relationships and interactions between females are as diverse and as complex as those already documented between males.

Summary

Data on female-female relationships and interactions from captive chimpanzees at the Arnhem Zoo, involving stable groups of familiar females, were compared with data on female-female relationships and interactions from the Detroit Zoo during the period of group formation when relationships were unstable. Compared with females in the stable Arnhem Zoo colony, females in the newly formed Detroit Zoo group were more likely to compete in contexts in which status alone appeared to be the contested resource. Females at Detroit Zoo reconciled after fights much more often than did Arnhem Zoo females, suggesting a strong motivation to repair relationships threatened by status competition. In addition, patterns of female agonistic support at Detroit Zoo indicated the use of coalitions as tools for improving competitive ability. These findings indicate that when females are first forming relationships with one another, they use a variety of complex competitive strategies reminiscent of the strategies documented for status-striving males. These results, which challenge previous characterizations of females as inherently less competitive than males, can be interpreted in terms of the payoffs associated with resource competition in the wild in different contexts and in different stages of female life history.

Acknowledgments

We are grateful to the Detroit Zoo for permission to conduct research on their captive chimpanzees and to Susan McDonald and Karen Pazol for help in data collection. Warren Holmes, John Mitani, and Richard Wrangham provided helpful feedback on this study. We thank David Gubernick, W.C. McGrew, and Frans de Waal for valuable comments on the manuscript and John Mitani for help with figures. This research was supported by NSF Grant BNS-8857969 to B. Smuts and by grants to K. Baker from the L.S.B. Leakey Foundation, the Nixon-Griffis Fund for Zoological Research, Sigma Xi, and the Evolution and Human Behavior program, University of Michigan.

References

Adang, O. M. J., J. A. B. Wensing, and J.A.R.A.M. van Hooff. 1987. The Arnhem colony of chimpanzees *(Pan troglodytes):* Development and management techniques. *Intern. Zoo Yearbook* 26:236–48.

Altmann, J. 1974. Observational study of behavior: Sampling methods. *Behav.* 49:227–65.

Bakeman, R., and J. Gottman. 1986. *Observing Interaction: An Introduction to Sequential Analysis.* Cambridge, England: Cambridge Univ. Press.

Baker, K. C. 1992. Hierarchy formation among captive female chimpanzees. Ph.D. diss., University of Michigan.

Bygott, D. B. 1974. Agonistic behavior in wild chimpanzee males. Ph.D. diss., Cambridge University

———. 1979. Agonistic behavior, dominance, and social structure in wild chimpanzees of the Gombe National Park. In D.A. Hamburg and E.R. McCown, eds., *The Great Apes,* pp. 405–27. Menlo Park, Calif.: Benjamin/ Cummings.

Chapais, B. 1992. The role of alliances in social inheritance of rank among female primates. In A.H. Harcourt and F.B.M. de Waal, eds., *Coalitions and Alliances in Humans and other Animals,* pp. 29–59. Oxford, England: Oxford Univ. Press.

Goodall, J. 1968. The behaviour of free-living chimpanzees in the Gombe Stream Reserve. *Anim. Behav. Monogr.* 1:165–311.

———. 1986. *The Chimpanzees of Gombe: Patterns of Behavior.* Cambridge, Mass.: Belknap Press.

Hasegawa, T., and M. Hiraiwa-Hasegawa. 1990. Sperm competition and mating behavior. In T. Nishida, ed., *The Chimpanzees of the Mahale Mountain: Sexual and Life History Strategies,* pp. 115–32. Tokyo: Univ. of Tokyo Press.

van Hooff, J.A.R.A.M. 1974. A structural analysis of the social behavior of a semi-captive group of chimpanzees. In M. von Cranach and I. Vine, eds., *Social Communication and Movement,* pp. 75-162. London: Academic Press.

McDonald, S. M., K. Pazol, K. C. Baker, and B. B. Smuts. In Prep. The integration of two infant chimpanzees *(Pan troglodytes)* into a newly established social group at Detroit Zoo.

McDonald Black, S. M. Heterosexual bonding and partner preference among captive chimpanzee. Ph.D. diss. University of Michigan.

Nishida, T. 1979. The social structure of chimpanzees of the Mahale Mountains. In D.A. Hamburg and E.R. McCown, eds., *The Great Apes,* pp. 73-121. Menlo Park, Calif.: Benjamin/ Cummings.

———. 1983 . Alpha status and agonistic alliance in wild chimpanzees *(Pan troglodytes schweinfurthii). Primates* 24:318–36.

———. 1989 . Social conflicts between resident and immigrant females. In P.G. Heltne and L.A. Marquardt, eds., *Understanding Chimpanzees,* pp. 68–89. Cambridge, Mass.: Harvard Univ. Press.

Nishida, T., and M. Hiraiwa-Hasegawa. 1987. Chimpanzees and bonobos: Cooperative relationships among males. In B.B. Smuts, D.L. Cheney, R.M. Seyfarth, R.W. Wrangham, and T.T. Struhsaker, eds., *Primate Societies,* pp. 165–77. Chicago: Univ. of Chicago Press.

Nishida, T., H. Takesaki, and Y. Takahata, 1990. Demography and reproductive profiles. In T. Nishida, ed., *The Chimpanzees of the Mahale Mountains: Sexual and Life History Strategies,* pp. 63–97. Tokyo: Univ. of Tokyo Press.

Noë, R., F.B.M. de Waal, and J.A.R.A.M. van Hooff. 1980. Types of dominance in a chimpanzee colony. *Folia Primatol.* 34:90–110.

Pusey, A. E. 1978. The physical and social development of wild adolescent chimpanzees *(Pan troglodytes schweinfurthii).* Ph.D. diss., Stanford University

———. 1980. Inbreeding avoidance in chimpanzees. *Anim. Behav.* 28:543–52.

Smuts, B.B. 1985. *Sex and Friendship in Baboons.* New York: Aldine Press.

Strum, S.C. 1982. Agonistic dominance in male baboons: An alternate view. *Am. J. Primatol.* 3:175–202.

de Waal, F.B.M. 1978. Exploitative and familiarity-dependent support strategies in a colony of semi-free-living chimpanzees. *Behav.* 66:268–312.

———. 1982. *Chimpanzee Politics: Power and Sex among Apes.* New York: Harper and Row.

———. 1984. Sex differences in the formation of coalitions among chimpanzees. *Ethol. and Sociobiol.* 5:239–55.

———. 1989. *Peacemaking among Primates.* Cambridge, Mass.: Harvard Univ. Press.

———. 1993. Sex differences in chimpanzee (and human) behavior: A matter of social values? In M. Hechter, L. Cooper, and L. Nadel, eds., *Towards a Scientific Understanding of Values.* Stanford, Calif.: Stanford Univ. Press.

de Waal, F.B.M., and A. van Roosmalen. 1979. Reconciliation and consolation among chimpanzees. *Behav. Ecol. and Sociobiol.* 5:55–66.

de Waal, F.B.M., and J. Hoekstra. 1980. Contexts and predictability of aggression in chimpanzees. *Anim. Behav.* 28:929–37.

de Waal, F.B.M., and D. Yoshihara. 1983. Reconciliation and redirected affection in rhesus monkeys. *Behav.* 85:224–41.

Wrangham, R.W., A.P. Clark, and G. Isabirye-Basuta. 1992. Female social relationships and social organization of Kibale Forest chimpanzees. In T. Nishida, W.C. McGrew, P. Marler, M. Pickford, and F.B.M. de Waal, eds., *Topics in Primatology.* Vol. 1, *Human Origins,* pp. 81–98. Tokyo: Univ. of Tokyo Press.

Wrangham, R.W., and B.B. Smuts. 1980. Sex differences in the behavioural ecology of chimpanzees in the Gombe National Park, Tanzania. *J. Reprod. Fert. (Suppl.)* 29:13–31.

Yerkes, R.M. 1943. *Chimpanzees: A Laboratory Colony.* New Haven, Conn.: Yale Univ. Press.

Chimpanzee's Adaptive Potential

A Comparison of Social Life under Captive and Wild Conditions

Frans B.M. de Waal

Enlightened Captive Environments

The chimpanzee is one of the very few nonhuman primates whose social behavior is better known in the natural habitat than in captivity. This situation is all the more surprising for a species that has been studied extensively in the home and laboratory with respect to symbolic communication, learning abilities, cognition, and tool use and that is maintained in considerable numbers by zoos and research institutions. The reason is that, until recently, captive chimpanzees were not housed in environments appropriate for investigations of social behavior.

Since the beginning of this century, zoos have displayed baboons and macaques in natural-sized groups on the ubiquitous monkey rocks, while housing their great apes singly or in very small groups. Moreover, as if the apes' resemblance to humans did not speak for itself, they were often featured in human clothing in so-called tea parties. When, in the 1970s, the social housing conditions of apes finally began to catch up with those of monkeys, chimpanzees were not among the first beneficiaries. Orangutans and gorillas were considered more impressive and attractive as well as less problematic in terms of escape tendency and proneness to aggression. What zoos did not seem to realize was that chimpanzees are more interesting to watch; the level and variety of social activity in a chimpanzee colony is considerably greater than in a similarly sized group of orangutans or gorillas. However, zoo managers seemed intolerant of the noisy quarrels typical of chimpanzees.

Presently, naturalistic enclosures for chimpanzees are on the increase in Western zoos (Coe 1992; Gold 1992), and a few biomedical institutions (such as the Yerkes Regional Primate Research Center in Atlanta, Georgia, and the Anderson Cancer Center Science Park in Bastrop, Texas) now have outdoor compounds for groups of up to 25 chimpanzees. As a result, prospects for research on the social behavior of chimpanzees in captivity are improving. To illustrate the benefits of such research, I will discuss studies at the Arnhem Zoo in the Netherlands and at the Yerkes Regional Primate Research Center. This work does more than complement field research; it has produced solid quantitative information on phenomena such as reconciliation behavior, political strategy, and social reciprocity that were either previously unknown in the field or had been described in qualitative terms only.

This does not necessarily reflect a discrepancy between chimpanzee behavior in captivity and in the field; the differences may be caused in at least two ways.

First, in captivity, continuous visibility of the subjects and human control over conditions make it possible to apply more comprehensive and often more detailed data collection techniques and, thus, to penetrate more deeply into certain aspects of chimpanzee social life than feasible in the field. The investigator at the zoo, for instance, can schedule a focal observation of a particular individual for the next day at a precise time, whereas the field observer is lucky to encounter every individual on a regular basis, and usually cannot predict when this will happen Riss and Busse describe a notable exception (1977). Also, complex social dynamics, such as a power struggle between adult males, can be followed closely for many consecutive days in captivity, whereas field-workers generally need to reconstruct such processes from observed bits and pieces. Hence, some of the differences in research results between captivity and the field are simply a product of observation conditions. These differences apply particularly to research on chimpanzees, who are partly arboreal and hard to follow in the forest, whereas some of the best-studied monkeys, such as baboons, are largely terrestrial.

Second, chimpanzee social life is no doubt modified by captivity. Even the most enlightened zoo conditions, such as those in the one-hectare island of the Arnhem Zoo (van Hooff 1973a: see figure 1), create behavioral constraints that the species does not encounter in nature. Here, again, chimpanzees are probably more affected than the best-studied monkey species, because in nature monkeys live in permanent groups whereas chimpanzees are characterized by temporary fission-fusion societies. Wild chimpanzees move through the forest alone or in small parties of a few individuals at a time. The composition of a party changes constantly; all associations, except the one between mother and dependent offspring, are of a temporary character.

Figure 1
The chimpanzee island at Burgers Dierenpark, Arnhem, the Netherlands, opened in 1971 and inspired similar moated exhibits elsewhere in the world. The Arnhem exhibit consists of an island of nearly one hectare, on which approximately 25 chimpanzees live seven months a year, and two large indoor halls for the five-month winter period. The building includes 10 night cages to which the chimpanzees return every evening. Nieuwenhuijsen and de Waal (1982) compared periods with the colony remained in a single winter hall with periods on the island. (Drawing by B. Willems.)

Initially, this flexibility made investigators wonder whether chimpanzee social groups have any stable membership at all. After years of documenting the composition of chimpanzee parties in the Mahale Mountains, Nishida (1968, 1979) was the first to crack the puzzle. He reported that chimpanzees form large *unit-groups,* also known as communities. The members of a single community mix freely in ever-changing parties, but members of different communities never gather. Goodall et al. (1979) added territoriality to this picture; not only do communities not mix, males of different communities may engage in lethal fights. Territoriality is, of course, another important aspect of chimpanzee social organization that is affected by captivity. At the zoo, there are rarely enemy groups and, if there are, it is impossible to do physical battle with them.

Inasmuch as enclosure forces chimpanzees into a mode of life that differs substantially from that in the wild, comparisons between captive and wild chimpanzees may yield greater differences than such comparisons for species that form cohesive groups. Whether a group of rhesus monkeys lives in a cage or around an Indian temple, its members face essentially the same internal problems of maintaining affiliative ties, supporting kin, confirming the hierarchy, and so on (though external problems, such as predation and migration, are of course quite different). An inspection of data on rhesus in a variety of environments (de Waal 1989a), from captive to free-ranging, reveals the same social organization and almost identical aggression rates, thus contradicting popular notions about how population density affects behavior. Chimpanzees, in contrast, face a different set of social problems in the wild and in captivity; only in captivity are they continually together, having lost the social fluidity and choice of companionship typical of their wild counterparts.

In sum, the student of chimpanzee behavior in captive settings enjoys a great advantage over the field-worker in terms of ease of data collection. But this student should also realize that limited space probably affects the social organization of this species more than that of other primates.

Adaptive Potentials

From an evolutionary perspective, the most relevant environment in which to study animals is the natural habitat, as this is the environment to which a species has been adapted through natural selection. Not that there is such a thing as *the* natural habitat for chimpanzees: environments in which these apes survive range from relatively open, wooded savannas to dense rain forests. Neither can current natural habitats be regarded as original or pristine, because most known chimpanzee populations have been affected by human activity, such as hunting, logging, or provisioning of artificial foods. Yet field-workers generally agree that the existing habitats are close enough to the hypothetical original to allow an understanding of the chimpanzee's evolutionary adaptations.

Power (1991) is the only author to question the efforts thus far to study chimpanzee adaptation. On the basis of her reading of the literature, Power has argued that the food provisioning by field-workers at some of the best-known field sites has turned the chimpanzees more violent and less egalitarian and, thus, has changed the tone of relationships both within and between communities. Power's analysis—a serious reexamination of available facts blended with nostalgia for the 1960s image of apes as noble savages—raises questions that will no doubt be settled in the near future by further research on unprovisioned wild chimpanzees. In the meantime, readers are referred to Asquith (1989) for a balanced review of the pros and cons of provisioning in the field.

In contrast to the study of evolutionary adaptation, the study of adaptive potential need not be limited to the natural habitat if *adaptive potential* is defined as the entire range of conditions to which a species can adjust without compromising its health, biological functions (such as reproduction), or major parts of its natural behavioral repertoire (such as species-typical communication). This range of conditions usually extends well beyond the various habitats occupied in nature. The experience of the last two decades demonstrates that chimpanzees are very well capable of living healthy lives, both physically and socially, in large enclosures even if this means forfeiting fission-fusion opportunities. We may conclude, therefore, that the chimpanzee's adaptive potential encompasses life in permanent associations.

To say that these chimpanzees do not behave naturally misses the point; following the same argument, very few humans behave naturally. People in modern societies and chimpanzees in zoos successfully adjust to novel circumstances due to their psychological and social plasticity, which in and of itself is a natural characteristic. Given the wholly artificial world in which many of us survive today, we may be pushing our adaptive potentials to the limit; for this reason, the potentials of our closest relatives deserve close attention. The chimpanzee's ability to live in stable groups is a case in point. Without a similar ability, it could be argued, our ancestors could never have made the step toward life in permanent settlements. As noted by Masters (1984:209), the study of captive chimpanzee colonies may "provide particularly valuable evidence of what happens to hominoids when they cannot easily leave their home community and, thus, like humans after the invention of agriculture, tend to substitute shifting social alignments for the group fission characteristic of hunter-gatherers and wild chimps."

A good indicator of the chimpanzee's capacity for adjustment to new conditions is the response to a sharp reduction in space, as investigated by comparing the Arnhem colony's behavior in a winter enclosure 20 times smaller than the colony's summer island (Nieuwenhuijsen and de Waal 1982). Novelty effects of the environment could be excluded as the apes were extensively familiar with both enclosures and were accustomed to being observed for many months. Under the crowded winter conditions, relaxing affiliative behaviors such as grooming increased the most (see table 1). Submissive greeting or pant-grunting is a vocalization commonly shown by subordinates when they encounter dominants (Bygott 1979; Noë et al. 1980). The observed increase in these behaviors may represent the apes'

Table 1
Effect of a 95% area reduction on the chimpanzee colony of the Arnhem Zoo; relative levels of behavior between outdoor (summer) and indoor (winter) periods.

Measure	Outdoor	Indoor
Aggression	1	1.7
Proportion of severe aggression	1	1.1
Submissive "greeting"	1	2.4
Social grooming	1	2.0
Social play	1.4	1

Notes: The relative rates of behavior seen in the table arise from weighted individual rates of behavior over two summer and three winter periods of observation. (Nieuwenhuijsen and de Waal 1982.)

attempt to reduce social tensions caused by crowding; both grooming and greeting appear to serve as appeasement. When indoors, the chimpanzees' aggressive behavior also increased in frequency, but the intensity of aggression did not change. The only behavior that dropped in frequency was social play.

The observed 1.7-fold increase in the rate of aggression is minimal compared to the results of crowding experiments on monkeys involving considerable smaller space reductions (Southwick 1967; Elton and Anderson 1977). To explain the response pattern of the Arnhem chimpanzees, I have proposed a *coping model.* According to this model, spatial crowding results in an increased risk of aggression, to which the apes react with calming gestures that serve to reduce this risk. In other words, a negative feedback system is in place that maintains peaceful coexistence (de Waal 1989a). For biologists, used to thinking in terms of homeostasis, this may seem an obvious proposition. Yet mechanisms of social adjustment are rarely discussed in the literature on crowding—including the literature on humans—which is still dominated by early reports of social pathologies in rodents due to high population densities (Calhoun 1962). Rodents, however, more than likely lack the powerful mechanisms of tension regulation and conflict resolution found in primates (de Waal 1989b).

Unchanged Male Bonds

Social organization is often summarized in terms of three relationship classes among adults: male-male, female-female, and male-female. This framework may prove useful to analyze the effect of the environment on the behavior of chimpanzees, as the two sexes appear to respond quite differently. For example, we don't know with certainty whether captive and wild males differ significantly in the way they deal with one another, except for the already mentioned absence of enemy males in captive settings. Female chimpanzees, on the other hand, seem an almost different species in captivity compared to what we know about them living in the wild in Gombe National Park and Mahale Mountains National Park in Tanzania.

Let us start with the males. Adult male chimpanzees form a distinct layer of society in Gombe and Mahale. They dominate all females, often travel in all-male parties, and compete fiercely over rank and sexual partners. Yet they also groom one another, hunt together, and raid neighboring territories (Goodall 1986; Nishida 1990). Because of this evolutionary heritage of bonding, association, and interdependency, the social problems posed by captivity may constitute less of a challenge to adult male chimpanzees than to other age and sex classes.

The Arnhem males often sat together under the highest oak tree in the middle of the island. On many days, toward the end of the afternoon, they would make one last round along the moat that surrounds the entire island, walking together in a manner somewhat reminiscent of the border patrols of wild chimpanzees (Goodall 1986). They showed a high degree of association and grooming despite a rate of male-male aggression that was 20 times higher than that among females (see figure 2). This aggression rate reflected the males' continual jockeying for dominance, which at times became so intense that they suffered serious, once even fatal, injuries (de Waal 1982, 1986a).

A combination of competition and bonding is less paradoxical than it may seem; male chimpanzees, captive or wild, possess effective mechanisms to keep social tensions in check. One of these is a *formalized hierarchy*, that is, a hierarchy communicated through status rituals involving bowing (pressing the whole body against the ground, often with repeated up-and-down movements) and pant-grunting by the subordinate (de Waal 1986b, 1989c: see figure 3). These rituals allow nonhostile confirmation of who is subordinate and who is dominant. Another tension-reduction mechanism is *reconciliation* defined as friendly reunions—typically a kiss and embrace—between former adversaries (de Waal and van Roosmalen 1979; de Waal 1989b: see figure 4). These interactions prevent permanent harm to relationships disturbed by aggression.

The probability of damaging fights between males increased by a factor of almost five in the absence of a formalized dominance relationship (during periods in which males failed to engage in status rituals). Re-establishment of formal dominance resulted in a sharp increase in grooming and reconciliation (de Waal 1986b). Thus, the male hierarchy appears to regulate conflict in such a manner that bonds are preserved, the net effect being that the hierarchy *unifies* the competitors. Nishida (1979:93) has expressed a similar view: "It is likely that complex sequences of threat-submission-reassurance may strengthen the male bond among chimpanzees."

Chimpanzee males may have evolved these unifying mechanisms in response to intercommunity aggression (Wrangham 1986). Male chimpanzees of a particular community operate with great solidarity against external enemies, yet within their own camp are divided into ever-changing partnerships (Riss and Goodall 1977; de Waal 1982, 1984; Goodall 1986; Nishida 1983). The flexible, or opportunistic, nature of intracommunity coalitions among males adds important reasons for effective conflict management. Coalition partners need to 'agree' on how to divide the payoffs of their cooperation, and even arch rivals cannot afford to hold grudges, as they may need one another against a third party in the future. In the meantime, all males in the community need to preserve a united front, or macrocoalition, against the males of neighboring territories. This need poses a special challenge to conflict-management skills, a challenge also known in human societies facing external threats (Boehm 1992).

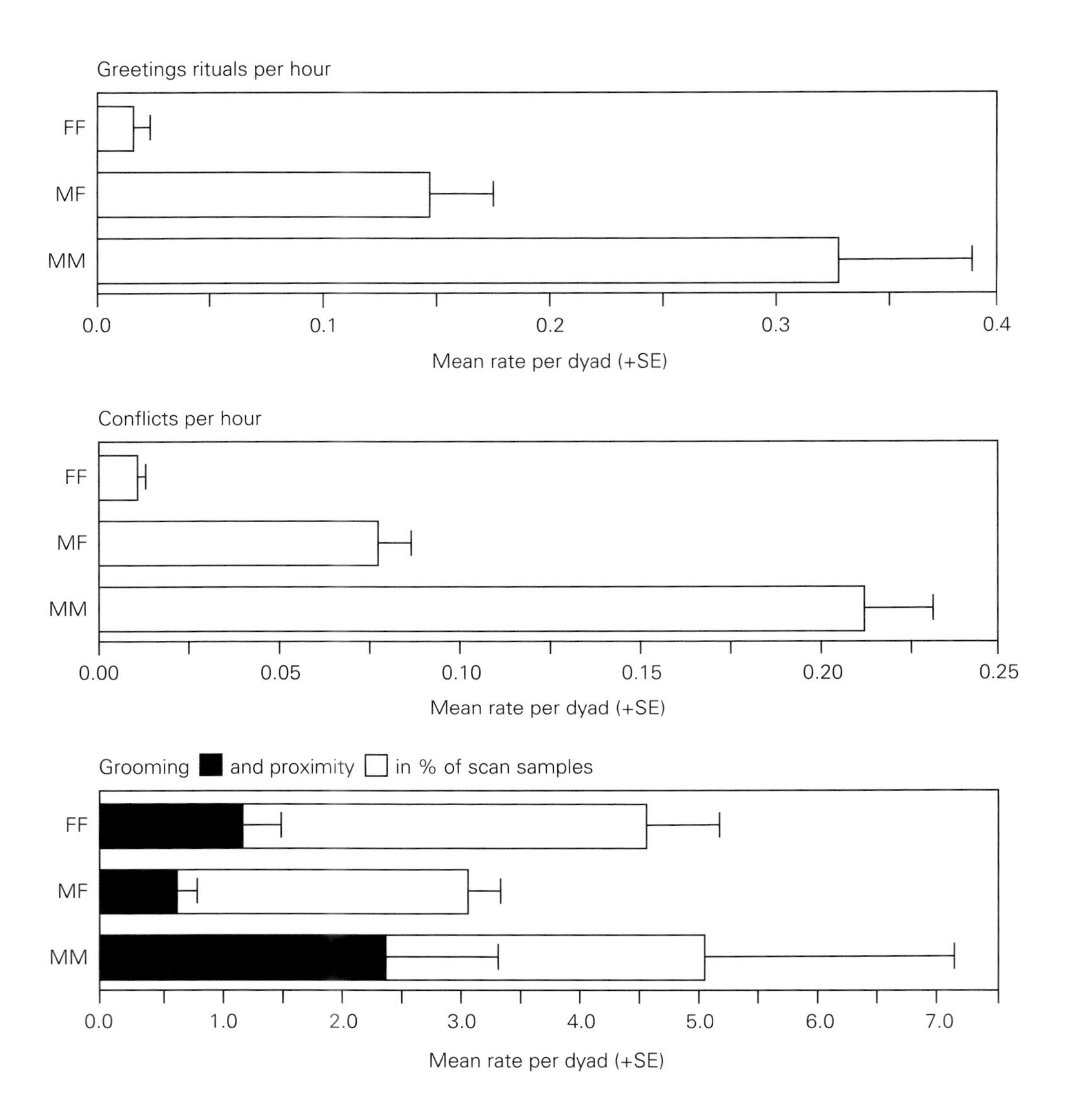

Figure 2

Relationships between adults in the Arnhem colony over five summer periods, from 1976 through 1980. Data were collected by pairs of simultaneous observers during a total of 334 hours, including 1,868 five-minute scan samples. SE represents standard error. The graphs provide the mean rate per dyad (+SE), for 36 female-female dyads (FF), 36 male-female dyads (MF), and 6 male-male dyads (MM). Submissive greeting rituals (i.e., pant-grunting) and agonistic conflicts are expressed in terms of frequency per hour. Proximity (i.e., sitting within arm's reach of another individual) has been divided into grooming and other proximity, both expressed as percentages of scan samples. (See de Waal 1986b for statistical details).

Figure 3
Status differences between male chimpanzees are communicated in ritual encounters. A dominant male (left) makes himself look as impressive as possible, walking bipedally and raising his hair, whereas the subordinate makes himself smaller and utters a series of pant-grunts. In reality, the two males are approximately the same size. (Photo by the author de Waal 1989c).

Figure 4
Reconcilation 10 minutes after a protracted, noisy conflict between two adult male chimpanzees. The challenged male (left) fled into the tree. His adversary now approaches stretching out a hand. Just after this photograph was taken, the two males kissed and embraced in the tree, then climbed down together to groom each other on the ground. (Photo by the author de Waal 1982).

The Flexible Female Domain

Because female chimpanzees live rather dispersed and solitary lives at the Gombe and Mahale field sites, we can expect females to be more affected than males by the forced proximity of captivity. Table 2 compares data on male and female relationships in the Arnhem colony with similar data from Gombe and the Mahale Mountains. Sex-related bonding patterns

Table 2
Social relationships of adult male and female chimpanzees in the Arnhem Zoo colony, the Netherlands, compared with wild populations in Gombe National Park and Mahale Mountains National Park in Tanzania (de Waal 1989c).

	Arnhem Zoo	Both	Natural Habitat
Hierarchy		Dominance is highly formalized among adult males, but infrequently expressed among females (Bygott 1979; Goodall 1986; Nishida 1989; de Waal 1986b).	
Coalitions	Females unite to defend against aggression and to influence status struggles among males (de Waal 1982, 1984).	Males follow flexible, opportunistic coalition strategies aimed at high status (Riss and Goodall 1977; de Waal 1982, 1984; Nishida 1983).	Female coalitions are virtually unknown.
Reconciliations		Male-male aggression is more often followed by reassurance behavior than is female-female aggression (de Waal 1986b; Goodall 1986).	
Social Bonding	Female bonding is as strong as male bonding; less association between the sexes (de Waal 1986b).	Strong association between males.	Females lead largely solitary lives with their dependent offspring (Nishida 1979; Goodall 1986).

similar to patterns observed at these sites have been reported for unprovisioned chimpanzees at the Kibale Forest in Uganda (Wrangham et al. 1992). As can be seen, the major difference between captivity and the wild occurs in female-female relationships: the Arnhem females clearly bond, as measured by the frequency of association and grooming (see figure 2), and they combine forces in aggressive coalitions to such an extent that they form a political block to be reckoned with by even the most dominant males (de Waal 1982, 1984). For example, male dominance struggles may be decided by female support for one male or the other. Another difference is that plant-food sharing between unrelated adult females—extremely rare in Gombe (McGrew 1975)—was regularly observed in the Arnhem colony and found to be reciprocally distributed in the colony of the Yerkes Field Station (de Waal 1989d).

Bonding with same-sexed partners, combined with the continuous presence of these partners within hearing distance, allows captive females to effectively curb male power by recruiting support and mounting massive defensive coalitions against attacks. Consequently, the single most conspicuous effect of captivity on chimpanzee social organization seems to be the increased sociability and social influence of females.

While relations among females in the Arnhem colony are not free of tensions, competition and aggression never reach the level typical of relations among males. Reconciliations are significantly less common among females than among males, and the female hierarchy is inconsistent and vague compared to the male hierarchy (see table 2). Marked asymmetries in the outcome of aggressive encounters do occur among females; what is lacking is conflict resolution through the ritualization of dominance relationships. In other words, female chimpanzees may have as agonistic hierarchy, but they do not have a formalized one. After six years of study of the Arnhem colony—with an estimated 6,000 hours of observation—there remained a dozen female-female pairs in which status rituals had *never* been observed.

Under certain circumstances, however, females may show more status competition, stronger ambitions, and a more clear-cut hierarchy. Table 2 characterizes the Arnhem colony when it was already well established, with virtually all females having one or several offspring. In its early history, however, the situation may have been more like that in the newly formed Detroit Zoo colony, with females engaging in male-like opportunistic strategies (Baker and Smuts this volume). Initially, it was the females who dominated the Arnhem colony, led by an older matriarch named Mama. Unfortunately, we have almost no systematic data on Mama's reign. The one thing we know for certain is that she did not give up her position willingly when adult males were introduced. After the males had taken over and Mama had given birth to the first offspring that she did not reject, her behavior changed dramatically. She still held an extremely influential position as leader of female alliances and mediator in disputes, but she ceased to actively engage others in fights of intimidation or for status improvement. In short, she lost the dominance orientation known from earlier years (de Waal 1982).

As for the absence of strong affiliations among female chimpanzees in the wild, female bonding may not have evolved because a solitary lifestyle suits foraging needs of females in the wild and, females may benefit little from each other's presence (Wrangham 1986). However, this argument fails to account for potential for female bonding so clearly in evidence when competition is reduced, as in the captive environment with its relative food abundance. As with most potentials realized in captivity, it is interesting and useful to look for possible parallels in the wild. For example, chimpanzees may have lived in permanent groups at one time or another in their evolutionary past, in the same way that it appears to be the current condition of small population trapped by agricultural encroachment in a forest of approximately six square kilometers on top of a mountain in

Figure 5
In its early years, the Arnhem colony was dominated by a female, Mama, who performed charging displays as spectacular as those of any male. Dethroned by newly introduced males, Mama reconciled herself with her new position only after Moniek was born. The photograph shows Moniek in the protective arms of the old matriarch. (Photo by the author de Waal 1982).

Bossou, Guinea. Although this population has 600 times more space available than the Arnhem colony, the two show striking similarities in social organization. Almost 50% of the parties encountered in the forest are made up of three quarters of the community members (Sugiyama 1984), and female-female grooming occurs at relatively high rates (Sugiyama 1988; Muroyama and Sugiyama this volume). Thus, the Bossou population shows some of the cohesiveness and close female ties typical of captive groups.

Similarly, in Taï Forest, Ivory Coast, female chimpanzees are more sociable than at other sites: they associate more, develop special friendships, share food, and support one another in coalitions. Boesch (1991, this volume) attributes the different grouping pattern of Taï chimpanzees to increased alertness for and cooperative defense against leopards. These observations suggest that the potential for female bonding observed under captive conditions may have survival value under particular conditions in nature. How common or rare these conditions have been in the chimpanzee's evolutionary history is unknown.

When female relations are close, females tend to support one another against males: not only do the intrasexual relations differ, but also male dominance is curbed or mitigated. In captivity, females gain an important voice in social matters, and males tend to rely on females for certain favors. Thus, the Arnhem males appeal to females for support during dominance struggles, rarely attack or intimidate females in the sexual context, and regularly relinquish food and desirable objects to the most important females, possibly in order to build up good relationships with them (Noë et al. 1980). Although the Arnhem males clearly formally dominate females, the dynamics of intersexual relationships appear different, more egalitarian, than in the wild.

In short, the greatest similarity between chimpanzees in the wild and in captivity can be found in male-male relations, and the greatest difference in female-female relations. Male-female relations may change secondarily in captivity as a result of the closer female-female relations that affect the power balance between the sexes, particularly since females tend to outnumber males in captive colonies.

Intergroup Diversity

The obvious similarities in the behavioral repertoires of captive and wild chimpanzees notwithstanding (compare the ethograms of van Lawick-Goodall 1968, and van Hooff 1973b), some behavior patterns have been observed under one condition but not the other. According to Kummer and Goodall (1985), a primate's behavioral repertoire may actually grow under captive conditions: the continuous proximity of conspecifics and the time freed up by the virtual elimination of foraging promote innovation and sophistication in the social sphere. Kummer and Goodall (1985) illustrated this point with examples of elaborate courtship displays, refined deception techniques, and mediated reconciliations in captive chimpanzees. Although my research on captive chimpanzees is partly responsible for the above examples, I am not convinced that chimpanzees in a large enclosure necessarily show increased social sophistication. The situation in a fission-fusion society being inherently unpredictable, wild chimpanzees face social problems of great complexity (such as one day A encounters B alone, the next day A encounters B in the presence of his closest ally, C) and their methods of solution may yet await discovery.

In addition to sharing certain common behaviors, all captive colonies have their distinctive characteristics and traditions in the same way that wild groups have distinctive characteristics and traditions—exhibiting intergroup differences in communication, choice of foods, and tool use (McGrew this volume; Huffman this volume; Mitani this volume). Often, these distinctive patterns relate to particular features of the environment. Thus, the Arnhem chimpanzees developed a rubber tire culture, carrying rubber tires filled with water around to drink from them in a favorite spot, to sit or sleep in them, to roll them in play, or to pile them up to reach elevated goals (see also McGrew 1992). The group at the Yerkes Field Station has integrated the substrate of its outdoor compound—pine-bark chips and dirt—into its daily communication. For example, it is common in this colony, and quite accepted, for a chimpanzee to attract the attention of another by throwing a handful of chips or sand (Tomasello this volume).

One of the best-documented traditions in captivity is the rhythmic hand clapping and foot clapping that occur during grooming in the bonobo colony of the San Diego Zoo. In 1984, 98.5% of the 673 observed instances of this behavior involved seven of the ten bonobos. Because the three nonperformers were the only individuals that had not been bottle-reared, I have speculated that this peculiar behavior might have originated at the zoo's nursery (de Waal 1989e). However, bonobo-to-bonobo transmission of the pattern must have taken place since this time: the same communication gestures have been adopted by several mother-reared individuals introduced as juveniles into the San Diego colony (Amy Parish pers. com.). Hand and foot clapping have never been reported for other bonobos, captive or wild, and the gestures appear to serve the same communicative function during grooming in the San Diego bonobos as the typical tooth clacking, sputtering, and lip smacking of grooming chimpanzees.

Research on captive chimpanzees may turn out to be crucial for our understanding of the origin of intergroup diversity, since the transmission of novel behavior in captivity can be followed in detail and can be manipulated by introducing novel situations or by moving individuals from one group to another. Sometimes we may be able to witness the gradual establishment of a new tradition. For example, in the chimpanzee colony at the Yerkes Field Station, we may have the beginning of a behavior pattern first described for wild chimpanzees by McGrew and Tutin (1978). A young adult female named Georgia occasionally initiates handclasp grooming with other adults (see figure 6). Over the last two years, this behavior has been observed only a dozen times, never lasting longer than a minute, but Georgia has shown it with a variety of adult partners, and it does require their cooperation. Although not yet a clear tradition, it will be of interest to chart further occurrences.

Figure 6
Two adult females, Georgia (right) and Peony, at the Field Station of the Yerkes Primate Center show handclasp grooming. (Photo by the author).

Summary

Captive environments challenge chimpanzees with new social problems, the most important one being life in permanent groups for a species that, in nature, forms fission-fusion societies. Chimpanzee group life is highly flexible, however, in that the entire social system may adjust to the environment. Research on chimpanzees in naturalistic captive environments provides information on the adaptive potential of the species, that is, the range of circumstances under which species-typical behavior, physical and psychological health, and biological functions can be maintained.

Of the various aspects of chimpanzee social organization, female-female relations are probably the most affected by captivity. The sociability of female chimpanzees increases dramatically; their tendency to bond and defend one another against male aggression makes for a powerful female alliance in captive colonies quite different from female dispersal and apparent lack of political influence in the best-known wild communities. Intergroup differences in social habits have been observed in captivity, and their study is all the more attractive because of opportunities to manipulate the transmission process.

Acknowledgments

I am grateful to the students who have contributed to the data presented in this paper and to Bill McGrew and Barbara Smuts for comments. Writing and part of the research were supported by a National Institutes of Health grant to the Yerkes Regional Primate Research Center (RR00165) and by a grant from the National Institutes of Mental Health (R03 MH49475).

References

Asquith, P. 1989. Provisioning and the study of free-ranging primates: History, effects, and prospects. *Yearb. Phys. Anthro.* 32:129–58.

Boehm, C. 1992. Segmentary 'warfare' and the management of conflict: Comparison of East African chimpanzees and patrilineal-patrilocal humans. In A.H. Harcourt and F.B.M. de Waal, eds., *Coalitions and Alliances in Humans and Other Animals,* pp. 137–74. Oxford, England: Oxford Univ. Press.

Boesch, C. 1991. The effects of leopard predation on grouping patterns in forest chimpanzees. *Behav.* 117:220–42.

Bygott, J. D. 1979. Agonistic behavior, dominance, and social structure in wild chimpanzees of the Gombe National Park. In D.A. Hamburg and E.R. McCown, eds., *The Great Apes,* pp. 405–28. Menlo Park, Calif.: Benjamin/ Cummings.

Calhoun, J.B. 1962. Population density and social pathology. *Sci. Am.* 206:139–48.

Coe, J.C. 1992. Advances in facility design for great apes in zoological gardens. In J. Erwin, ed., *Chimpanzee Conservation and Public Health: Environments for the Future,* pp. 95–102. Rockville, Md: Diagnon.

Elton, R.H., and B.V. Anderson. 1977. The social behavior of a group of baboons *(Papio anubis)* under artificial crowding. *Primates* 18:225–34.

Gold, K.C. 1992. Chimpanzee exhibits in zoological parks. In J. Erwin, ed., *Chimpanzee Conservation and Public Health: Environments for the Future,* pp. 103–11. Rockville, Md.: Diagnon.

Goodall, J., and van Lawick. 1968. The behaviour of free-living chimpanzees in the Gombe Stream Reserve. *Anim. Behav. Monogr.* 1:161–311.

Goodall, J. 1986. *The Chimpanzees of Gombe: Patterns of Behavior.* Cambridge, Mass.: Belknap Press.

Goodall, J., A. Bandora, E. Bergman, C. Busse, H. Matama, E. Mpongo, A. Pierce, and D. Riss. 1979. Intercommunity interactions in the chimpanzee population of the Gombe National Park. In D.A. Hamburg and E.R. McCown, eds., *The Great Apes,* pp. 13–53. Menlo Park, Calif.: Benjamin/ Cummings.

van Hooff, J.A.R.A.M. 1973a. The Arnhem Zoo chimpanzee consortium: An attempt to create an ecologically and socially acceptable habitat. *Intern. Zoo Yearbook* 13:195–205.

———. 1973b. A structural analysis of the social behaviour of a semi-captive group of chimpanzees. In M. von Cranach and I. Vine, eds., *Expressive Movement and Non-verbal Communication,* pp. 75–162. London: Academic Press.

Kummer, H., and J. Goodall. 1985. Conditions of innovative behaviour in primates. *Phil. Trans. R. Soc. London B* 308:203–14.

Masters, R. 1984. Review of "Chimpanzee Politics." *Politics and the Life Sci.* 2:208–9.

McGrew, W.C. 1975. Patterns of plant food sharing in wild chimpanzees. In *Contemporary Primatology,* 304–9. 5th Cong. Intern. Primatol. Soc., Nagoya 1974. Basel: Karger.

———. 1992. *Chimpanzee Material Culture: Implications for Human Evolution.* Cambridge, Mass.: Cambridge Univ. Press.

McGrew, W.C., and C.E.G. Tutin. 1978. Evidence for a social custom in wild chimpanzees? *Man* 13:234–51.

Nieuwenhuijsen, K., and F.B.M. de Waal. 1982. Effects of spatial crowding on social behavior in a chimpanzee colony. *Zoo Biol.* 1:5–28.

Nishida, T. 1968. The social group of wild chimpanzees in the Mahale Mountains. *Primates* 9:167–224.

———. 1979. The social structure of chimpanzees in the Mahale Mountains. In D.A. Hamburg and E.R. McCown, eds., *The Great Apes,* pp. 73–121. Menlo Park, Calif.: Benjamin/Cummings.

———. 1983. Alpha status and agonistic alliance in wild chimpanzees. *Primates* 24:318–36.

———. 1989. Social interactions between resident and immigrant female chimpanzees. In P. Heltne and L. Marquardt, eds., *Understanding Chimpanzees,* pp. 68–89. Cambridge, Mass.: Harvard Univ. Press.

———. 1990. *The Chimpanzees of the Mahale Mountains: Sexual and Life History Strategies.* Tokyo: Univ. of Tokyo Press.

Noë, R., F.B.M. de Waal, and J.A.R.A.M. van Hooff. 1980. Types of dominance in a chimpanzee colony. *Folia primatol.* 34:90–110.

Power, M. 1991. *The Egalitarians: Human and Chimpanzee.* Cambridge, Mass.: Cambridge Univ. Press.

Riss, D.C., and C.D. Busse. 1977. Fifty-day observation of a free-ranging adult male chimpanzee. *Folia Primatol.* 28:283–97.

Riss, D., and J. Goodall. 1977. The recent rise to the alpha-rank in a population of free-living chimpanzees. *Folia Primatol.* 27:134–51.

Southwick, C.H. 1967. An experimental study of intragroup agonistic behavior in rhesus monkeys *(Macaca mulatta). Behav.* 28:182–209.

Sugiyama, Y. 1984. Population dynamics of wild chimpanzees at Bossou, Guinea, between 1976 and 1983. *Primates* 25:391–400.

———. 1988. Grooming interactions among adult chimpanzees at Bossou, Guinea, with special reference to social structure. *Intern. J. Primatol.* 9:393–408.

de Waal, F.B.M. 1982. *Chimpanzee Politics.* London: Cape.

———. 1984. Sex-differences in the formation of coalitions among chimpanzees. *Ethol. and Sociobiol.* 5:239–55.

———. 1986a. The brutal elimination of a rival among captive male chimpanzees. *Ethol. and Sociobiol.* 7:237–51.

———. 1986b. Integration of dominance and social bonding in primates. *Q. Rev. Biol.* 61:459–79.

———. 1989a. The myth of a simple relation between space and aggression in captive primates. *Zoo Biol. Suppl.* 1:141–48.

———. 1989b. *Peacemaking among Primates.* Cambridge, Mass.: Harvard Univ. Press.

———. 1989c. Dominance 'style' and primate social organization. In V. Standen and R. Foley, eds., *Comparative Socioecology: The Behavioral Ecology of Humans and Other Mammals,* pp. 243–63. Oxford England: Blackwell.

———. 1989d. Food sharing and reciprocal obligations among chimpanzees. *J. Human Evol.* 18:433–59.

———. 1989e. Behavioral contrasts between bonobo and chimpanzee. In P. Heltne and L. Marquardt, eds., *Understanding Chimpanzees,* pp. 154–75. Cambridge, Mass.: Harvard Univ. Press.

de Waal, F.B.M., and A. van Roosmalen. 1979. Reconciliation and consolation among chimpanzees. *Behav. Ecol. Sociobiol.* 5:55–66.

Wrangham, R. 1986. Ecology and social relationships in two species of chimpanzee. In D. Rubenstein and R. Wrangham, eds., *Ecological Aspects of Social Evolution: Birds and Mammals,* pp. 352–78. Princeton, New Jersey: Princeton Univ. Press.

Wrangham, R.W., A.P. Clark, and G. Isabirye-Basuta. 1992. Female social relationships and social organization of Kibale Forest chimpanzees. In T. Nishida, W.C. McGrew, P. Marler, M. Pickford, and F.B.M. de Waal, eds., *Topics in Primatology.* Vol. 1: *Human Origins,* pp. 81–98. Tokyo: Univ. of Tokyo Press.

Section 3

Cognition

Fifi fishing for termites.

Overview—Culture and Cognition

Frans B.M. de Waal

The term *culture* evokes images of art and music (as in the "cultural" life of our town), of symbols and language (as in the "cultural" anthropologist's claim that all human behavior is symbolically mediated), and of a heritage that needs to be protected against the eroding effects of the mass-consumption society. A so-called cultured person has achieved a refinement of tastes, a well-developed intellect, and a particular set of values and moral principles. Given these connotations, it is no wonder that culture is often presented as the antithesis of nature, a tradition going back to Thomas Henry Huxley, who compared the relation between culture and nature to that between a gardener and his garden: nature is something that needs to be kept under control, and culture allows us to do so (Nitecki and Nitecki 1993). Sigmund Freud (1930) made a similar connection between civic order and our brutish ancestry the centerpiece of *Civilization and Its Discontents,* in which he argued that people need to renounce their baser instincts before they can build a modern society. Needless to say, the views of Huxley and Freud are extraordinarily pessimistic about the nature of human nature.

Indications that our closest primate relatives, the chimpanzees and bonobos, exhibit a degree of behavioral diversity that borders on cultural variation thoroughly upset a neat dichotomy between human culture and animal nature. In an attempt to maintain a distinction one could, of course, argue that the behavioral diversity of apes does not necessarily reflect processes of cultural transmission; after all, it is unlikely that the behavioral diversity of apes depends on symbols and language. Maintaining such a distinction, however, assumes that all human cultural variation requires symbols and language and that a cognitive gap exists between humans and other extant hominoids that permits us to adopt different explanations for variation within the group.

First, although it is true that human cultural variation is closely tied to and expressed in language and that many aspects of human culture are unthinkable without this connection, some aspects of variation may reflect

socialization practices or processes of observational learning that involve but do not necessarily require language. Second, increasing evidence exists for cognitive continuity between humans and great apes. Admittedly, substantial differences remain, but more and more scientists feel that most of these are differences in degree. Accordingly, we cannot exclude the possibility that certain processes of cultural transmission take place in species other than our own.

The following section reviews the cognitive abilities of chimpanzees and bonobos relevant to the question of how behavioral diversity may arise. Most chapters do not directly connect cognitive ability and behavioral diversity for the simple reason that study of cognitive abilities often takes place in the laboratory where variables can be most effectively controlled, while the study of intergroup variability takes place mainly in the wild or in large captive groups outside the laboratory. Increasingly, laboratory workers find inspiration for their experiments in reports about ape behavior under natural or naturalistic conditions, whereas field-workers are paying attention to capacities documented in captivity so as to determine the functions those capacities might serve in the natural habitat. Further integration of laboratory and field research will be necessary in order to disentangle the factors and abilities underlying behavioral diversity.

To begin this section, van Hooff outlines the various cognitive domains—such as self-awareness, deception, attribution of intentions to others, imitative learning, active teaching, and so on—in which chimpanzees and bonobos seem to excel. The evidence is stronger for some of these capacities than for others—a few are positively problematic. Nonetheless, the overall picture emerging from research on social cognition is that members of the genus *Pan* occupy a special position with respect to social cognition, perhaps closer to humans than to monkeys, a not entirely unexpected outcome given their taxonomic position.

In the next two chapters, both Povinelli and Tomasello explore specific cognitive abilities that bear directly on culture and on the process of knowledge transmission. Povinelli focuses on the cognitive ability of attribution of knowledge and intentions to others. This ability may be instrumental both in the evolution of moral systems (moral judgment depends heavily on perceived intentions) (Povinelli and Godfrey 1993) and in the capacity for observational learning and teaching. An individual is probably more motivated to pay attention to another whom it assumes to possess special knowledge than to another without such knowledge. Conversely, a chimpanzee is observed to demonstrate special techniques to another or give other kinds of instruction if it understands that this may improve the other's knowledge and skill. Povinelli assumes that processes of attribution depend on self-awareness and explains how chimpanzees differ from monkeys in this regard.

Tomasello discusses observational learning and the ability for imitation. Many people recognize the ability to imitate to be present in apes; hence, the verb *to ape* and its equivalent in a number of languages. However, actually demonstrating aping in apes turns out to be quite a challenge, especially with respect to complex tool-using tasks and interactional skills, such as gestural communication. Thus far, controlled experiments have failed to support claims regarding full-blown imitation of tool use in nonhuman primates.

Learning processes in the natural habitat may differ from those in the laboratory. In addition, observational learning covers a wide range of processes that cannot easily be disentangled in the field. For this reason, field reports suggest but cannot demonstrate conclusively the presence of particular cognitive capacities relating to knowledge transmission.

Rumbaugh, Savage-Rumbaugh, and Sevcik as well as Boysen report research on the acquisition of symbolic communication, numerical competence, categorization, language comprehension, and complex learning tasks. Whether the language label applies to the ape-human communication documented at the Rumbaughs' laboratory may not be the most central issue (although it certainly is the issue that has attracted most attention). Far more significantly, their research indicates how apes order the world around them and provides an important window into their minds. Boysen avoids the language issue altogether by testing a subjects' understanding of the environment with different means, such as heart-rate responses to photographs or ability to match visual and vocal cues of familiar conspecifics. Her studies confirm the incredible individual variability found in apes and warn against generalizations from one individual to the rest of his or her species.

Individual variation is also reflected in Matsuzawa's observations of nut cracking by chimpanzees at Bossou, Guinea. A few chimpanzees in this community failed to use tools; they had to content themselves with scraps left behind by nut-cracking group mates. Reporting three instances of a tool arrangement more complex than the usual hammer-anvil combination, Matsuzawa compares these instances to the hierarchical ordering of lexigrams constructed in the laboratory.

References

Freud, S. 1961 (1930). *Civilization and Its Discontents.* Translated by J. Strachey. New York: Norton.

Nitecki, M.H., and D.V. Nitecki 1993. *Evolutionary Ethics.* New York: State Univ. of New York Press.

Povinelli, D.J., and L.R. Godfrey. 1993. The chimpanzee's mind: How noble in reason? How absent in ethics? In M.H. Nitecki and D.V. Nitecki eds., *Evolutionary Ethics,* pp. 277–324. New York: State Univ. of New York Press.

Carmen with a rose in her mouth.

Photo by J. Goodall.

Understanding Chimpanzee Understanding

Jan A.R.A.M. van Hooff

Reacting Versus Understanding and Designing

The last two decades have witnessed a paradigmatic change in our thinking about the behavior of animals. Specifically, an emphasis on cognitive aspects of behavior has become prominent (Cheney and Seyfarth 1990b). *Cognition* may be defined as an animal's ability to represent relevant aspects of the world in which it lives in mental schemes that can be manipulated in simulation types of procedures. On the basis of this ability the animal can design its behavior and make choices about its course of actions; in other words, it can behave intentionally (Dickinson 1988). This creative aspect in the management and processing of information distinguishes the cognitive mode from the reflexive mode, in which even comparatively complex mental schemes are restricted in their functioning to the selective perception of matching stimulus configurations (Walker 1983; McFarland 1985).

The questions of interest now are, first, what types of information management and processing can be distinguished? Specifically, what are the mechanisms of "animal thought," and what principles of reasoning emerge from this—such as analogical reasoning (Gillan 1981; Gillan et al. 1981)? Second, what features of the world are comprised and accounted for in these internal schemes? That is, what sorts of information are specified in these schemes, not only about environmental features but also about the properties and functioning of the animal itself and about the relationship of these properties with relevant environmental variables? Such internal schemes would allow, if perhaps only primitively, reflection on the conditions of the animal's own functioning, a kind of meta-knowledge, that has been postulated even for a bird such as the pigeon (Shimp 1982, 1983).

The Genus *Pan:* A Special Case?

Recent research on primate cognition has drawn attention to the chimpanzee. The early language studies of chimpanzees raised high expectations about these animals' linguistic competence. In retrospect the expectations may have been too high (Snowdon 1990), yet these studies have been of tremendous interest, opening up ways to explore the nature and content of the cognitive domain of these species by exploiting their competence to attach labels to elements of cognition. This possibility was examined spectacularly not only in the case of the primates (Premack 1983), but also in the case of birds (Pepperberg 1990, 1992).

These and other experiments gradually yield the suggestion that *Pan,* and perhaps the other great apes, occupy a special position among the nonhuman primates. This special position concerns their faculties of self-recognition, imitation, and attribution.

Meta-Knowledge

Meta-Knowledge About the Self

Gallup's experiments on the recognition of mirror images by primates were ground breaking because Gallup realized the theoretical importance of self-recognition (Gallup 1970, 1982). Recognizing a mirror image as oneself means that an animal is able to detect the synchrony of the mirror image's movements with its own movements by relating the perception of its own movements, as seen in the mirror, to some internal representation of their execution. Such representations exist in the form of sequential motor command schemes, or efference copies (Roitblat 1985; von Holst and Mittelstaedt 1950).

Information of this kind must be available in a cognitively accessible form if an animal is to be able to perceive the congruence of its behavior with that of a mirrored image, particularly if the animal sees only movements of parts of its body that it cannot normally see directly; such as when an animal experiments with mimicking itself in a mirror, or on a TV screen (Savage-Rumbaugh 1986). Such knowledge is about information in the system being used to achieve the desired effect and thus qualifies for the designation of meta-knowledge.

This form of self-recognition has not been found in cercopithecoid monkeys (Anderson 1984). Experiments with other higher mammals were negative as well. Thus, self-recognition could not be shown in the elephant (Povinelli 1989), where an awareness of the trunk movements and positions would not be astonishing. More remarkable is the difference that seems to exist within the great apes. Whereas orangutans displayed self-recognition

(Lethmate and Dücker 1973), its occurrence in the gorilla is controversial (Ledbetter and Basen 1982; Suarez and Gallup 1981; Patterson 1984).

Various behaviors specific to the chimpanzee are explained most easily on the basis of knowledge of the nature and effects of self-manifestation: for example, the ways in which chimpanzees extend the signal value of their displays by incorporating instruments such as branches and sticks, resounding objects, and the like (Kortlandt 1968; Goodall 1986). Also playful self-adornment (van Hooff 1973) can be understood as an exploratory fascination with one's own appearance. An example of a possible expression of self-knowledge is given by de Waal (1982): a subordinate male chimpanzee of the Arnhem colony who was displaying his sexual interest to a female suddenly hid his erect penis behind his hands when he noticed that a dominant male was about to pass.

Ecological Pressures for Meta-Knowledge of Self

Meta-knowledge about the state and the coordination of bodily structures may be the basis of self-awareness. But, why would a capability for this knowledge have developed? Do some special ecological conditions and matching life modes offer the incentive for the evolution of motor awareness? Kortlandt (1968) drew attention to the extreme multifunctional flexibility in use of limb and hand by the four-handed chimpanzee, and he pointed out the important evolutionary consequences of that flexibility. Povinelli and Cant (1992) claim that the occurrence of advanced mental capabilities such as self-conception and mental-state attribution in only chimpanzees and orangutans is not satisfactorily explained by social hypotheses. They noted that these two species are the heaviest of the arboreal primates and that both have adopted a highly flexible and nonstereotyped form of locomotion called *orthograde clambering.* Choosing the appropriate action in conjunction with choosing a route, circumventing and crossing gaps, as well as anticipating the stratum deformations caused by the movements of the animal itself, as in tree swaying (Rijksen 1978) require that the animals know about their body movements and the functional consequences of these. The question is whether these apes differ essentially in these behaviors from other heavy arboreal primates, such as the South American muriquis and spider monkeys, which seem to display insightful arboreal behavior as well.

The locomotion-flexibility hypothesis at least adds to hypotheses that seek to explain the origin of meta-knowledge about the self in terms of the demands of an increasingly complex social organization (Humphrey 1976; Whiten and Byrne 1988). There can be little doubt that the intelligence of nonhuman primates manifests itself primarily in the social realm. They know about social relations in a much more refined way than they know about their non-social environment (Cheney and Seyfarth 1990). The question to be investigated further is whether monkeys and apes differ essentially in social cognitive skills in ways that explain the differences in body awareness. The locomotion-flexibility hypothesis at least circumvents the assertion that a social explanation does not account for the lack of self-recognition in certain primates, including macaques, baboons, and vervets, even though these animals live in complex and cohesive societies where acquiring and maintaining a favored social position requires complex social strategies such as coalition formation. In a form that suggests basic elements of negotiation, coalitions have, for instance, been demonstrated in baboons (Bercovitch 1988; Noë 1990, 1992). On the other hand, mirror recognition has been demonstrated in orangutans (Lethmate and Dücker 1973), even though their present natural life appears to offer little need for social sophistication.

Attribution

Representing the Notion and Intentions of Others

Meta-knowledge about the self is supposed to be a prerequisite for the capacity to attribute notions and intentions to others (Bischof-Köhler 1989; Gallup 1991). Social attribution implies knowledge about the behavioral dispositions and intentions of others and, also, of oneself. In the case of self-attribution, it means that an individual has a mental representation of some of its own characteristics such as its behavioral inclinations, of relevant aspects of its social and nonsocial environment, and of the relationships between these—that is, the way its own functioning may affect and modify this relationship and, thus, release reactions from the environment.

The penis-hiding incident, recounted by de Waal, suggests the possibility of such self-knowledge and how self-knowledge can generate a smart solution for a social problem. The experiments of Premack and Woodruff (1978) strongly suggest attributive abilities in the chimpanzee (Premack 1988).

Recently Povinelli et al. (1990, this volume) have again addressed the problem of attribution experimentally. Can chimpanzees and macaques attribute knowledge to another individual? Povinelli et al. trained a chimpanzee to distinguish between two persons. The chimpanzee could see that one person (the knower) was watching where a food reward, not visible to the chimpanzee, was hidden. The other person (the guesser) could not see it either. When, subsequently, both persons pointed to a possible hiding place, the chimpanzee followed the indications of the knower. The same was accomplished by children after they had reached the age of four (Povinelli and deBois 1992). A similar experiment with rhesus monkeys yielded negative results (Povinelli et al. 1991).

Another experiment, on spontaneous role change, showed that chimpanzees were able to take the perspective of a partner, whereas macaques were not (Povinelli et al. 1992). Extensive observations and experiments on feral vervet monkeys led Cheney and Seyfarth (1990a and b) to conclude that these monkeys lacked attribution skills.

Planning, Calculation, and Reciprocity in Social Manipulation

Attributive competence makes biological sense only if it is integrated into the programming of social strategies. All kind of animals demonstrate anticipation by extrapolation on the basis of schemes such as mental maps of past experience. Observations of Savage-Rumbaugh and MacDonald (1988) unequivocally indicate that, at least in chimpanzees, the goal is actively represented in the process that leads the animal to adopt a plan (Walker 1983; Roitblat 1985). Both bonobo and chimpanzee could use a learned, computer symbol system to indicate beforehand the destination of their travel and the things of interest to be found at that destination.

In his *Chimpanzee Politics,* de Waal (1982) expressed his conviction that such planning plays a great role in the social strategies that we recognize in chimpanzee interaction. He has built a picture of their intelligent social functioning on the basis of detailed, long-term observations of a chimpanzee colony. Moreover, since Trivers (1971) drew attention to reciprocity as a principle of organizing cooperative and seemingly altruistic relationships, he and others have studied not only the occurrence of patterns of reciprocity in primates but also the implications of these patterns with respect to the proximate mechanisms and to cognition (Seyfarth and Cheney 1984; de Waal and Luttrell 1988; de Waal 1989b, 1991a; Hemelrijk et al. 1992).

Increasingly, evidence points to the existence of true systems of reciprocity, as revealed, for instance, by the study of de Waal and Luttrell (1988) on intervention in conflicts by macaques and chimpanzees. De Waal and Luttrell found a striking difference, however, between these taxa. Both macaques and chimpanzees intervened to return help to those who had helped them. Chimpanzees, in addition, chose sides against those who had intervened against them; in other words, they showed a revenge system. However, this difference need not be ascribed to a difference in cognitive ability: the comparatively strict dominance system of macaques may simply inhibit aggression and interventions against the formal hierarchy. Aureli et al. (1989), found that after losing a conflict, Japanese macaques turned on the weaker relatives of their opponent; thus striking back indirectly at a high-ranking opponent.

These studies suggest a calculating mind, keeping a record of how much was given and how much was received. Of course, even animals whom we expect to be functioning at modest cognitive levels appear to value relevant aspects of their interaction with the environment quantitatively . For example, animals often find near-optimal solutions for dilemmas, suggesting processes in which costs and benefits are weighed against one another. They adjust their attitude and the associated behavior on the basis of the negative (investments, risks, and punishments) and positive (rewards) effects of their behaviors (Roitblat 1985).

Our analysis tells us that such processes can occur at an emotional, intuitive level, where cognitive contents are little specified and not very explicit. The calculations involved can clearly be executed partly if not wholly at an unconscious level (Jackendorf 1987). The outcomes we become aware of are the *emotive connotations* of likes and dislikes and the *normative connotations* of gratitude, guilt, spite, and the like, and these outcomes are formed, built up, or diminished as a function of the valuing of experiences.

The question is this: should all forms of reciprocity be explained as processes that occur on the unconscious level? Or do certain forms of reciprocity require more explicit, calculative reasoning of cognition, that is, where the animals have some level of awareness of the ins and outs of their own calculatory processes. From the laboratory have come remarkable results showing that chimpanzees have a certain numerical and arithmetic competence (Boysen and Berntson 1990; Matsuzawa 1990). Have these competences proved their adaptive value in tasks required in natural life? Or are they the artificially generated by-products of a more general competence?

Deceptive Manipulation

Deception is an especially fascinating phenomenon in connection with the discussion of cognitive capacities. In a comprehensive historical review, Mitchell (1986) discusses four levels of deception with the following definition as a starting point: *Deception* is a behavior programmed to achieve a beneficial result for the deceiver as a consequence of a mistaken registration of the behavior by the receiver. The level of deception depends on the openness of the programming and on how the programming has developed.

It is Mitchell's levels three and four that are interesting in connection with the issue of cognition. At level three the animal can adjust the programming of its behavior on the basis of its own experience. The animal thus has the freedom to manipulate its signals. At level three the animal may regulate its action without any reference to the receiver's beliefs. Thus, the action is intended, though not intended to be deceptive.

At level four the deceiver programs and reprograms its behavior on the basis of knowledge about the receiver's beliefs. Mitchell concluded that only chimpanzees behave in ways that suggest deception at level four. However, the evidence for such behavior consists almost exclusively of single records or anecdotes. Whiten and Byrne (1988) have justified collecting such rare anecdotal accounts. They have reasoned that deception can only be effective if it would not occur sufficiently frequently or systematically to enable receivers to detect it. A certain rarity of incidents of deceiving is to be expected. This rationale has been made most explicit in a recent report on a new inquiry among primatologists (Byrne and Whiten 1992): If a sufficient number of careful descriptions of possible deceptive behaviors by experienced observers can be accumulated; then, in spite of the rarity of the incidents, certain regularities may become visible. If independent accounts of well-studied species reveal species-specific variations, then these variations add conviction to a conclusion of the presence of deceptive behavior.

Byrne and Whiten's levels one-and-a-half (visual perspective-taking) and two (mental perspective-taking) correspond with Mitchell's level four. Their inventory has yielded many records for these two levels for all the pongids, especially for the chimpanzee. By contrast, there were no records for the other monkeys (Cheney and Seyfarth 1990a) with one remarkable exception: for baboons *(Papio)* there were five records, one at level two, and four at level one-and-one-half, suggesting special cognitive capacities.

Kummer et al. (1990) have taken a critical stance. They do recognize the importance of investigating cognitive phenomena and the role of cognitive phenomena in shaping purposeful social strategies. However, these researchers fear a swing of the pendulum toward easy, unjustified, unparsimonious, and anthropomorphic conclusions of high levels of cognition and conscious intentionality if anecdotes are awarded the power of proof. Thus, Kummer et al. view anecdotes as suggestive at best and request that the principle of parsimony be adhered to with traditional vigor and that experimental testing be given a crucial role in deciding between alternative explanations.

De Waal (1991b) has reacted to this conflict about methodology by taking issue with the customary interpretation of parsimony. He argued that, in view of evolutionary continuity, the acceptance of cognitive operations and a conscious intentionality in chimpanzees analogous to our own can be regarded as the parsimonious position. Interpreting the behavior of our nearest relatives in terms of cognition and consciousness may make us sensitive to peculiarities in unique occurrences, which we might otherwise fail to recognize. A precise, sensitive, and at the same time conscientious description of single incidents can make visible the temporal coincidences that are very unlikely to occur by chance. Such a description has a high convincing power, as the accounts of deceptive behaviors by de Waal (1986) demonstrate. Still the anecdotal approach cannot be the only method. The ideas generated by the anecdotal approach must be tested against alternatives in more controlled observations or experiments.

The ability to attribute beliefs and intentions to others certainly makes for more efficient social manipulation. The descriptions of chimpanzee intrigues, notably by de Waal (1982, 1986), are impressively suggestive in this respect. If macaques and baboons do not possess this faculty, then complex social processes such as alliance formation and alleged tactical deception have to be explained by more parsimonious mechanisms in these species. This observation leads to the question of whether there is evidence of behavior in the wild, other than the manipulation of social relationships, for which the faculty of attribution is necessary. The systematic, cooperative hunting behavior in chimpanzees might represent an example.

Social Cooperation

Chimpanzee Hunting Behavior

The first reports on hunting by chimpanzees came from the Gombe study site in Tanzania (Teleki 1973; McGrew 1979). Hunting at Gombe was described as an opportunistic behavior, occurring when the prey animals—baboons, colobus monkeys, or other small animals—are encountered in a position that make them an accessible prey. Only the male chimpanzees hunted and, remarkably, they often acted in what seemed a coordinated fashion. They encircled their victim shutting off its possible escape routes from the tree. Similar behavior has been observed elsewhere.

In contrast, in a population in the Taï forest in Ivory Coast, hunting has been reported as occurring almost daily (Boesch and Boesch 1989). The males trek through the forest in a regular fashion looking for red colobus monkeys. Once they have located a group, they fan out silently and surround the victim by taking positions that seem to acknowledge the flight options of the prey depending on the topography of trees and branches and depending on the approach options of their fellow hunters. The degree of coordination shown during collaborative hunting is remarkable.

This behavior sets the chimpanzee clearly apart from the other higher primates. Some of the other higher primates also capture animal prey on occasions—however, in a more erratic and opportunistic pattern. For example, a hunting tradition was developed in a population of baboons in Kenya in an area where the baboons' natural predators had largely disappeared. The baboons captured young antelopes, rodents, and the like, as they stumbled upon them (Harding 1975). Gradually, this pattern developed into a more systematic searching for prey, during which a small group of baboons went scouting on the savanna (Strum 1981). However, whereas the West African chimpanzees seemed to be directed by a complementary anticipation of the movements of their hunting partners, there was never any indication of this in the baboons.

This contrast between chimpanzees and baboons raises the interesting question of whether the interspecific differences found in the coordination of hunting behavior are in any way related to the interspecific differences with respect to self-recognition and attribution. Do these behavioral differences simply reflect differences in the environment? Or do they reflect a hard-wired difference in attributive competence?

Complementary Roles in Social Cooperation

Do we have evidence that cooperative hunting is possible only because one individual is able to value and interpret the situation from the perspective of a partner (Povinelli et al. 1990)? In other words, is one individual able to recognize the intention of another individual and able to survey his own options, so that he can extrapolate about the other's course of actions and can adjust his own choices of action to complement the most effective joint strategy? Since two or even more partners are involved, as well as a prey animal striving to make the hunters' choices unsuccessful, choosing the appropriate complementary actions is complex—involving a mutual, iterative approximation of the successful joint strategy.

Clearly, we need precise analyses of the tempero-spatial coordination of the course of action taken with regard to the options available in a given topographical situation. These analyses should reveal the extent to which the participants in a hunt do base their own behavior on an anticipatory interpretation of the behaviors of other partners. And, to exclude a more parsimonious explanation, we should establish whether other primate species and other social carnivores do the same.

The traditional view holds that cooperative hunting and concomitant developments, such as communication and tool use, were stimulated when our ancestors moved into the savanna. However, Boesch and Boesch (1989) cast doubt, to say the least, on this view. The complex arboreal environment may well have provided the selective challenges ultimately responsible for the development of the necessary cognitive precursors.

Imitation and Instruction

In searching for behavioral manifestations in the natural environment that could rest on capacities similar to attribution, certain phenomena—namely, imitation, instruction, and consolation—come to mind that may exist exclusively in the chimpanzee or, perhaps, in the other pongids.

Imitation

For a long time it was assumed that animals can imitate certain behavior patterns when they observe the occurrence of these patterns in others. Gradually, however, two things have become clear. First, the mechanisms that permit animals to engage in behavior similar to that of others can vary considerably. Second, of the acquired animal behaviors long considered to be the product of imitation, most can be, and most likely should be, explained in a more parsimonious way (Whiten 1989; Tomasello 1990; Visalberghi and Fragaszy 1990; Whiten and Ham 1992).

True imitation is the modeling of a behavior pattern by copying the example set by another (Thorpe 1956). Not only does the observer respond selectively to certain circumstances (stimuli), but this observational learning also shapes the motor response. This motor response presupposes the kind of motor awareness that recently has been found only in the apes. Thorpe indeed argued that simpler processes can lead to a similar result: for example, the activity of one animal or even the mere presence of that animal can direct the attention of another animal of the same species to particular features of its environment. As a result of such local enhancement and/or stimulus enhancement, the conspecific more readily experiences the same contingencies of trial-and-error reinforcement and discovers the same successful routine. The potato washing and the grain rinsing described for Japanese macaques (Kawai 1965), often cited as an example of cultural transmission by imitation, are better explained by local enhancement and stimulus enhancement (Galef 1990).

In recent reviews, both Visalberghi and Fragaszy (1990) and Whiten and Ham (1992) conclude that in practically all reported cases for monkeys, processes simpler than true imitation can indeed explain observed behavioral similarities equally well or even better. Both reviews take a critical stance with respect to the easily accepted assumption that true imitation determines the cultural transmission of tool use—such as ant and termite fishing, sponge dipping, and nut cracking—in chimpanzees (Nishida 1987). Instead, the slow and gradual development of these behaviors accords with a shaping process by operant conditioning facilitated by stimulus enhancement.

However, Tomasello et al. (1987) and Tomasello (1990) showed experimentally that in chimpanzees something more elaborate is occurring mentally. Their chimpanzees learned a task because these chimpanzees understood, from watching a demonstrator, the instrumental connection between a simple procedure and its meaningful consequences. These chimpanzees then learned the simple procedure for themselves. This process of acquisition is called *emulation.* A more focused attending to the details of the modeled routine might occur if the emulator was unable to discover the adequate procedure quickly. However, in these experiments the emulators did not imitate the actions of the demonstrator. A transition from emulation to more precise attending leading to real imitation may have taken place in the teaching of the nut-cracking procedure described by Boesch (1991). Recent work of Tomasello and Savage-Rumbaugh suggests that chimpanzees indeed have the competence to be educated to attend to the precise actions of a demonstrator (Tomasello 1990).

Instruction

Instruction, or teaching, is the logical complement to imitation. In human teaching, the intentional orientation of a teacher to a disciple is implied. The teacher monitors the performance of a disciple in a certain task and adjusts his or her own behavior so that the disciple learns to distinguish the appropriate perceptive categories and/or comes to master the appropriate motor programs. Such teaching clearly derives its adaptive flexibility from the ability of teachers to interpret the knowledge, beliefs, attitudes, and motivations of their disciples.

Nonetheless, such attribution is not a necessary condition for teaching, and the teacher need not even intend to teach the pupil. Caro and Hauser (1992) have distinguished different forms of teaching and have explored what is known about the occurrence of these forms in different animal taxa. Various species of birds such as raptors and mammals such as felids merely bring conspecifics (as a rule, their offspring) into situations where the young can acquire a competence through trial and error. For example, the parents offer prey animals to their offspring in a way that incites excitation of the prey-catching routine. The teaching process could then be, but need not be, a partly fixed and partly hard-wired routine involving little if any attributive insight from the teacher.

In contrast, true teaching involves active molding by the teacher, who guides the performance of the disciple with well-tuned encouraging and discouraging feedback. So far, except for chimpanzees, we have no evidence for the occurrence of such active molding (Caro and Hauser 1992). These authors refer to the observations of Fouts et al. (1989). Loulis, a young chimpanzee in their sign-language project, gradually learned signs even though she never received any instruction by humans. On a few occasions, these investigators saw how the elder female Washoe spontaneously took Loulis's hand and molded it in the configurational position of a required sign.

Evidence for similar processes in the natural situation is still limited. Boesch (1991) has studied the manner in which chimpanzee mothers in the Taï forest influence the attempts of their infants to crack nuts. Much of this process involves active, stimulus enhancement. In a few cases, however, Boesch observed clear, active, teaching, as when a mother gave a demonstration in emphatic slow motion of how the hammer should be used after the child had unsuccessfully struggled with the tool. Subsequently, the child used the hammer in the correct way. Even more so than in the case of true imitation, it seems difficult to explain these admittedly rare patterns of teaching behavior without assuming attribution and self-awareness in the sense of Gallup.

Reconciliation and Consolation

Living in a society means striking the right balance between affiliation and collaboration on the one hand and competition and conflict on the other hand. *Reconciliation,* a process that serves the restoration of a disturbed social balance, has been described in recent years for a great many species. In a number of primate species, former opponents may show post-conflict affiliative behaviors toward one another to reduce the tension in their strained relations (Aureli et al. 1989; Aureli 1992; de Waal 1989c).

In chimpanzees there is evidence for other forms of post-conflict behaviors (de Waal and van Hooff 1981), in particular for consolation (de Waal 1989c). The function implied is the attempt at restoration of the emotional balance in a disturbed animal—for example, the victim of a conflict—by an outsider. The sympathetic attitude of the outsider can be understood in terms of an attributional process. A more parsimonious, social-releaser explanation is certainly possible: signals of distress by young and helpless individuals release caring behavior from the parent in many animal species. The releasing value of distress signals is generalized to other caring group members in a number of cooperative species (van Hooff 1987). Nevertheless, it is striking that the chimpanzee is the only species for which post-conflict consolation has been reported so far. Systematic studies have failed to demonstrate consolation in monkeys (Aureli et al. 1989).

Summary

Clearly there are different cognitive realms that are spread in different ways over animal taxa. The ability to attach symbolic labels to conceptual categories is clearly not restricted to our nearest relatives. The ability may occur not only in other mammals (Schusterman et al. 1986; Gisimer and Schusterman 1992) but also in birds (Pepperberg 1990).

Other cognitive capacity may be much more limited. Only the chimpanzee, bonobo and orangutan (White Miles 1990) show clear evidence of the abilities of perspective taking and attribution. Moreover, it is only these species that have convincingly demonstrated the ability to form and to handle meta-knowledge about their own appearance and functioning in relation to the environment. Similarly, the abilities to imitate motor patterns and to perform intentional instruction may be restricted to one or more of the great apes.

A dichotomy appears to exist between the genus *Pan* and its relatives, on one hand, and the monkeys, on the other hand. Some researchers have been inclined to translate this dichotomy into an anthropo- and chimpocentric position (Beck 1982) and to conclude that these species have some fundamental difference in cognitive disposition, that is, in capacity for consciousness. Before accepting this position, however, we should consider several alternatives. Do behavioral differences that suggest this conclusion reflect species-specific differences or environmentally determined differences in the readiness and the opportunity to engage in certain forms of interaction? Do motivational constraints affect the expression of certain competencies? And, can we exclude the possibility that the social characteristics of a species either preclude or do not stimulate the ontogenetic development and manifestation of certain cognitive capacities, even though the genetic disposition is present?

Various social characteristics of a species can be related to the regime of competition between its members (van Schaik 1989; van Hooff and van Schaik 1992). Furthermore, social organizations of different species can differ in being either more hierarchical or more egalitarian (Vehrencamp 1983) as evidenced by dominance styles and patterns of agonistic interaction (Thierry 1985; de Waal 1989a). The more a society rests on cooperation and on an egalitarian negotiation about benefits (Hand 1986; Noë et al. 1991), the more sensitive the individuals have to be to each others' wants and needs. To what extent, therefore, can such differences be responsible for the expression of certain cognitive competencies? For example, processes that might be based on attribution could be facilitated in a system of social relationships in which cooperative exchange is important. Stolba (1979) has described the processes by which hamadryas baboon males—both harem owners and follower males—come to agree on route and destination choices. It is suggested that the baboons investigate the behavior tendencies (such as probe intentions of their partners). At the same time, they invoke representations of their own evaluations of the available options in the minds of their partners (Kummer 1992). Clearly, we should investigate the attributive competence of the baboons.

In conclusion, the plea here is to select species and tasks for comparative investigation in which the natural cooperative tendencies are most clearly expressed. Moreover, we should be critical of accepting high-level explanations for the processes we encounter. While accepting the outstanding position of *Pan,* we should be careful not to create a new dichotomy in the animal kingdom.

References

Anderson, J.R. 1984. Development of self-recognition: A review. *Developmental Psychobiol.* 17:35–49.

Aureli, F. 1992. Reconciliation, redirection and the regulation of social tension in macaques. Ph.D. diss., Universiteit Utrecht, Utrecht, Germany.

Aureli, F., C.P. van Schaik, and J.A.R.A.M. van Hooff. 1989. Functional aspects of reconciliation among captive long-tailed macaques *(Macaca fascicularis). Am. J. Primatol.* 19:39–51.

Aureli, F., H.C. Veenema, C.J. van Panthaleon van Eck, and J.A.R.A.M. van Hooff. In Press. Reconciliation, consolation, and redirection in Japanese macaques. *Behav.*

Beck, B. 1982. Chimpocentrism: Bias in cognitive ethology. *J. Human Evol.* 11:3–17.

Bercovitch, F.B. 1988. Coalitions, cooperation and reproductive tactics among adult male baboons. *Anim. Behav.* 36:119–209.

Bischof-Köhler, D. 1989. *Spiegelbild und Empathie: Die Anfänge der sozialen Kognition.* Bern, Switzerland: Huber.

Boesch, C. 1991. Teaching among wild chimpanzees. *Anim. Behav.* 41:530–2.

Boesch, C., and H. Boesch. 1989. Hunting behavior of wild chimpanzees in the Taï National Park. *Am. J. Phys. Anthro.* 78:547–73.

Boysen, S.T., and G.C. Berntson. 1990. The development of numerical skills in the chimpanzee *(Pan troglodytes).* In S. Taylor Parker and K.R. Gibson, eds., *"Language" and Intelligence in Monkeys and Apes,* pp. 435–50. Cambridge, England: Cambridge Univ. Press.

Byrne, R., and A. Whiten. 1992. Cognitive evolution in primates: Evidence from tactical deception. *Man N.S.* 27:609–27.

———. eds. 1988. *Machiavellian Intelligence: Social Expertise and the Evolution of Intellect in Monkeys, Apes and Humans.* Oxford, England: Clarendon.

Caro, T.M., and M.D. Hauser. 1992. Is there teaching in nonhuman animals? *Q. Rev. Biol.* 67:151–74.

Cheney, D.L., and R.M. Seyfarth. 1990a. *How Monkeys See the World.* Chicago: Univ. of Chicago Press.

———. 1990b. Attending to behavior versus attending to knowledge: Examining monkeys' attribution of mental states. *Anim. Behav.* 40:742–53.

Dickinson, A. 1988. Intentionality in animal conditioning. In L. Weiskrantz, ed., *Thought Without Language,* pp. 305–25. Oxford, England: Clarendon.

Fouts, R., D.H. Fouts, and T.E. van Cantfort. 1989. The infant Loulis learns signs from cross-fostered chimpanzees. In R.A. Gardner, B.T. Gardner, and T.E. van Cantfort, eds., *Teaching Sign Language to Chimpanzees,* pp. 280–92. Albany: State Univ. of New York Press.

Galef, B.G. 1990. Tradition in animals: Field observations and laboratory analyses. In M. Bekoff and D. Jamieson, eds., *Interpretations and Explanations in the Study of Behavior,* pp. 74–95. Boulder, Colorado: Westview Press.

Gallup, G.G. 1970. Chimpanzees: Self-recognition. *Science* 167:86–7.

———. 1982. Self-awareness and the emergence of mind in primates. *Am. J. Primatol.* 2:237–48.

———. 1991. Toward a comparative psychology of self-awareness: Species limitations and cognitive consequences. In G.R. Hoethals and J. Strauss, eds., *The Self: An Interdisciplinary Approach,* pp. 121–35. New York: Springer.

Gillan, D.J. 1981. Reasoning in the chimpanzee: II. Transitive inference. *J. Exp. Psychol.: Anim. Behav. Processes* 7:52–77.

Gillan, D.J., D. Premack, and G. Woodruff. 1981. Reasoning in the chimpanzee: I. Analogical reasoning. *J. Exp. Psychol.: Anim. Behav. Processes* 7:1–17.

Gisimer, R., and R.J. Schusterman. 1992. Sequence, syntax and semantics: Responses of a language-trained sea lion *(Zalophus californianus)* to novel sign combinations. *J. Comp. Psychol.* 106:78–91.

Goodall, J. 1986. *The Chimpanzees of Gombe: Patterns of Behavior.* Cambridge, Mass.: Belknap Press.

Hand, J.L. 1986. Resolution of social conflicts: Dominance, egalitarianism, spheres of dominance and game theory. *Q. Rev. Biol.* 61:201–20.

Harding, R.S.O. 1975. Meat-eating and hunting in baboons. In R.H. Tuttle, ed., *Socioecology and Psychology of Primates,* pp. 247–57. Den Haag, Netherlands: Mouton.

Hemelrijk, C.K., G.J. van Laere, and J.A.R.A.M. van Hooff. 1992. Sexual exchange relationships in captive chimpanzees. *Behav. Ecol. Sociobiol.* 30:269–75.

von Holst, E., and H. Mittelstaedt. 1950. Das Reafferenzprinzip: Wechselwirkungen zwischen Zentralnervensystem und Peripherie. *Naturwiss.* 37:464–7.

van Hooff, J.A.R.A.M. 1973. The Arnhem Zoo chimpanzee consortium: An attempt to create an ecologically and socially acceptable habitat. *Internat. Zoo. Yearbook,* 13:195–205.

———. 1987. On the ethology of pain, its experience and expression. In A.C. Beynen and H.A. Solleveld, eds., *New Developments in Biosciences: Their Implication for Laboratory Animal Science,* pp. 41–6. Dordrecht: Nijhoff.

van Hooff, J.A.R.A.M., and C.P. van Schaik. 1992. Cooperation in competition: The ecology of primate bonds. In S. Harcourt and F.B.M. de Waal, eds., *Cooperation in Conflict: Coalitions and Alliances in Humans and Other Animals,* pp. 356–90. Oxford, England: Oxford Univ. Press.

Humphrey, N.K. 1976. The social function of intellect. In P.P.G. Bateson and R.A. Hinde, eds., *Growing Points in Ethology,* pp. 303–17. Cambridge, England: Cambridge Univ. Press.

Jackendorf. R. 1987. *Consciousness and the Computational Mind.* Cambridge, Mass.: MIT Press.

Kawai, M. 1965. Newly-acquired pre-cultural behavior of the natural troop of Japanese monkeys on Koshima Islet. *Primates* 6:1–30.

Kortlandt, A. 1968. Handgebrauch bei freilebenden Schimpansen. In B. Rensch, ed., *Handgebrauch und Verständigung bei Affen und Frühmenschen,* pp. 59–102. Bern, Switzerland: Hans Huber.

Kummer, H. 1992. *Weiße Affen am Roten Meer: Das soziale Leben der Wüstenpaviane.* Munich, Germany: Piper.

Kummer, H., V. Dasser, and P. Hoyningen Huene. 1990. Exploring primate social cognition: Some critical remarks. *Behav.* 112:84–98.

Ledbetter, D.H., and J.A. Basen. 1982. Failure to demonstrate self-recognition in gorillas. *Am. J. Primatol.* 2:307–10.

Lethmate, J., and G. Dücker. 1973. Untersuchungen zum Selbsterkennen im Spiegel bei Orangutans und einigen anderen Affenarten. *Z. Tierpsychol.* 33:248–69.

Matsuzawa, T. 1990. Spontaneous sorting in human and chimpanzee. In S. Taylor Parker and K.R. Gibson, eds., *"Language" and Intelligence in Monkeys and Apes,* pp. 451–68. Cambridge, England: Cambridge Univ. Press.

McFarland, D. 1985. *Animal Behaviour.* London: Pitman.

McGrew, W.C. 1979. Evolutionary implication of sex differences in chimpanzee predation and tool use. In D. Hamburg and E.M. McCown, eds., *The Great Apes,* pp. 441–64. Menlo Park, Calif.: Benjamin/Cummings

Mitchell, R.W. 1986. A framework for discussing deception. In R.W. Mitchell and N.S. Thompson, eds., *Deception: Perspectives on Human and Nonhuman Deceit,* pp. 3–39. New York: State Univ. of New York Press.

Nishida, T. 1987. Local traditions and cultural transmission. In B.B. Smuts, D.L. Cheney, R.M. Seyfarth, R.W. Wrangham, and T.T. Struhsaker, eds., *Primate Societies,* pp. 462–74. Chicago: Univ. of Chicago Press.

Noë, R. 1990. A veto game played by baboons: A challenge to the use of the prisoners' dilemma as a paradigm for reciprocity and cooperation. *Anim. Behav.* 39:78–90.

———. 1992. Alliance formation among male baboons: Shopping for profitable partners. In A.H. Harcourt and F.B.M. de Waal, eds., *Cooperation in Conflict: Coalitions and Alliances in Animals and Humans,* pp. 285–322. Oxford, England: Oxford Univ. Press.

Noë, R., C.P. van Schaik, and J.A.R.A.M. van Hooff. 1991. The market effect: An explanation for pay-off asymmetries among collaborating animals. *Ethology* 87:97–118.

Patterson, F. 1984. Self-recognition by *Gorilla gorilla gorilla. J. Gorilla Found.* 7:2–3.

Pepperberg, I.M. 1990. Conceptual abilities of some nonprimate species, with an emphasis on an African Grey parrot. In S. Taylor Parker and K.R. Gibson, eds., *"Language" and Intelligence in Monkeys and Apes.* pp. 469–507. Cambridge, England: Cambridge Univ. Press.

———. 1992. Proficient performance of a conjunctive, recursive task by an African gray parrot *(Psittacus erithacus). J. Comp. Psychol.* 106:295–305.

Povinelli, D.J. 1989. Failure to find self-recognition in Asian elephants *(Elephas maximus)* in contrast to their use of mirror cues to discover hidden food. *J. Comp. Psychol.* 103:122–32.

Povinelli, D.J., and S. deBois. 1992. Young children's *(Homo sapiens)* understanding of knowledge formation in themselves and others. *J. Comp. Psychol.* 106:228–30.

Povinelli, D.J., and J.G.H. Cant. 1992. Orangutan clambering and the evolutionary origins of self-conception. *Abstr. 14th Cong. Internat. Primatol. Soc.,* p. 46. Strasbourg, Germany: Société Francophone de Primatologie.

Povinelli, D.J., K.E. Nelson, and S.T. Boysen. 1990. Inferences about guessing and knowing in chimpanzees: Evidence of empathy. *J. Comp. Psychol.* 104:203–10.

Povinelli, D.J., K.A. Parks, and M.A. Novak. 1991. Do rhesus monkeys *(Macaca mulatta)* attribute knowledge and ignorance to others? *J. Comp. Psychol.* 105:318–25.

———. 1992. Role reversal by rhesus monkeys, but no evidence of empathy. *Anim. Behav.* 43:269–81.

Premack, D. 1983. Animal cognition. *Ann. Rev. Psychol.* 34:351–62.
———. 1988. "Does the chimpanzee have a theory of mind?" revisited. In R. Byrne and A. Whiten, eds., *Machiavellian Intelligence,* pp. 160–79. New York: Oxford Univ. Press.
Premack, D., and G. Woodruff. 1978. Does the chimpanzee have a theory of mind? *Behav. and Brain Sci.* 1:515–26.
Rijksen, H.D. 1978. A field study of Sumatran orangutans *(Pongo pygmaeus abelii,* Lesson 1827): Ecology, behavior and conservation. *Meded. Landbouwhogesch. Wageningen* 78 (2):1–420.
Roitblat, H.L. 1985. *Introduction to Comparative Cognition.* New York: Freeman.
Savage-Rumbaugh, E.S. 1986. *Ape Language: From Conditional Response to Symbol.* New York: Columbia Univ. Press.
Savage-Rumbaugh, E.S., and K. MacDonald. 1988. Deception and social manipulation in symbol-using apes. In R.W. Byrne and A. Whiten, eds., *Machiavellian Intelligence,* pp. 224–37, Oxford, England: Oxford Univ. Press.
van Schaik, C.P. 1989. The ecology of social relationships amongst female primates. In V. Standen and G.R.A. Foley, eds., *Comparative Socioecology, the Behavioural Ecology of Humans and Other Mammals,* pp. 195–218. Oxford, England: Blackwell Scientific Publications.
Schusterman, R.J., J.A. Thomas, and F.G. Wood, eds., 1986. *Dolphin Cognition and Behavior: A Comparative Approach.* Hillsdale, N.J.: Lawrence Erlbaum.
Seyfarth, R.M., and D.L. Cheney. 1984. Grooming, alliances and reciprocal altruism in vervet monkeys. *Nature* 308:541–3.
Shimp, C.P. 1982. Meta-knowledge in the pigeon: An organism's knowledge about its own adaptive behavior. *Anim. Learn. Behav.* 10:358–64.
———. 1983. The local organization of behavior: A dissociation between a pigeon's behavior and a self-report of that behavior. *J. Exp. Anal. Behav.* 39:61–8.
Snowdon, C.T. 1990. Language capacities of nonhuman animals. *Yearb. Phys. Anthro.* 33:215–43.
Stolba, A. 1979. *Entscheidungsfindung in Verbänden von Papio hamadryas.* Ph.D. Diss., Universität Zürich, Zürich.
Strum, S. 1981. Processes and products of change: Baboon predatory behavior at Gilgil, Kenya. In R.S. Harding and G. Teleki, eds., *Omnivorous primates: Gathering and hunting in human evolution,* pp. 255–302. New York: Columbia Univ. Press.
Suarez, S.D., and G.C. Gallup. 1981. Self-recognition in chimpanzees and orangutans, but not gorillas. *J. Human. Evol.* 12:175–88.
Teleki, G. 1973. *The Predatory Behavior of Wild Chimpanzees.* East Brunswick, N.J.: Bucknell Univ. Press.
Thierry, B. 1985. Patterns of agonistic interactions in three species of macaque *(Macaca mulatta, M. fascicularis, M. tonkeana). Aggressive Behav.* 11:223–33.
Thorpe, W.H. 1956. *Learning and Instinct in Animals.* London: Methuen.
Tomasello, M. 1990. Cultural transmission in the tool use and communicatory signaling of chimpanzees? In S. Taylor Parker and K.R. Gibson, eds., *"Language" and Intelligence in Monkeys and Apes,* pp. 274–311. Cambridge, England: Cambridge Univ. Press.
Tomasello, M., M. Davis-Dasilva, L. Camak, and K. Bard. 1987. Observational learning of tool use by young chimpanzees. *Hum. Evol.* 2:175–83.
Trivers, R.L. 1971. The evolution of reciprocal altruism. *Q. Rev. Biol.* 46:35–57.

Vehrencamp, S. 1983. A model for the evolution of despotic versus egalitarian societies. *Anim. Behav.* 31:667–82.

Visalberghi, E., and D.M. Fragaszy. 1990. Do monkeys ape? In S. Taylor Parker and K.R. Gibson, eds., *"Language" and Intelligence in Monkeys and Apes.* pp. 247–73. Cambridge, England: Cambridge Univ. Press.

de Waal, F.B.M. 1982. *Chimpanzee Politics.* London: Jonathan Cape.

———. 1986. Deception in the natural communication of chimpanzees. In R.W. Mitchell and N.S. Thompson, eds., *Deception: Perspectives in Human and Nonhuman Deceit,* pp. 221–44. Albany, N.Y.: State Univ. of New York Press.

———. 1989a. Dominance "style" and primate social organization. In V. Standen and R.A. Foley, eds., *Comparative Socioecology,* pp. 243–64. Oxford, England: Blackwell.

———. 1989b. Food sharing and reciprocal obligations among chimpanzees. *J. Human. Evol.* 18:433–549.

———. 1989c. *Peacemaking among Primates.* Cambridge, Mass.: Harvard Univ. Press.

———. 1991a. The chimpanzee's sense of social regularity and its relation to the human sense of justice. *Am. Behav. Scientist* 34:335–49.

———. 1991b. Complementary methods and convergent evidence in the study of primate social cognition. *Behav.* 118:297–320.

de Waal, F.B.M., and J.A.R.A.M. van Hooff. 1981. Side-directed communication and agonistic interactions in chimpanzees. *Behav.* 77:164–98.

de Waal, F.B.M., and L.M. Luttrell. 1988. Mechanisms of social reciprocity in three primate species: Symmetrical relationship characteristics or cognition? *Ethol. and Sociobiol.* 9:101–18.

Walker, S. 1983. *Animal Thought.* London: Routledge and Kegan Paul.

White Miles, H.L. 1990. The cognitive foundations for reference in a signing orangutan. In S. Taylor Parker and K.R. Gibson, eds., *"Language" and Intelligence in Monkeys and Apes,* pp. 511–39. Cambridge, England: Cambridge Univ. Press.

Whiten, A. 1989. Transmission mechanisms in primate cultural evolution. *Trends Evol. Ecol.* 4:61–2.

Whiten, A., and R.W. Byrne. 1988. Tactical deception in primates. *Behav. and Brain Sci.* 11:233–44.

Whiten, A., and R. Ham. 1992. On the nature and evolution of imitation in the animal kingdom: Reappraisal of a century of research. *Adv. Study Behav.* 21:239–84.

What Chimpanzees (Might) Know about the Mind

Daniel J. Povinelli

In this chapter I review the results of an ongoing comparative research program designed to determine whether primates differ phylogenetically in their ability to attribute mental states to others. I focus on chimpanzees, *Pan troglodytes,* for two reasons. First, they are the only species of great apes that have undergone systematic experimental tests for these abilities (Premack 1988a; Povinelli 1993; Whiten 1993). Second, a theoretical reason exists for this focus. Over a decade ago, for reasons that will be described later, Gallup (1982) predicted that chimpanzees (and orangutans) ought to differ markedly from other nonhuman primates in their natural inclinations to attribute mental states such as intention and knowledge to other organisms. In addition to attempting to reconstruct the evolution of mental-state attribution in general, much of my own research has focused on testing Gallup's idea. Although our results are somewhat ambiguous, it is fair to conclude that chimpanzees, but no monkey species tested to date, have provided intriguing evidence that they may be reasoning about more than just the observable world.

Monkeys, Apes, Mirrors, and Minds

Gallup (1970) reported that after several days of exposure to mirrors, chimpanzees show evidence that they successfully identify the sources of their mirror images. Initially, the four chimpanzees he tested (separately) behaved as if they were in the presence of other chimpanzees. By day two or three, however, these social behaviors declined and were replaced by efforts to use the mirror as a tool to explore themselves in ways they had never done before, such as making exaggerated facial expressions and using

their fingers to inspect their teeth, eyes, and anal-genital region (see figure 1). After 10 days of such exposure, Gallup anesthetized each subject and applied an odorless, red dye that could not be felt to an upper eyebrow ridge and the opposite ear. Upon recovery, the subjects were observed for 30 minutes without the mirror present. The chimpanzees made virtually no attempts to touch the surreptitiously applied marks, presumably because they were unaware of them. The mirror was then reintroduced, and the subjects made a number of attempts to touch the marks that they could now see on themselves with the aid of the mirror. Gallup (1970) concluded that the chimpanzees had recognized themselves and that this recognition seemed to imply the presence of at least a rudimentary self-concept. Several years later the phenomenon was extended to include orangutans (Lethmate and Dücker 1973; Suarez and Gallup 1981).

Compared with these findings on chimpanzee and orangutan self-recognition, other species tested have provided little in the way of convincing evidence for that capacity. To date, studies of nearly 20 species of Old and New World monkeys, lesser apes, and even gorillas[1] have failed to show compelling evidence of the emergence of spontaneous self-exploratory behaviors using a mirror as a guide. Nor have these subjects detected the presence of experimentally placed marks on their faces with the aid of a mirror (Anderson 1984; Gallup 1991). Although it is tempting to dismiss these results as reflecting species differences in domains other than

Figure 1
A four-year-old chimpanzee displays signs of self-recognition by using a mirror to explore parts of herself which are otherwise invisible. Photo by D. Bierschwale.

cognition (i.e., motivation), these kinds of interpretations have not fared well against the often ingenious (but unsuccessful) efforts to prompt species other than chimpanzees and orangutans to display evidence of self-recognition (Gallup 1991).

Reviewing early findings, Gallup (1982, 1983) developed a model to further test his interpretation of the chimpanzee's and orangutan's capacity for self-recognition. Adopting a Cartesian epistemological argument, Gallup reasoned that if chimpanzees possessed some knowledge of self (as deduced from their ability to recognize themselves), then they might be able to infer analogous knowledge in others. Although the extent of self-knowledge possessed by chimpanzees was unclear from his work on self-recognition, Gallup speculated that chimpanzees might use their own experiences as a means of modeling the likely experiences of others, and thus be able to attribute intentions, desires, plans, and knowledge to other individuals.

In search of evidence in support of his theory, Gallup (1982) turned to the landmark set of investigations of a chimpanzee theory of mind conducted by Premack and Woodruff (1978). Their experimental evidence that an adult chimpanzee might be reasoning about unobservable intentions of human actors represented exactly the kind of ability he suspected was widespread in a species showing evidence of self-recognition. But far more startling than Gallup's predictions about chimpanzees were his companion predictions about lesser apes and monkeys. The inability of these primates to recognize themselves suggested to Gallup that they had no access to their own minds and, therefore, had no access to the minds of others. His model thus predicted that chimpanzees and orangutans should succeed on tests designed to tap into an understanding of the mental world of desires, intentions, and beliefs, whereas other species should not succeed.

Initial Tests of the Model

Our initial strategy to test Gallup's model was to compare two species using tests related to mental state attribution. We chose chimpanzees because of their widely replicated ability to recognize themselves in mirrors and rhesus monkeys because of their similarly widely replicated inability to do so. We thus set out to test the null hypothesis that there were no species differences between chimpanzees and rhesus monkeys in their abilities to pass tests of mental state attribution.

Role Reversal as a Measure of Understanding Personal Agency

In our first experiments, we tested four chimpanzees (8, 9, 10 and 28 years old) and four rhesus monkeys (14 to 20 years old) to determine whether they could understand other individuals and themselves as intentional agents.

In humans, this requires the ability to understand that others are animate beings who can control their own behavior. This ability may develop from early forms of role playing and empathy and begins to emerge in infants at an explicit level by about 16–18 months and then undergoes further refinement and elaboration (Poulin-DuBois and Schultz 1988).

We reasoned that we could test for this limited understanding of the mental world by pairing each chimpanzee or rhesus monkey subject with a human partner and creating two distinct roles, one an operator role and the other an informant role. The subject and the human partner were seated opposite each other with a cooperative apparatus between them. All the subjects were initially trained to pull handles that controlled movement of food trays, some of which contained a small food reward. Once the subjects could accurately pull the handle corresponding to the baited tray, the chimpanzee and rhesus monkey subjects were divided into two groups (operators and informants). The apparatus was then modified by placing opaque shields that prevented the operators from seeing which trays contained food. From the informant side, the baited trays were perfectly visible, but the informant had no handles to bring the trays within reach. The operators needed to respond to the pointing of their human partner in order to obtain the food reward. Gradually the operators learned to do so with near-perfect accuracy, thus delivering a food reward to both participants. The informants, likewise, gradually learned to produce a pointing or reaching gesture in front of the correct food trays, thus enabling their naive human partners to pull the correct handle. We continued training the subjects (chimpanzee and rhesus monkey alike) until they were performing their operator or informant roles at near-perfect levels.

Up to this point our data could tell us very little. Although members of both species were cooperating with a human partner, we had no evidence that they understood anything at all about what they were doing. Did the chimpanzees and rhesus monkeys understand that they and their partners were acting as autonomous agents, directing and coordinating their behavior to achieve a desirable outcome? Or were they were merely responding to the imposed reward contingency of the setup we had devised? We asked them this question by staging two critical sessions of *role reversal:* operators were required to become informants and informants to become operators. The apparatus was set up as usual, but then slowly turned around as the subjects watched; thus, each subject was now confronted with the task from the opposite perspective. Gallup's model predicted that the chimpanzees would understand the reversal and realize that it was now their turn to perform the role originally performed by the human agent. In contrast, the model predicted that the rhesus monkeys would initially respond at chance levels and only gradually learn the new role as the reward contingencies became explicit. These are exactly the results we obtained. Three of the four chimpanzees showed clear evidence of understanding their new role, whereas none of the rhesus monkeys did (Povinelli, Nelson, and Boysen 1992; Povinelli, Parks, and Novak 1992; Hess et al. 1993). We were thus able to reject the null hypothesis—at least as far as this test was concerned.

Understanding That Seeing Leads to Knowing

The same subjects were next tested to determine whether they understood that the observable behavioral act of *seeing* creates an internal, unobservable mental-state of *knowing*. We reasoned that we could have the subjects observe two human actors, one of whom (the knower) would hide food under one of several cups, and one of whom (the guesser) would be outside the room during the hiding procedure. The subjects would see that the knower had food and was placing it in one of several cups, but a screen would prevent them from seeing exactly which cup held the food; they would also see that the guesser had left the room. The guesser would return and the screen would be removed. Then, the guesser and knower would simultaneously offer the chimpanzees "advice" (by pointing) at a cup. The knower would point to the correct cup; the guesser would point to an incorrect cup (see figure 2a–c). If subjects understood that seeing results in knowing, they would follow the advice of the knower.

Our subjects were administered 10 trials a day over several weeks in which the roles of the guesser and knower were randomly altered.

Figure 2
This procedure is used to determine if chimpanzees understand the connection between seeing and knowing:

A. chimpanzee sees guesser leave the room while knower hides food, or

B. chimpanzee observes as knower watches third experimenter hide food while guesser's vision is obstructed, and

C. guesser and knower offer advice to chimpanzee.

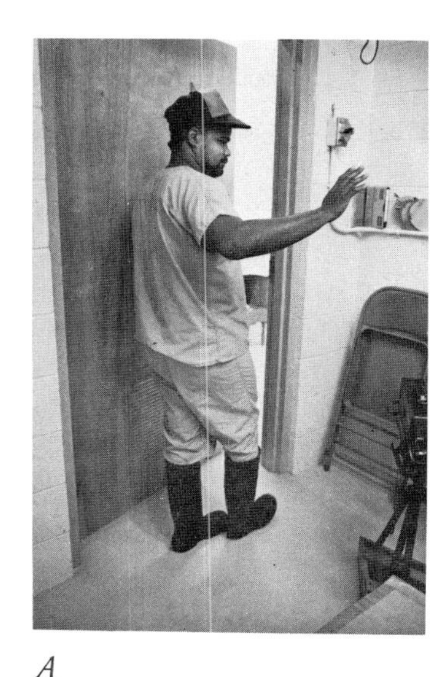

A

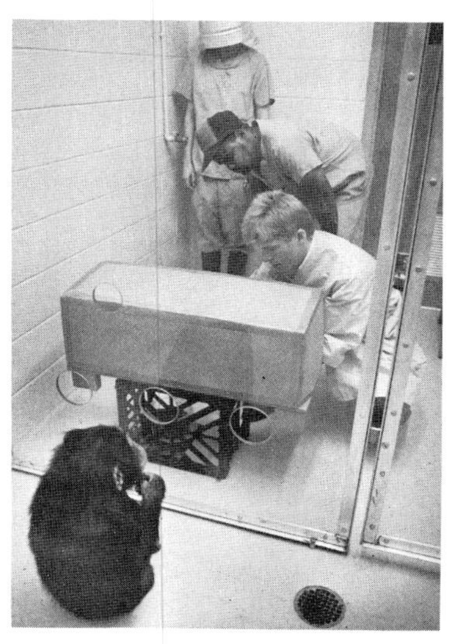

B

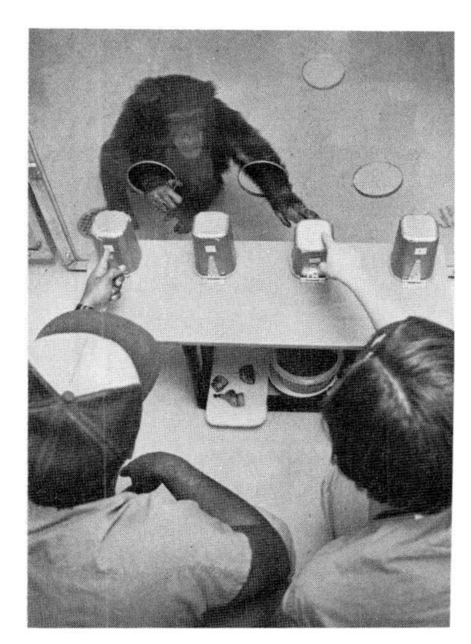

C

At least three of the chimpanzees showed a fairly stable selection of knower's advice during the initial several weeks of testing. However, there was some reason to think that they may have been solving the task another way. Instead of realizing that the knower knew something that the guesser did not, the chimpanzees may have been learning a behavioral role such as "pick the person who stays in the room." To probe their understanding further, we gave them three test sessions in which we altered the relevant behavioral variables but kept the epistemological relations intact: instead of hiding the food, the knower merely watched as a third person hid the food; instead of leaving the room, the guesser stayed inside, next to the knower, but placed a paper bag over his or her head while the food was being hidden (see figure 2b). The chimpanzees who had made the earlier discrimination showed reasonably good transfer into this procedure (Povinelli et al. 1990). Premack (1988a) has briefly reported analogous experimental results with four six- to seven-year-old chimpanzees.

The rhesus monkeys, on the other hand, never made a discrimination between the guesser and knower despite our best efforts to make the cues as obvious as possible. They did try a number of strategies, such as picking the person on the right or left, or picking the experimenter wearing a glove on and so on. But the manipulations that resulted in denying one person visual access to the hiding procedure did not seem to have any significance in terms of guiding their searches for the food (Povinelli et al. 1991). Cheney and Seyfarth (1990), working with Japanese macaques, have reported similar findings using different experimental approaches.

In an effort to understand whether our task was really tapping into an ability to understand the seeing-knowing relationship, we gave the same test to three- and four-year-old human children (Povinelli and deBlois 1992). Previous research by other investigators using analogous linguistic measures had found that young children did not appreciate the seeing-knowing connection until about four years of age (Wimmer et al. 1988; Gopnik and Graf 1988; O'Neill and Gopnik 1991). When we administered our task, the three-year-old performed like the rhesus monkeys described above. In contrast, the four-year-old performed at near-perfect levels from the first trial forward. Thus, we felt reasonably confident that the nonverbal test given to the chimpanzees and rhesus monkeys was measuring the seeing-knowing relationship. On the other hand, our chimpanzees never performed as well as the four-year-olds. Thus, determining whether the chimpanzees had attributed the mental states of knowledge and ignorance to the actors or simply, had learned the difference using behavioral cues, must await experiments with additional controls designed to tease these issues apart.

The Ontogeny of Mental State Attribution

As I was conducting these original tests of a chimpanzee theory of mind, I began to reconsider questions concerning the evolutionary history of mental-state attribution. Just as physical anthropologists attempt to determine the timing and order of the development of key morphological traits (brain size, bipedalism, reduced prognathism) important in human evolution, I realized that ultimately our research could allow us to reconstruct (through cladistic inference) the likely psychology of mental state attribution possessed by the ancestor of the great ape-human clade.

In considering how to achieve this broader goal, I realized that our approach (and Gallup's model) suffered from some very serious limitations that needed to be addressed. The picture from developmental psychology strongly suggests that the emergence of mental state attribution in human children does not appear as a uniform package. Rather, individual aspects related to intention and to knowledge may develop at different rates. For instance, although young two-year-olds display some understanding of the mental world—especially as related to desires and, later, intentions—they still do not have a very good understanding of knowledge or belief. [For good overviews of the development of the child's theory of desire and belief, see Astington and Gopnik (1991); Perner (1991); Wellman (1990).]

The fact that components of mental state attributions can be dissociated during development raises several distinct possibilities with regard to their evolution. First, chimpanzees and humans may share developmental pathways related to mental state attribution only up to the point of self-recognition. In other words, human psychology may have been modified from an ancestor that reached a terminal point of conceptual development akin to that possessed by chimpanzees and orangutans.[2] Of course, this possibility assumes that, despite a good deal of morphological evolution, chimpanzees and orangutans have undergone relatively little psychological evolution since their initial divergence. This need not be true. A second possibility is that chimpanzees and orangutans, as well as humans, have undergone unique psychological evolution within their respective lineages, thus reaching derived (and different) psychological terminal points in the different species. A final possibility is that the ancestral condition of the last common ancestor of the great ape-human lineage was more derived than that displayed by children, chimpanzees, or orangutans at the point that they display self recognition. In other words, humans, orangutans, and chimpanzees may share an extensive overlap in their development of mental state attribution as the result of the inheritance of an extensive pleisiomorphic condition.

The consideration of these three possibilities has direct and important implications for research into a chimpanzee's theory of mind. It means that comparative ontogeny must become as important a tool in psychological evolutionary reconstructions as it has been in morphological evolutionary reconstructions (Parker and Gibson 1979; Parker 1990). These considerations argue in favor of following cohorts of relevant species (for example, chimpanzees) through critical landmarks in psychological development. If the onset of robust ontogenetic landmarks could be mapped, then tests of mental state attribution could be administered both before and after their emergence. For example, by testing young chimpanzees for mental state attribution both before and after they develop the capacity for mirror self-recognition, we could perhaps determine in a more straightforward manner whether Gallup's interpretation of mirror self-recognition is of heuristic import. In addition, and quite independent of Gallup's model, such tests (if conducted comparatively with young children) could produce a fine-grained picture of the ontogeny of these abilities, as well as a clearer picture of which abilities are shared, ancestral character states and which are exclusively derived in the various lineages.

Theory of Mind Ontogeny in Chimpanzees and Children

Recently, we have adapted the strategy described above in a further effort to understand exactly what chimpanzees know about the mind. First, we are utilizing cross-sectional methods with both young children and mature and immature chimpanzees in order to develop additional and more refined measures of mental state attribution. Second, we have been following a cohort of seven young chimpanzees from about 36 months of age.

Parameters of Self-Recognition

To begin, we have tried to determine the ontogenetic parameters of self-recognition in chimpanzees. As we have discussed, we seek these parameters for both theoretical and practical reasons. First, at what age do most chimpanzees develop the capacity to recognize themselves in a mirror? Second, do all chimpanzees develop the capacity, or just some? Third, how long does it take for mature, mirror-naive chimpanzees to recognize themselves in mirrors? In order to answer these questions, we have adopted both cross-sectional and longitudinal methods.

The results of our initial, cross-sectional research program, involving over 100 socially housed chimpanzees ranging in age from 10 months to nearly 40 years of age, revealed some surprising patterns (Povinelli et al.

1993). First, in a cross-sectional study of almost 50 animals younger than 6 years of age, only a few (less than 10%) showed compelling evidence of self-recognition. In contrast, nearly 80% of the subjects between 7 and 15 years of age showed very clear signs of self-recognition. Although this finding does not fit with previous results using much smaller sample sizes (Lin et al. 1992), we were able to replicate our findings by intensively studying six three- and four-year-old chimpanzees who (with one precocious exception—a three-year-old female) initially tested negative for self-recognition. Despite weeks of mirror exposure in a variety of settings, the five negative animals displayed no signs of recognition nor did they pass a controlled mark test similar to that described earlier in this paper. In contrast, the single positive (control) subject showed clear self-exploratory behaviors using the mirror and passed the mark test. A longitudinal research program is currently underway, and with only a couple of notable exceptions, the ontogenetic patterns detected in our cross-sectional research program appear fairly robust (Povinelli et al. 1993). It is still too early to definitively state the typical age at which the capacity for self-recognition develops in chimpanzees. Povinelli et al. (1993) discusses factors that may affect the rate of development of this capacity. Nonetheless, it now appears that the onset of the capacity is markedly later than in humans, who show evidence of self-recognition by about 18 to 24 months of age (Lewis and Brooks-Gunn 1979).

In addition, our results reveal that, contrary to the findings of Gallup and his colleagues, chimpanzees do not require two or three days of mirror exposure before they recognize themselves. Most of our chimpanzees that showed clear behavioral patterns indicating self-recognition, did so within 30 to 40 minutes of mirror presentation. This difference may be the result of methodology; our chimpanzees were tested while in their social groups, whereas Gallup's were not. Also, contrary to the data presented by Swartz and Evans (1991), our cross-sectional results suggest that most chimpanzees do display evidence of self-recognition, but that this capacity (or the expression of the capacity) declines in midadulthood. The extent to which this decline can be written off as a motivational difference depends on certain other factors. For instance, one could imagine that adults might simply be less interested in their images in mirrors. However, adults who do not show signs of self-recognition spend as much or more time staring into the mirror as do the younger animals (or even their same-age peers) who do show evidence of self-recognition (Povinelli et al. 1993).

Understanding the Perceptual Sources of Knowledge

Age differences in self-recognition allowed us to test a prediction, derived from the developmental framework, that young chimpanzees who did not test positive for self-recognition would likewise not show signs of understanding the perceptual sources of knowledge. We thus tested six three- and four-year-old chimpanzees on the seeing-knowing paradigm; these subjects were the five who had previously tested negative for self-recognition and the one precociously positive female. Our predictions for the precocious subject were uncertain because, as discussed earlier, young children do not appear to understand the seeing-knowing relationship until about four years of age, or roughly two years after the onset of self-recognition. Thus, although this young chimpanzee was capable of self-recognition, she may well have developed only marginally (or not at all) past the point of children that are 18- to 24-months old.

This investigation revealed that none of the six subjects discriminated between the guesser and knower in any of the variations we presented to them (Povinelli et al. 1994). Indeed, the overall pattern of results mirrored the pattern we had obtained with rhesus monkeys several years earlier. Thus, consistent with both Gallup's model and with the developmental pattern in children, the absence of self-recognition was associated with an undeveloped understanding of the perceptual sources of knowledge. This finding will achieve far greater importance if, as these subjects mature and become capable of self-recognition, we subsequently obtain data that they are able to understand the epistemological distinction between the guesser and the knower.

Assessing One's Own Knowledge

Although we have yet to obtain definitive evidence that chimpanzees understand *how* knowledge states arise (an ability that would be required for success on the seeing-knowing task), it has occurred to us that chimpanzees (and young children) may first understand less sophisticated aspects of knowledge. For example, even though they may not possess an accurate understanding of how knowledge states arise, they may at least understand that such states exist (Premack 1988b). How might we go about testing them for such ability? Imagine that a subject observes an experimenter hiding food under one of several cups. The subject can see that the food is being hidden but cannot determine exactly which cup is baited because of a screen obscuring the cups. The screen is then removed, revealing the experimenter pointing to the correct cup. The response latency of the subject—that is, the time (in seconds) it takes the subject to look under a cup after the screen is removed—can be measured. We predicted that at some point after young children display evidence of self-recognition, but before they pass the seeing-knowing test, young children (and perhaps chimpanzees) would begin to hesitate on probe trials where the experimenter refrained from pointing. In other words, at some point the young subjects should realize that they are uncertain of the location of the reward and, hence, should show a marked hesitation (latency) before looking under a cup, or perhaps, simply refuse to search altogether. That is, the young child or young chimpanzee will show an ability to assess the state of its own knowledge and act accordingly.

We have tested both young children and young chimpanzees using this paradigm (Povinelli et al. unpubl. manu.). Initially, we tested 48 children ranging in age from 23 to 48 months. Consistent with our prediction, before about 28 months of age, the children did not discriminate between trials in which the experimenter pointed and trials in which he did not. After that age, children began to show much longer response latencies on average, sometimes coupled with verbal inquiries such as, "Where is it?" or "I don't know!" Figure 3 displays the critical age transition—somewhere between 23 and 34 months of age. We have undertaken additional tests to rule out less demanding interpretations of these response latencies.

Thus far young chimpanzees show no evidence of discriminating between conditions in which information is provided about which cup contains food and conditions in which such information is not provided. For example, just like the younger children, chimpanzees showed no increased hesitation on trials in which no pointing was provided to guide

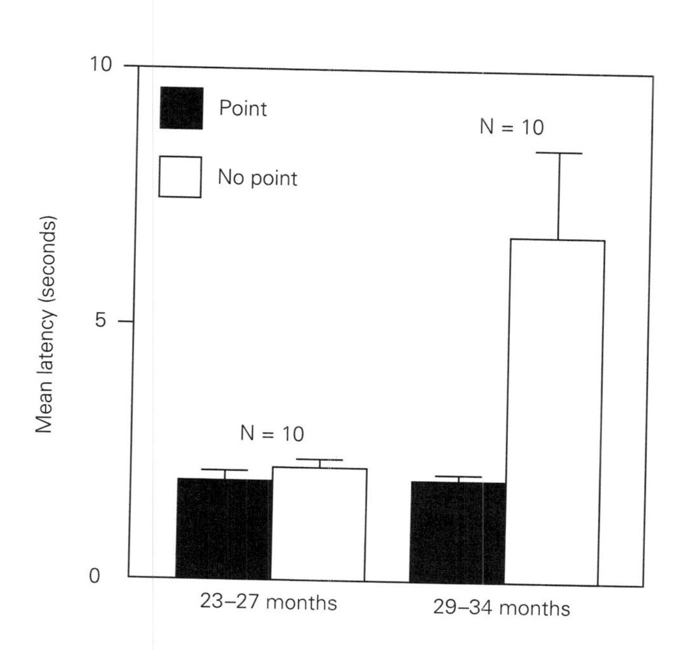

Figure 3
Self-knowledge assessment tasks show the development of response latency in young children. Data show the development of hesitation on trials when experimenter refrains from pointing to a cup, "no point trials", thus depriving a child of knowledge as to the location of a reward. Data represent mean period of hesitation (latency) on "point trials", (±SE of the mean) compared to "no point trials". Children appear to begin to discriminate between the conditions by about 28 months of age. N refers to number of subjects.

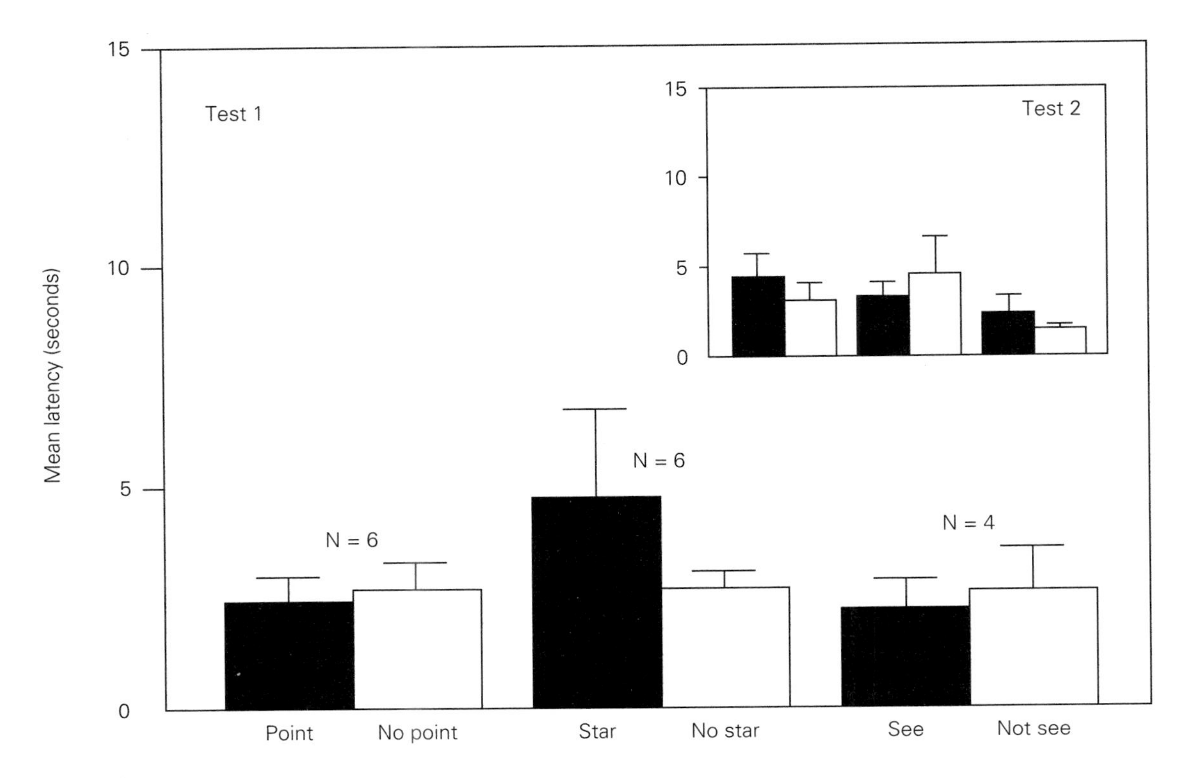

Figure 4
Self-knowledge assessment is absent in young chimpanzees. Subjects did not hesitate longer on probe trials where information necessary to solve the task was eliminated (no point, no star, not see) versus baseline trials when information was provided (point, star, see). N refers to number of subjects.

their searches. Likewise the young chimpanzees did not hesitate longer on trials in which a green star (that they had been reliably using to find the correct cup) was removed from the cups. Finally, they did not hesitate longer on trials when they did not see which cup was baited as opposed to trials when they directly witnessed the experimenter's placement of the reward (see figure 4). We obtained essentially the same results in a second administration of these proceeds. This is not to say that they never will make such discriminations but at this point in their development, chimpanzees do not yet appear to discriminate between situations in which they know the answer and situations in which they do not. This finding is consistent with the ontogenetic pattern detected in our cross-sectional research with young children; for instance even children who recognize themselves may still be too young to assess their own knowledge in this rudimentary manner.

Mature Versus Immature Self-Recognition

Finally, although chimpanzees may have some awareness of mental states in themselves and in others, do they conceive of themselves (and others) as enduring entities through space and time? Although they may well possess some kind of on-line, or moment-by-moment, representation of certain aspects of themselves (for example, of themselves as intentional agents), do they, like humans, conceive of themselves as enduring and changing through time? Consider the following paradigm as one method of determining this. Imagine that a chimpanzee, sophisticated in recognizing and exploring herself with a mirror, were to witness a trainer approaching her cage for an annual physical examination. And imagine that the event was unusual in some fashion; perhaps the trainer is wearing a bright orange costume. Imagine further that the trainer sedates the animal for the examination, and while the animal is under anesthesia, the animal's hair is dyed pink. Finally, the animal recovers, none the wiser about her new hair color. Suppose now that the chimpanzee is given the opportunity to view a short videotape of the entire procedure, beginning with the trainer's approach to the cage in the strange costume and ending with the hair dye. How would the chimpanzee react to witnessing such a video later that same day? Upon observing the result of the hair dye, would she reach up to inspect her head? Would she realize that what she was witnessing on tape happened just that morning and that her hair might still be pink?

Although we have begun to pilot aspects of this procedure with chimpanzees, we realized that first we needed to know how young children would react to such a manipulation. The children, of course, were not sedated, but an analogous paradigm was used. Forty-two two-, three-, and four-year-old children were videotaped while playing an out-of-the-ordinary game with an adult. During the game, the experimenter surreptitiously placed a large sticker in each child's hair, just above the forehead, in the context of praising them. After an observer made certain the children were not aware of the sticker, they were invited to watch a videotape of what they had just been doing. Despite their capacity for mirror self-recognition, none of the two-year-olds, and only about 25% of the three-year-olds (the older ones) reached up to search for the sticker when they saw it being placed in their hair, despite the fact that the whole scene occurred just several minutes earlier. In contrast, nearly all the four-year-olds reached up

Table 1
Results of mature self-recognition with young children.

	Children who passed test			
	Videotape		Photo	
Age	%	N	%	N
2	0	(0/10)	—	—
3	25	(4/16)	25	(4/16)
4	75	(12/16)	100	12/12)

Notes: Age ranges: 2-year-olds = 22–34 months; 3-year-olds = 35–46 months; 4-year-olds = 47–60 months. Age transition is highly significant for both tasks (videotape task, Fisher's exact test, $p < .001$; photo task, $\chi^2 = 16.56$, $df = 2$, $p < .001$).

within seconds of seeing the marking event. We have replicated this same effect using Polaroid pictures.

Although it may be tempting to think that children simply cannot recognize things in photos or on videotape, all the children were capable of identifying themselves in the photos or videotapes when the experimenter asked, "Who is that?" Do the children simply not understand the physical technology of these media? Probably not, but they do not understand the physical properties of mirrors either, yet this does not prevent them from using them appropriately. We speculate that the difference is a conceptual one, related to other broader shifts in their representational systems. The result is that children younger than about three-and-a-half have a very good on-line representation of themselves, and they also have access to previous memories of themselves, but they are very poor at integrating the two. As a profound corollary, they may not yet understand their own ontogeny. Obviously, testing chimpanzees of various ages on this task is a high priority because a subject's performance appears to be quite a robust marker of a more mature representation of self than mirror self-recognition alone.

Summary: Do Chimpanzees *Really* Attribute Mental States?

At present it is impossible to say with much definitiveness what chimpanzees really know about the mind. They clearly act *as if* they understand intention, knowledge, and possibly belief; but the extent to which that behavior is supported by learned or inferred behavioral rules remains uncertain (Cheney and Seyfarth 1990; Premack 1988a; Whiten 1993; Povinelli 1993). Data collected from spontaneous social interactions in captivity and in the field suggest that chimpanzees as well as other primates possess a very complex psychology that may take into account the motives, intentions, and the knowledge possessed by competitors and allies (de Waal 1982; Whiten and Bryne 1988). Ultimately, experimental approaches of the type pioneered by David Premack and Gordon Gallup, which we have been attempting to elaborate and extend, will help us to place constraints upon the kinds of interpretations that make sense in such contexts. These investigations, coupled with parallel research using young children and other primates, will provide a more meaningful answer to the question implicit in this chapter's title.

Endnotes

1. Gorillas may require special explanation. Povinelli (1993) has argued that gorillas may have undergone a secondary reversal in important aspects of their cognitive development as the result of selective pressure favoring early, rapid

physical maturation. This shift in development may have important implications for their typical failure to express evidence of self-recognition.

2. This does not assume that such modifications were tacked on to the end of the existing developmental program of the last common ancestor of the greater ape-human clade. Human language, for instance, was clearly not simply a terminal addition to the primitive, cognitive developmental program of humans and apes, because children begin to utter simple words and sentences before they display evidence of self-recognition.

References

Anderson, J.R. 1984. The development of self-recognition: A review. *Developmental Psychobiol.* 17:35–49.

Astington, J.W., and A. Gopnik. 1991. Theoretical explanations of children's understanding of the mind. *British J. Developmental Psychol.* 9:7–31.

Cheney, D.L., and R.M. Seyfarth. 1990. Attending to behaviour versus attending to knowledge: Examining monkeys attribution of mental states. *Anim. Behav.* 40:742–53.

Gallup, G.G., Jr. 1970. Chimpanzees: Self-recognition. *Science* 167:86–7.

———. 1982. Self-awareness and the emergence of mind in primates. *Am. J. Primatol.* 2:237–48.

———. 1983. Toward a comparative psychology of mind. In R.E. Mellgren, ed., *Animal Cognition and Behavior*, pp. 473–510. New York: North-Holland.

———. 1991. Toward a comparative psychology of self-awareness: Species limitations and cognitive consequences. In G.R. Goethals and J. Strauss, eds., *The Self: An Interdisciplinary Approach.* New York: Springer-Verlag.

Gopnik, A., and P. Graf. 1988. Knowing how you know: Young children's ability to identify and remember the sources of their beliefs. *Child Development* 59:1366–71.

Hess, J., M.A. Novak, and D.J. Povinelli. 1993. 'Natural pointing' in a rhesus monkey, but no evidence of empathy. *Anim. Behav.* 46:1023–5

Lethmate, J., and G. Dücker. 1973. Untersuchungen am sebsterkennen im spiegel bei Orangutans einigen anderen Affenarten (Self-recognition by orangutans and some other primates.) *Z. Tierpsychol.* 33:248–69.

Lewis, M., and J. Brooks-Gunn. 1979. *Social Cognition and the Acquisition of the Self.* New York: Plenum Press.

Lin, A.C., K.A. Bard, and J.R. Anderson. 1992. Development of self-recognition in chimpanzees *(Pan troglodytes). J. Comp. Psychol.* 106:120–7.

O'Neill, D.K., and A. Gopnik. 1991. Young children's ability to identify the sources of their beliefs. *Developmental Psychol.* 27:390–7.

Perner, J. 1991. *Understanding the representational mind.* Cambridge, Mass.: MIT.

Parker, S.T. 1990. The origins of comparative evolutionary developmental studies of primate mental abilities. In S.T. Parker and K.R. Gibson, eds., *"Language" and Intelligence in Monkeys and Apes.* pp. 3–63. New York: Cambridge Univ. Press.

Parker, S.T., and K.R. Gibson. 1979. A developmental model for the evolution of language and intelligence in early hominids. *Behav. and Brain Sci.* 2:367–408.

Poulin-DuBois, D., and T.R. Schultz. 1988. The development of the understanding of human behavior: From agency to intentionality. In J.W. Astington, P.L. Harris, and D.R. Olson, eds., *Developing theories of mind,* pp. 109–25. New York: Cambridge Univ. Press.

Povinelli, D.J. 1993. Reconstructing the evolution of mind. *Am. Psychol.* 48:493–509.

Povinelli, D.J., and S. deBlois. 1992. Young children's *(Homo sapiens)* understanding of knowledge formation in themselves and others. *J. Comp. Psychol.* 106:228–38.

Povinelli, D.J., K.E. Nelson, and S.T. Boysen. 1990. Inferences about guessing and knowing by chimpanzees *(Pan troglodytes). J. Comp. Psychol.* 104:203–10.

———. 1992. Comprehension of social role reversal by chimpanzees: Evidence of empathy? *Anim. Behav.* 43:633–40.

Povinelli, D.J., K.A. Parks, and M.A. Novak. 1991. Do rhesus monkeys *(Macaca mulatta)* attribute knowledge and ignorance to others? *J. Comp. Psychol.* 105:318–25.

———. 1992. Role reversal by rhesus monkeys, but no evidence of empathy. *Anim. Behav.* 44:269–81.

Povinelli, D.J., A.B. Rulf, and D. Bierschwale. 1994. Absence of knowledge attribution and self-recognition in young chimpanzees *(Pan troglodytes). J. Comp. Psychol.* 108:74–80

Povinelli, D.J., A.B. Rulf, K. Landau, and D. Bierschwale. 1993. Self-recognition in chimpanzees *(Pan troglodytes):* Distribution, ontogeny, and patterns of emergence. *J. Comp. Psychol.* pp. 347–72.

Premack, D. 1988a. "Does the chimpanzee have a theory of mind?" revisited. In R. Byrne and A. Whiten, eds., *Machiavellian Intelligence,* pp. 160–79. New York: Oxford Univ. Press

———. 1988b. Minds with and without language. In L. Weiskrantz, ed., *Thought Without Language,* pp. 46–65. Oxford: Clarendon Press.

Premack, D., and G. Woodruff. 1978. Does the chimpanzee have a theory of mind? *Behav. and Brain Sci.* 1:515–26.

Suarez, S.D., and G.G. Gallup, Jr. 1981. Self-recognition in chimpanzees and orangutans, but not gorillas. *J. Human Evol.* 10:175–88.

Swartz, K.B., and S. Evans. 1991. Not all chimpanzees show self-recognition. *Primates* 32:483–96.

de Waal, F.B.M. 1982. *Chimpanzee Politics.* Baltimore: Johns Hopkins Univ. Press.

Wellman, H.M. 1990. *The Child's Theory of Mind.* Cambridge, Mass.: Bradford.

Whiten, A. 1993. Evolving theories of mind: The nature of non-verbal mentalism in other primates. In S. Baron-Cohen, H. Tager-Flusberg, and F. Volkmar, eds., *Understanding Other Minds: Perspectives From Autism,* pp. 367–96. Oxford, England: Oxford Univ. Press.

Whiten, A., and R.W. Byrne. 1988. Tactical deception in primates. *Behav. and Brain Sci.* 11:233–44.

Wimmer, H., G.J. Hogrefe, and J. Perner. 1988. Children's understanding of informational access as a source of knowledge. *Child Development.* 59:386–96.

The Question of Chimpanzee Culture

Michael Tomasello

In the early days of chimpanzee fieldwork, researchers talked of *precultural* or *protocultural* behaviors (Menzel 1973b). More recently the prefixes have been dropped, and talk has turned to *culture, cultural traditions,* and *cultural transmission* (Goodall 1986; Nishida 1987; McGrew 1992; Boesch 1993). The terminology has changed, at least partly in response to the growing list of ontogenetically acquired, population-specific behavioral traditions observed in different chimpanzee communities. Many of these behaviors are documented in this volume. It is also likely that the meanings of culture and cultural transmission have been broadened over the past two decades, inspired by the writings of evolutionary biologists who see evidence of culture in a variety of social activities in all kinds of animal species (Bonner 1980; Boyd and Richerson 1985). Unfortunately, this shift in meaning has led to some unproductive debates in which different researchers use the same words to talk about different things.

I have argued (1990) that the most productive way to approach the question of chimpanzee culture is to investigate the processes of social learning. Different animal species engage in different processes of social learning. Regardless of which of these learning processes we choose to call *cultural,* we must distinguish them if we hope to understand the way particular animal species organize their social activities and the way the young of the species enter into these activities. The tools for making such distinctions have only recently become available. A very large body of research on the social learning of all kinds of animal species is now accumulating and, as a result, we have both a wider range of empirical observations and more detailed theoretical analyses of the various ways that animals may learn from one another (Zentall and Galef 1988). Of special interest in the current context are studies of the social learning of primates, including monkeys of various types (Cheney and Seyfarth 1990; Visalberghi and Fragaszy 1990), gorillas (Byrne 1992), orangutans (Russon and Galdikas 1993), chimpanzees (Tomasello 1990), and human beings of

various ages (Meltzoff 1988). The key theoretical advance made by this group of researchers is the analysis of social learning into a number of related but distinct processes, including social facilitation, response facilitation, matched-dependent learning, response matching, local enhancement, stimulus enhancement, mimicking, emulation learning, action-outcome contingency learning, imitation, program-level imitation, movement imitation, and imitative learning (Whiten and Ham 1992.) Although much work remains to be done, this analysis now makes available to researchers a variety of distinctions and hypotheses that must be tested before precise learning processes may be attributed to a particular animal species or group.

Within this emerging paradigm, the question of whether chimpanzees have cultures may be reformulated initially as a straightforward question about which of the various underlying processes of social learning chimpanzees employ in the acquisition and maintenance of population-specific behavioral traditions. In addition, recent approaches to the study of social learning have found that, to identify specific types of social learning, it is useful, if not necessary, to make cross-species comparisons. When the issue is cultural transmission, a particularly important comparison is, of course, to humans. The concept of culture was specifically formulated to describe group differences in human behavior and, thus, behavioral traditions of humans provide the prototypical case of cultural transmission. In this comparative perspective, posing the question of chimpanzee culture in terms of underlying processes of social learning is facilitated and enlightened by an explicit comparison to the processes of social learning underlying human cultural traditions. In any case, my investigation of chimpanzee culture employs both a focus on social learning and an explicit comparison to humans.

Given this overall formulation, my strategy in this chapter is to address the question of chimpanzee culture in three ways. First, I review recent research on the types of social learning that chimpanzees and human children employ in two domains of particular importance to discussions of chimpanzee culture: tool use and communicatory gesturing. Second, I attempt to clarify some of the issues involved in the study of chimpanzee social learning and cultural traditions through a brief look at research using captive chimpanzees raised in humanlike cultural environments. Third, I identify some key characteristics of human cultural traditions and ask if chimpanzee behavioral traditions display these same characteristics.

Tool Use

A number of population-specific traditions of tool use have been documented for either one or a few chimpanzee communities. These traditions include termite fishing, ant fishing, ant dipping, nut cracking, and leaf sponging (McGrew 1992, this volume.) Researchers such as Goodall (1986), Nishida (1987), Boesch (in press), and McGrew (1992) claim that specific tool-use practices are *culturally transmitted* between the individuals within the various communities; Goodall, Nishida, and Boesch use the process of imitation as a key part of their explanation. However, these pioneering field researchers have not had available for their analyses the broad range of hypothesized learning processes that are now employed by a new wave of social-learning researchers. Thus, it is unclear from the perspective of the new and more differentiated point of view what specific processes of social learning are at work in the chimpanzee communities that traditionally use tools.

Animals may learn in social situations in a number of ways. Especially important for current purposes is *local enhancement* or *stimulus enhancement,* in which one individual draws the attention of another to a locale or to a stimulus and, thus, increases the probability of successful discovery and learning by the naive individual. For example, one chimpanzee might use a stone hammer and anvil to crack open nuts and then leave the stone tools, and some nuts, and some broken nut shells all in one place. This behavior might facilitate individual learning of tool use by group mates, especially in concert with local enhancement. In fact, based on detailed observations of individual learning, this is precisely the hypothesis of Sumita et al. (1985) concerning the limited spread of nut cracking in their captive

Figure 1
Juvenile male chimpanzee fishing an object from the moat at the Arnhem Zoo. Photo by Frans de Waal.

group. This finding of the limited spread of nut cracking is especially important given the many laboratory studies showing that chimpanzees *can* learn to use many types of tools individually, without any observation of others using tools (Beck 1980). Thus it is possible in all of the reported cases of the rapid spread of a tool-use behavior (Menzel 1973a; Hannah and McGrew 1987) that one creative individual has made a discovery and, as a result, propitious learning conditions and stimulus enhancement facilitate the individual discovery of others. This process is different from true *imitative learning,* in which the learner actually copies another's behavior or behavioral strategy. What is needed to tease apart the different possibilities is experimental manipulation.

Tomasello et al. (1987) trained an adult chimpanzee (the demonstrator) to rake food items into her cage with a metal T-bar. When a food item was in the center of the serving platform, she learned simply to sweep, or rake, the item within reach. When the food was located either along the side or against the back of the serving platform's raised edges, she had to employ more complex, two-step procedures to be successful. Several young chimpanzees from four to six years old (the experimental group) were then exposed to this adult demonstrator as she employed all three of her feeding strategies. Several other chimpanzees in this same age range (the control group) were exposed to the demonstrator in an unoccupied state. Results showed that experimental subjects learned to use the tool after only a few trials in most cases, while most control subjects did not. It is important to note, however, that the experimental subjects employed a wide variety of raking-in procedures and that none of them learned either of the demonstrator's more complex, two-step procedures. Even though they were trying to learn the two-step procedure, they failed more than 75% of the time.

We interpreted these results as demonstrating that the chimpanzees in this study benefited from observation in learning to use the tool but that the form of this learning was not imitative learning, because the experimental subjects did not learn to copy either the demonstrator's precise behaviors or behavioral strategies. Observation of the demonstrator, in our interpretation, drew the attention of the experimental subjects to the functional significance of the tool in obtaining the out-of-reach food—a kind of function-based stimulus enhancement. The experimental group then used this knowledge of the tool's function in their subsequent individual attempts. We proposed that the social learning in this case might best be called *emulation learning,* because subjects attempted to reproduce the observed end result (obtaining food), without copying the behavioral methods of the demonstrator.

In a second experimental study, Nagell et al. (1993) attempted to investigate the emulation hypothesis more directly. A human demonstrator presented chimpanzees (both young and mature) and two-year-old human children with a rakelike tool and a desirable but out-of-reach object. The tool was such that it could be used in either of two ways leading to the same end result. For each species, one group of subjects observed one method of tool use, and another group of subjects observed another method of tool use. The point of this design was that the stimulus enhancement of the tool was the same in both experimental conditions; only the precise method of use was different. What we found was that human children in general copied the method of the demonstrator in each of the two experimental conditions. Chimpanzees, in contrast, used the same method or methods no matter which demonstration they observed. This result occurred despite generally equal levels of tool use and of success with tools for the different experimental groups. Moreover, the children persisted in reproducing adult demonstrator behavior, even when it meant that they would be less successful than if they had simply used individual learning strategies, as the chimpanzees apparently did. We concluded from this pattern of results, once again, that chimpanzees in this task were paying attention to the general functional relations of tool and food and to the end results obtained by the demonstrator, but they were not attending to the methods of tool use actually demonstrated; chimpanzees were engaged in emulation learning. In contrast, human children focused on the demonstrator's actual methods of tool use (her behavior) and, thus, children were engaged in imitative learning.

I believe these studies suggest that chimpanzees and human children understand the tool-using behavior of conspecifics in different ways. For human children, the goal or intention of the demonstrator is a central part of what they perceive and, thus, her actual methods of tool use—the details of the way she is attempting to accomplish that goal—become salient. For chimpanzees, the tool, the food, and their physical relation are salient; the intentional states of the demonstrator and her precise methods, on the other hand, are either not perceived or seem less relevant.

A major criticism of these experiments is that captive chimpanzees are not representative of wild chimpanzees. Captive chimpanzees may be impoverished in various ways relative to wild chimpanzees and, thus, they tell us little about what is going on in the wild. One possibility is that captive chimpanzees may be impoverished with regard to tool use and, thus, they may not have the ability to understand through observation how certain tools work: imitative learning in all species relies on the ability to understand the task at hand. However, in the Nagell et al. (1993) study, chimpanzees and children did have the same levels of skill with the tool; the only difference was in how they used the tool.

Another possibility is that captive chimpanzees may be socially deficient compared with wild chimpanzees in ways that affect social learning. However, in the Nagell et al. (1993) study, mother-reared and nursery-reared chimpanzees performed in equivalent fashion. Although neither of these rearing conditions is the same as being raised in the wild, this equivalence shows that the lack of imitative learning was not a result of some simple factor, such as lack of adequate exposure to conspecifics. A related possibility is that chimpanzees do not attend to a human demonstrator (Nagell et al. 1993) as they would to a conspecific demonstrator. But the Tomasello et al. (1987) study did use a chimpanzee demonstrator. While it is true that the chimpanzee demonstrator was not closely related to the chimpanzee subjects, it is important to note that there was a strong stimulus enhancement for chimpanzees in both our studies: chimpanzees were attracted to the tool if a conspecific or human manipulated it in any way. This attraction suggests that the chimpanzee subjects were influenced in a positive way by demonstrators of either species. Despite all of this, it is still possible that the captive subjects in these two studies were socially deficient in precisely the dimension of interest. That is, because of their species-atypical social experiences during ontogeny, they had not been exposed to the kinds of interactions necessary to develop skills of imitative learning. This possibility cannot be ruled out at this time, but such atypicality seems to be much less of a factor for subjects in the gestural communication studies that I will now report.

Gestural Communication

In the debate over chimpanzee culture and imitation, the focus has been on tool use and other behaviors involving the use of instruments; but gestural communication may be just as important a focus. Population differences in gestural signaling of chimpanzees are well documented and provide evidence that at least some gestures are learned (Goodall 1986; Tomasello 1990). The best known of these gestures are leaf clipping in the Mahale K and Bossou communities and the grooming handclasp in the Mahale K, Kibale, and Yerkes communities. Inspection of data for individuals in these communities shows that, in addition to differences between populations, marked individual differences exist within communities in the use (intended effect) of the signal. Although data relevant to the learning processes by which individuals might acquire these gestures have not been collected, McGrew and Tutin (1978) and Nishida (1987) have speculated that young chimpanzees might learn these gestures by means of some form of cultural transmission such as imitation.

For the past eight years we have been conducting a longitudinal investigation of the gestural signals of a group of chimpanzees at the Yerkes Regional Primate Research Center Field Station in ways that allow for some inferences about learning processes. In a first study, we observed infants and juveniles from one to four years old, with special emphasis on how they used their signals (Tomasello et al. 1985). Looking only for intentional signals accompanied by an indication that a response was expected, as in gaze alternation or response waiting, we found a number of striking developmental patterns. For example, we found that juveniles used many gestures not used by adults, that adults used some gestures not used by juveniles, and that some juvenile gestures for particular functions were replaced by more adultlike forms in later developmental periods. None of these developmental patterns supported the idea that gestural signals are culturally transmitted from adults to juveniles.

In a longitudinal follow-up to these observations, Tomasello et al. (1989) observed the same juvenile chimpanzees of the Yerkes group four years later (at five to nine years old). In contrast to the hypothesis that imitative learning is the way young chimpanzees acquire their intentional gestures, we proposed an alternative hypothesis involving individual conventionalization (Smith 1977). In *conventionalization*, a communicatory signal is created by two organisms who shape each others' behavior in repeated instances of social interaction. For example, an infant may initiate nursing by going directly for the mother's nipple, perhaps grabbing and

Figure 2
Juvenile male chimpanzee holding out his hand to beg for the return of an object just taken from him. Photo by Frans de Waal.

moving her arm in the process. In some future encounter the mother might anticipate the infant's desire at the first touch on her arm, and so become receptive at that point. This leads the infant to abbreviate its behavior to touch the mother's arm, which becomes the so-called intention movement. In attempting to test this hypothesis, we noted two main pieces of evidence. First, in some cases, a number of idiosyncratic signals were used by only one individual. These signals could not have been learned by imitative processes and so must have been individually invented and conventionalized. Second, in other cases, the young chimpanzees used a number of signals in their interactions with adults where the youngsters had not been recipients of the actions; for example, no youngsters ever had been begged for food or been solicited for tickling or been requested to allow nursing. The youngsters' signals for these actions thus could not have been a product of imitation of signals directed to them (so-called second-person imitation). Furthermore, in many cases the youngsters were extremely unlikely to have imitated other youngsters' gestures (so-called third-person imitation, or "eavesdropping") because many of these gestures were produced in close quarters between mother and child with little opportunity for others to observe.

We have recently completed observations representing a third longitudinal time point on the Yerkes group (Tomasello et al. 1994). In this study a completely new generation of one- to four-year-old youngsters was the object of study. In order to investigate the question of potential learning processes, we systematically compared the concordance rates of each individual with all other individuals across the three points in time over the eight years. This analysis revealed quite clearly, by both qualitative and quantitative comparisons, that much individuality existed in the use of gestures, especially nonplay gestures, with much individual variability both within and across generations. As before, single individuals used a number of idiosyncratic gestures at each point in time. Also notable is the fairly large gap of four years between the two generations of this study, so that many of the gestures learned by the younger generation were ones that the older generation might have used only infrequently. Furthermore, many of the gestures common to most or all individuals of the studies were gestures that are learned also by captive youngsters raised elsewhere in peer groups with no opportunity to observe older conspecifics (Berdecio and Nash 1981). Commonality is presumably explained by commonalties in learning conditions. For example, all infant chimpanzees desire to nurse and must do so in the same basic way, so the gestures they might conventionalize in that context are limited in number, much as human children have only a few basic ways for gesturing to be picked up.

We concluded from this pattern of results that the young chimpanzees in the Yerkes group were not learning their communicatory gestures through imitation; instead, they were individually conventionalizing them with one another. The explanation for this conclusion is analogous to the explanation for emulation learning in the case of tool use. Conventionalization is like emulation learning in that neither process requires understanding the intentions of others in the same way that imitative learning requires understanding the intentions of others. For example, imitatively learning to solicit play requires that an individual understand the intentions of a conspecific when it raises its arm; in contrast, conventionalizing the arm-raise gesture requires only an anticipation of the future behavior of a conspecific and the ability to translate this anticipation into an instrumental behavior. The capacity for simple anticipation contrasts sharply with the imitative learning skills that human children use in acquiring linguistic conventions. These skills require, in a very specific way, an understanding of the intentions of others (Tomasello 1992). My conclusion is, thus, that chimpanzees are not employing skills of imitative learning as human children typically do, although they are displaying much individual inventiveness, behavioral flexibility, and creative intelligence in this domain. At these stages, chimpanzees do not imitatively learn gestures from one another because they either do not perceive or do not attend to the intentions of others, in contrast to human children who typically do.

The inferences about learning processes made on the basis of these studies of chimpanzee gestural communication are admittedly indirect; experimental studies are planned. However, compensating for the lack of directness is the fact that the subjects of these studies have led lives that resemble the lives of chimpanzees in the wild to a much greater degree than the subjects of the tool-use experiments. The Yerkes Field Station group has been relatively stable over many years. The physical setting is relatively diverse so that the environment of developing chimpanzees in the Yerkes group has been similar to the environment of developing chimpanzees in the wild. This similarity is especially true with regard to peer interaction and play, which seem to occur very much as they do in the wild and which were the contexts for the majority of the signals we observed and analyzed.

Social Cognition and the Role of Human Interaction

One may point out that a number of very convincing observations of chimpanzee imitation exist in the literature. However, almost all the clearest examples of imitation in the exhaustive review of Whiten and Ham (1992) concern chimpanzees that have had extensive amounts of human contact. The major exception is the instance reported by de Waal (1982), in which one individual walked with a limp, appearing to mimic the walk of an injured group mate. In many cases, observed imitation has resulted from intentional instruction involving human attention, encouragement, and direct reinforcement for imitation, that is, involving social interactions of the type human children are exposed to routinely (Hayes and Hayes 1952). This raises the possibility that development of imitative learning skills may be influenced, or even enabled, by certain kinds of social interaction during early development.

Confirmation of this point of view is provided by Tomasello, Savage-Rumbaugh, and Kruger (Tomasello et al. 1993) using *enculturated* chimpanzees, who are raised like human children and exposed to a languagelike system of communication. In this study, we compared the imitative learning abilities of mother-reared captive chimpanzees, enculturated chimpanzees who were raised like human children and exposed to a languagelike system of communication, and two-year-old human children. Each subject was shown 24 different, novel actions. Each subject's behavior was scored as to whether or not the subject successfully reproduced the demonstrated action and/or the behavioral means used by the demonstrator. The major result was that the mother-reared chimpanzees hardly ever reproduced the novel actions. In contrast, the enculturated chimpanzees and the human children imitatively learned the novel actions much more frequently, and they did not differ from one another in this learning. Corroborating this latter finding is the fact that earlier in their ontogeny, these same enculturated chimpanzees seemed to learn many of their humanlike symbols by means of imitative learning (Savage-Rumbaugh 1990).

For the issue of chimpanzee culture, these results raise a very important question. Which group of captive chimpanzees is more representative of chimpanzees in their natural habitats: mother-reared or enculturated? Are enculturated chimpanzees simply displaying *species-typical* imitative learning skills because their enriched rearing conditions, as compared with the impoverished rearing conditions of other captive chimpanzees, more closely resemble the rearing conditions of chimpanzees in the wild? Or might it be that the humanlike socialization process experienced by enculturated chimpanzees differs significantly from the natural state and, in effect, helps to create *species-atypical* imitative learning skills, which are similar to those of humans? We have no definitive answer to these questions at this time, but it is my hypothesis that a humanlike sociocultural environment is essential to the development of humanlike social-cognitive and imitative learning skills. This hypothesis might hold true not only for chimpanzees but also for human beings. Thus, a human child raised in an environment lacking intentional interactions and other cultural characteristics might not develop humanlike skills of imitative learning either.

The current hypothesis is, thus, that understanding the intentions of others is necessary for reproducing another's behavioral strategies and that this understanding develops in, and only in, the context of certain kinds of interactions with others (Tomasello et al. 1993). More specifically, for a learner to understand intentions of another individual requires that the learner be treated as an intentional agent: the other individual encourages attention to and specific behaviors toward some object of mutual interest on the part of the learner and, then, reinforces the learner's successful attempts in this direction. Direct instruction is not sufficient of course; many animals are subjected to all kinds of human interaction, including the array outlined here, without developing humanlike skills of imitative learning. The same is true of some human children, such as those with autism. The important point for current purposes is that, in terms of recognition, direction, encouragement, and reinforcement of social interaction, captive chimpanzees raised by conspecifics may be a better model for chimpanzees in the wild than are chimpanzees raised in humanlike cultural environments, because chimpanzees in the wild receive little in the way of direct instruction from conspecifics. Boesch (1991) reports some possible exceptions. Further, I hypothesize that the learning skills developed in the wild in the absence of human interaction (that is, skills involving individual learning supplemented by local enhancement, emulation learning, and conventionalization) are sufficient to create and maintain population-specific behavioral traditions, but these skills are not sufficient to create and maintain the behavioral traditions that display key characteristics of human culture. It is to these key characteristics to which I now turn.

Human and Chimpanzee Behavioral Traditions

Documenting differences in social learning and social cognition in chimpanzees and humans does not directly address questions of the nature, extent, and mechanisms of chimpanzee culture. But, I would contend that there are desireable differences in the population-specific, behavioral traditions of humans and chimpanzees and these may be tied to differences in social learning and social cognition. This tie makes all the foregoing experimental work relevant to discussions of chimpanzee culture in the wild.

As a subtype of population-specific behavioral traditions, human cultural tradition may be said to have at least three characteristics (Tomasello et al. 1993). First, in all human societies, some traditions are practiced by virtually everyone in the society; any child who does not learn these traditions simply would not be considered a normal member of the group. This is true of language and religious rituals in many cultures, as well as more mundane subsistence behaviors having to do with food, dress, and the like. We may call this characteristic *universality.* Second, with many human behavioral traditions, the methods employed by different persons—both within and across generations—show a high degree of similarity. This is true to some degree for concrete tasks such as using a hoe or weaving cloth, although there are often individual idiosyncrasies. In the case of social-conventional behaviors such as the use of linguistic symbols or participation in religious rituals, individual discovery and idiosyncratic use are not viable options; these behaviors would not be functional unless the methods of the mature users were reproduced faithfully. This second characteristic, thus may be called *uniformity.* A third characteristic of human cultural traditions derives from the second and is, perhaps the most telling. Human cultural traditions often show the ratchet effect, an accumulation of modifications over generations. For example, hammerlike tools and the way they are used show a gradual increase in complexity over time as they are modified again and again to meet novel exigencies (Basalla 1988). Thus, at the same time that cultural traditions are passed on rather faithfully from one generation to the next, if a modification is made, that modified version is what is passed on to the next generation. As a result, many human traditions have what may be called a *history.*

We have some evidence for each of these three characteristics in the behavioral traditions of chimpanzees although the evidence is in many cases not completely convincing and, in all cases, the candidate behaviors are very few in number and confined to only one or a few communities. With respect to universality, only a few communities have been observed long enough to come to reasonable conclusions. McGrew (pers. com.) reports that virtually all the noninfant members of the Kasakela community at Gombe fish for termites. It is also likely that virtually all the noninfant individuals in the Mahale K Group fish for ants (Nishida pers. com.). And virtually all the physically capable members of the Taï forest chimpanzee group engage in nut cracking (Boesch 1993). However, only some members of the Mahale K Group practice grooming, handclasp and leaf-clipping behaviors (Nishida pers. com.). Based on their survey of the innovative behaviors of chimpanzees and other primate species Kummer and Goodall conclude that "only a few will be passed on to other individuals, and seldom will they spread through the whole troop" (1985: 213). My overall conclusion is that evidence for universality of some behavioral traditions in some chimpanzee communities is strong. In any given chimpanzee community, however, all members engage in only one or just a few population-specific behavioral traditions. This differs, at least quantitatively, for humans who have many human cultural traditions that display universality.

With respect to uniformity, a number of keen observers have remarked that individual chimpanzees often apply their own creative techniques to all kinds of concrete tasks, from termite fishing to nut cracking (Goodall 1986; Hannah and McGrew 1987; Sumita et al. 1985). This innovation is as it should be; slavishly copying the techniques of others is not always the best strategy (Byrne 1992). More telling for current purposes, however, is chimpanzees' use of idiosyncratic techniques in social-conventional behaviors for which humans show such marked uniformity as, for example, in humans' choice and use of particular linguistic symbols. Much individuality among chimpanzees has been reported, for example, in the communicatory gesturing observed in wild populations (Goodall 1986); our own studies of the Yerkes group confirm the prevalence of marked individuality. These idiosyncrasies involve not only the way in which a particular signal is expressed by individuals but also the use or nonuse of particular signals in particular contexts. Of course, no simple measure exists by which we might compare the degree of uniformity of behavior within groups and across species. However, I would argue that only a limited number of learned behaviors seem to be expressed with a high degree of uniformity by all the chimpanzees of a community. This is true especially for the strict uniformities of the type we see in human languages and other social-conventional behaviors.

Finally, in the case of behaviors with a history, Boesch (1993) reports two examples he believes show the ratchet effect in the chimpanzee group of the Taï Forest. His first and clearest example concerns leaf clipping, which is a noise-making gesture. For some time adult males used leaf clipping only in the drumming context (in group movement) to indicate something about their target location; then, they began using it in the resting context. Some other group members then began using leaf clipping during the resting context, some for the first time and only in this context. In this observation, however, what changed was the context of the behavior. There was no change in the leaf-clipping behavior itself. This behavior seems to be a very weak candidate for the ratchet effect. Of course, we simply may not have observed chimpanzees in their natural habitats long enough to know whether some of their practices show the ratchet effect; many human cultural traditions remain unchanged for long periods of time. Nevertheless, it seems that only a very few chimpanzee behaviors have a history in the sense of an accumulation of modifications over time. It may be that there are none.

The differences observed in human and chimpanzee behavioral traditions, especially along the three characteristics just outlined, may very plausibly be attributed to differences in the processes of social learning employed by the two species, especially processes documented by myself and others in experimental contexts (Heyes in press). It is my hypothesis that human cultural traditions have universality, uniformity, and history because human beings learn from one another in ways that lead to a high degree of fidelity of transmission. For example, human beings learn to use tools and symbols through imitative learning and instruction (Tomasello et al. 1993). In contrast, chimpanzee societies have a looser structure because, when they learn from one another, they employ processes of socially enhanced, individual learning as, for example, emulation learning and conventionalization (Tomasello 1990). My view is that attributing culture to chimpanzees has some basis in empirical fact: chimpanzees rely on social learning in some behavioral domains and, in those domains, the behavioral traditions show some sign of universality and uniformity. Nevertheless, the processes of social learning and the characteristics of each species' behavioral traditions differ in important ways, notably in terms of their histories. These differences would lead to the unique cultural products of human societies such as languages, complex tool use, and other cultural traditions and institutions.

The Evolution of Culture

I am perfectly happy to say that chimpanzees have cultural traditions, as long as we specify how these traditions are similar to and different from the cultural traditions of other animal species, especially humans. I prefer, however, to use other terms because human cultural traditions seem to have unique characteristics. Human cultural traditions show universality, uniformity, and history in a manner and to a degree that makes these traditions seem qualitatively different from the behavioral traditions of other species. It is very likely, in fact, that many species have behavioral or cultural traditions that show their own unique stamp. How the behavioral and cultural traditions of chimpanzees are similar to and different from those of other primate species is a question we have just begun to investigate systematically.

Following Galef (1992)—from whom I borrowed the title of this paper—I also think we can pose the question of chimpanzee culture in a way that avoids much of this definitional indeterminacy. Quite simply, we may ask whether the behavioral traditions of various animal species are phylogenetically analogous or homologous. In the current context, the question is whether the common ancestors of chimpanzees and humans possessed cultural traditions displaying universality, uniformity, and histories five to ten million years ago. I believe that they did not. In fact, much evidence from the study of human prehistory indicates that cultural traditions of the clearest kind—a full-blown language and other means of symbolic expression—may be of relatively recent origin, emerging with *Homo sapiens sapiens* less than 100,000 years ago (Mellars and Stringer 1989; Davidson and Noble 1989). Even if human cultural traditions as I have defined them emerged in earlier periods of human prehistory, as some researchers believe, then that development still occurred well after the divergence of chimpanzees and humans. Furthermore, because current experimental and observational evidence suggests that the social learning and social cognition of chimpanzees and humans are different, I believe that the evolutionary evidence supports the view that human and chimpanzee behavioral traditions are only analogous. The social-cognitive adaptations on which human culture and cultural learning depend came only after the differentiation of the two species.

Acknowledgments

Portions of the research reported in this paper were supported by NIH Grant RR-00165 from the Division of Research Resources to the Yerkes Regional Primate Research Center. The Yerkes Center is fully accredited by the American Association for the Accreditation of Animal Care. Thanks to Joseph Call, Bill McGrew, and Frans de Waal for helpful comments on an earlier version of the manuscript.

References

Basalla, G. 1988. *The Evolution of Technology.* Cambridge, England: Cambridge Univ. Press.

Beck, B. 1980. *Animal Tool Behavior.* New York: Garland.

Berdecio, S., and A. Nash. 1981. Chimpanzee visual communication: Facial, gestural, and postural expressive movements in young, captive chimpanzees. Arizona State Research Papers, no. 26.

Boesch, C. 1991. Teaching among wild chimpanzees. *Anim. Behav.* 41:530–2.

———. 1993. Toward a new image of culture in wild chimpanzees? *Behav. and Brain Sci.* 16:514.

———. In press. Transmission aspects of tool use in wild chimpanzees. In T. Ingold and K. Gibson eds., *Tool, Language, and Intelligence: An Evolutionary Perspective.* Cambridge, England: Cambridge Univ. Press.

Bonner, J. 1980. *The Evolution of Culture in Animals.* Princeton, N.J.: Princeton Univ. Press.

Boyd, R. and P. Richerson. 1985. *Culture and the Evolutionary Process.* Chicago: Univ. of Chicago Press.

Byrne, R. 1992. The evolution of intelligence. In P. Slater and T. Halliday, eds., *Behavior and Evolution.* Cambridge, England: Cambridge Univ. Press.

Cheney, D., and R. Seyfarth. 1990. *How Monkeys See the World.* Chicago: Univ. of Chicago Press.

Davidson, I., and W. Noble. 1989. The archaeology of perception: Traces of depiction and language. *Current Anthro.* 30:125–55.

Galef, B. 1992. The question of animal culture. *Human Nature* 3:157–78.

Goodall, J. 1986. *The Chimpanzees of Gombe.* Cambridge, Mass.: Harvard Univ. Press.

Hannah, A., and W. McGrew. 1987. Chimpanzees using stones to crack open oil palm nuts in Liberia. *Primates* 28:31–46.

Hayes, K., and C. Hayes. 1952. Imitation in a home-raised chimpanzee. *J. Comp. and Physiological Psychol.* 45:450–9.
Heyes, C. In Press. Imitation, cognition, and culture. *Anim. Behav.*
Kummer, H., and J. Goodall. 1985. Conditions of innovative behavior in primates. *Phil. Trans. R. Soc. London* 308:203–14.
McGrew, W. 1992. *Chimpanzee Material Culture.* Cambridge, England: Cambridge Univ. Press.
McGrew, W., and C. Tutin. 1978. Evidence for a social custom in wild chimpanzees? *Man* 13:234–51.
Mellars, P., and C. Stringer. 1989. *The Human Revolution.* Oxford, England: Oxford Univ. Press.
Meltzoff, A. 1988. The human infant as "homo imitans." In T. Zentall and B. Galef, eds., *Social Learning: Psychological and Biological Perspectives.* Hillsdale, N.J.: Erlbaum.
Menzel. E., ed. 1973a. Further observations on the use of ladders in a group of young chimpanzees. *Folia Primatol.* 19:450–7.
———. 1973b. *Precultural Primate Behavior.* Vol. 1. Berlin: Karger.
Nagell, K., R. Olguin, and M. Tomasello. 1993. Processes of social learning in the imitative learning of chimpanzees and human children. *J. Comp. Psychol.* 107:174–86.
Nishida, T. 1987. Local traditions and cultural transmission. In B. Smuts, D. Cheney, R. Seyfarth, R. Wrangham, and T. Struhsaker, eds., *Primate Societies.* Chicago: Univ. of Chicago Press.
Russon, A., and B. Galdikas. 1993. Imitation in free-ranging rehabilitant orangutans. *J. Comp. Psychol.* 107: 155–73.
Savage-Rumbaugh, S. 1990. Language as a causeffect communication system. *Philosophical Psychol.* 3:55–76.
Smith, J. 1977. *The Behavior of Communicating.* Cambridge, Mass.: Harvard Univ. Press.
Sumita, K., J. Kitahar-Frisch, and K. Norikoshi. 1985. The acquisition of stone tool use in captive chimpanzees. *Primates* 26:168–81.
Tomasello, M. 1990. Cultural transmission in the tool use and communicatory signaling of chimpanzees? In S. Parker and K. Gibson, eds., *Language and Intelligence in Monkeys and Apes: Comparative Developmental Perspectives.* Cambridge, England: Cambridge Univ. Press.
———. 1992. The social bases of language acquisition. *Social Development* 1:67–87.
Tomasello, M., J. Call, C. Nagell, R. Olguin, and M. Carpenter. 1994. The learning and use of gestural signals by young chimpanzees: A trans-generational study. *Primates.* 35:137–45
Tomasello, M., M. Davis-Dasilva, L. Camak, and K. Bard. 1987. Observational learning of tool use by young chimpanzees. *Hum. Evol.* 2:175–83.
Tomasello, M., B. George, A. Kruger, J. Farrar, and E. Evans. 1985. The development of gestural communication in young chimpanzees. *J. Hum. Evol.* 14:175–86.
Tomasello, M., D. Gust, and T. Frost. 1989. A longitudinal investigation of gestural communication in young chimpanzees. *Primates* 30:35–50.
Tomasello, M., A. Kruger, and H. Ratner. 1993. Cultural learning. *Behav. and Brain Sci.* 16:495–552.
Tomasello, M., S. Savage-Rumbaugh, and A. Kruger. 1993. Imitative learning of actions on objects by chimpanzees, enculturated chimpanzees, and human children. *Child Development* 64:1688-1705.

Visalberghi, E., and D. Fragaszy. 1990. Do monkeys ape? In S. Parker and K. Gibson, eds., *Language and Intelligence in Monkeys and Apes: Comparative Developmental Perspectives.* Cambridge, England: Cambridge Univ. Press.

de Waal, F. 1982. *Chimpanzee Politics.* New York: Harper and Row.

Whiten, A., and R. Ham. 1992. On the nature and evolution of imitation in the animal kingdom: Reappraisal of a century of research. In P. Slater and J. Rosenblatt, eds., *Advances in the Study of Behavior.* New York: Academic Press.

Zentall, T., and B. Galef, eds. 1988. *Social Learning: Psychological and Biological Perspectives.* Hillsdale, N.J.: Erlbaum.

Chimpanzee dressed for a show.
Photo by J. Goodall.

Biobehavioral Roots of Language

A Comparative Perspective of Chimpanzee, Child, and Culture

Duane M. Rumbaugh, E. Sue Savage-Rumbaugh, and Rose A. Sevcik

Humans have long been anxious about their identity as a species. Although everyone can note the similarity between the structure and function of body parts common to animals and to us, not everyone has been comfortable about it. Thus, some persons would seemingly hope that if distance from animals cannot be attained on the basis of anatomical uniqueness, it surely can be defined on dimensions of psychological uniqueness—reason and language.

Quest for Uniqueness

Unencumbered and frequently impatient with the results of scientific inquiry, historical and even contemporary records are replete with proclamations about our species' distinguishing attributes. One by one, however, these proclamations have been found to be wanting and without empirical corroboration (Heltne and Marquardt 1989). Uniqueness is, indeed, the hallmark of taxonomic classification, and each species has its own constellation of defining attributes. However, the attributes that define various species are, in themselves, rarely unique.

Nevertheless, one point of apparent distinction many have looked to as seemingly the ultimate, unyielding line that demarcates humans from all other animals is *language.* In this chapter, the argument will be made that systematic data, now in hand from controlled experiments, are sufficient to conclude that the chimpanzee and the bonobo are, indeed, capable of acquiring a range of language skills.

Is Language Unique to Humans?

To many, it seems clear that language, as commonly viewed, is unique to humans in that they and they alone develop speech and attendant language skills with apparent spontaneity during early development. Yet, on the other hand, research with apes during the past 25 years, particularly during the past 10 years, has served to formulate a new question, one that has brought forth a new perspective of language: Are the behavioral requisites to language unique to humans?

We will argue that chimpanzees *(Pan troglodytes)* and bonobos *(Pan paniscus)* can acquire an impressive array of language skills, including the ability to understand the syntax, or way words are put together in sentences, of requests spoken to them. But *Pan's* competence for such language is extraordinarily dependent upon its being reared from birth in a language-saturated environment—one in which members of a social group coordinate their activities via language throughout the course of each day. Being reared thus, any member of the species *Pan* acquires language without special training. It comes to comprehend language well before it is used productively—in the same pattern that characterizes language development in the human child.

History of Thought Regarding Language

Hewes (1973, 1977) presents a historical review of theories of glottogenesis, the origin of language, and perspectives regarding the possibility that apes might be capable of language and even speech. He recounts de la Mettrie's 1748 anticipation that if one were to select an ape with "the most intelligent face" and were to have it taught by Amman, a gifted teacher of the deaf, that one might just succeed in teaching an ape to talk. Interestingly, de la Mettrie anticipated that as the ape acquired language, it would no longer be "a wild man, nor a defective man, but he would be a perfect man, a little gentleman . . .".

Although we no longer view it probable that an ape will literally learn to talk, the bonobo Kanzi has had his vocal repertoire modified, apparently as a function of being reared in an environment in which people talked a great deal (Hopkins and Savage-Rumbaugh 1991). And those of us who have had the privilege to work with apes and have succeeded in demonstrating their abilities for representational language would agree in principle with de la Mettrie that somehow the apes' psychology is basically transformed as they acquire representational language. Apes they remain, but they become apes with an attentiveness, a sensitivity, a gentleness, and even a civility not otherwise common to them. To the degree they acquire such a psychology, apes become enabled to achieve a remarkable social commerce with humans they trust. As we come to understand in detail the essence of these changes, we will surely come to have an enlightened perspective regarding what it means both *to be human* and *to be chimpanzee.*

Complex Learning

Research of the past 75 years has made clear that primates share with us the abilities not just to learn simple responses, but to acquire concepts and generate hypotheses and strategies (Tuttle 1986; Rumbaugh and Pate 1984; Goodall 1989). In our view, the perspective advanced by Descartes (1956) that animals are *beast machines* with no capacity for pain and reason stands as one of the most tragic conclusions of scholarship. As we will argue both implicitly and explicitly in this chapter, Darwin (1859) was correct: there is substantial continuity in the psychology as well as in the biology of animal and human. Regrettably, the Cartesian perspective still permeates the majority of world societies today.

Ape Language Research

Research interests in the chimpanzee's potential for language were revived in the mid-1960's as studies by the Gardners (Premack 1989) of the University of Nevada and the Premacks (1975), then at the University of California, Santa Barbara, were initiated. The Gardners used American Sign Language for the deaf to establish two-way communication; the Premacks used plastic tokens of different shapes and colors as words in their experimental analysis of language functions.

The early 1970's saw the launching of the LANA Project by Rumbaugh (1977) of Georgia State University in collaboration with the Yerkes Regional Primate Research Center and colleagues from the University of Georgia. A computer-based keyboard system was created that provided the chimpanzee Lana (see figure 1a) with a large number of keys, each having a distinctive geometric symbol called a *lexigram* embossed on its surface. The lexigrams functioned as words, and the *lexigram grammar,* the rules by which the lexigrams were joined, was monitored for correctness by the computer.

Correct sentences activated a bank of devices for the selective dispensing of various foods and drinks, movies, slides, and music. Lana readily learned her lexigrams to request things and to give the names and colors of objects. Tests revealed that she saw Munsell color chips (Essock 1977) in a manner that resembles humans, an observation recently reaffirmed by Matsuzawa (1990, this volume) in Japan with his chimpanzee, Ai.

Within limits, Lana demonstrated her ability to modify her sentences so as to achieve ends other than the specific ends for which they were designed. For example, she used sentences to attract attention to malfunctioning food vendors and to the units that produced slides and music. She innovatively called a cucumber, "banana which-is green"; an overly ripe banana, "banana which-is black"; and the citrus orange, "apple which-is orange (colored)." Comprehensive analyses clearly indicated that Lana's productions could not be satisfactorily attributed either to rote learning of sequences or to imitation (Pate and Rumbaugh 1983).

Project LANA affirmed the ability of the chimpanzee to learn large numbers of symbols, to use these symbols in prescribed sequences, to alter use of those sequences creatively, and to use symbols to facilitate perceptions of sameness and difference between items when one item was presented *visually,* to the sense of sight and the other item presented *haptically,* to the sense of touch. For the past 24 years, the technology and methods devised for Project LANA have served our research into the parameters of the apes' abilities for learning language. Moreover, we have extended these techniques to the benefit of children and young adults who, by reason of mental retardation, have profound deficiencies both in language and speech (Romski 1989; Romski and Sevcik 1991).

The early 1970s also saw the launching of Project Nim Chimpsky at Columbia University by Terrace (1979). Although Nim learned manual signs as did the Gardners' Washoe, Terrace reached the conclusion that Nim's signs and Washoe's signs, were, for the most part, due to the result of imitating what others working with their subjects had recently signed. Regrettably, many readers interpreted Terrace's report to mean that, because neither Nim nor Washoe had manifested language, no ape either had or could have language. Such a conclusion was patently unwarranted because it is tantamount to concluding that the null hypothesis (i.e., chimpanzees cannot acquire language) had been proved–which is impossible to do. Although the null hypothesis may be retained, it can never be proven.

Also in the 1970s, Lynn Miles of the University of Tennessee at Chattanooga began Project Chantek with an orangutan from the Yerkes Center. Chantek was taught manual signs, but Miles' analyses revealed no major indication that Chantek's signs were formed in imitation of others (Miles 1990).

Research of the mid to late 1970s by Savage-Rumbaugh and Rumbaugh with the chimpanzees Sherman and Austin used the computer-based keyboard developed for the LANA Project. Initially, it proved relatively difficult for Sherman and Austin to acquire symbols. The chimpanzees even had difficulty discriminating reliably between symbols and, as they first

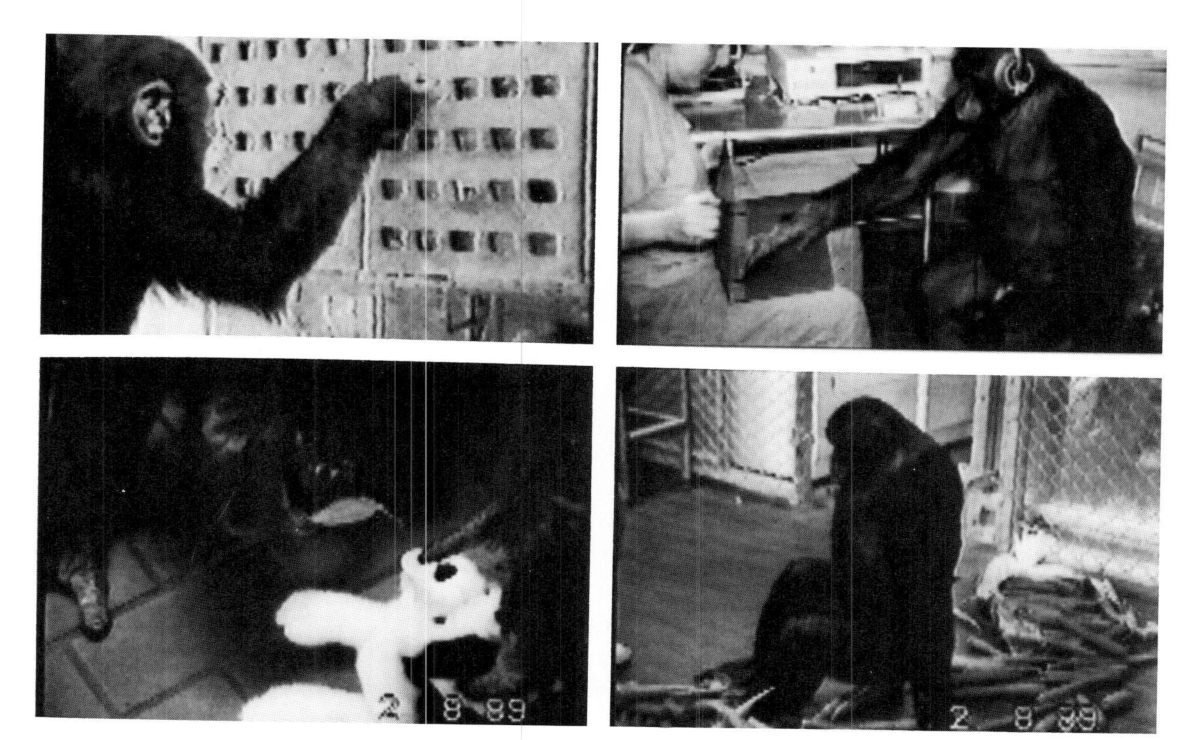

Figure 1

a. Upper left.
Lana chimpanzee at her keyboard

b. Upper right.
Kanzi selects the lexigram that corresponds to a spoken English word (Sherman) heard via the headphones.

c. Lower left.
Kanzi was instructed to make the doggie bite the snake, and he prepares to do so.

d. Lower right.
Kanzi responds to the request, "Give a carrot to Rose".

learned symbols, they tended to recall them by position on the keyboard rather than by their visual configuration. Initially as well, Sherman and Austin limited their symbol usage to requests, that is, they used symbols accurately only when they wanted to obtain something. When asked to show or to give a specific object, the two chimpanzees behaved as if they could no longer remember which lexigram went with a given item. They were also unable to demonstrate a reliable correspondence between what they said and what they did. For example, they might say they wanted an apple but, when taken to the refrigerator, they would select a banana instead. After learning to ask for things, they then required further training before they became able to label things. And they required yet additional training to respond appropriately when others used symbols to communicate with them.

However, after receiving subsequent training that emphasized communication with one another or with an experimenter, Sherman and Austin began to understand that symbols could be used to represent things. For example, they could announce that they needed a key to open a locked box and then proceed to search through a bunch of objects and find a key. Or they could say that they wanted an apple and, when sent to the refrigerator, they would return with an apple, resisting all other options available in the refrigerator's contents. Moreover, they were able to use these skills to communicate specific information about absent objects.

Of critical significance, Sherman and Austin categorized 17 word-lexigrams where each was the name of a specific food or tool that they had learned in earlier training. The chimpanzees categorized these 17 lexigrams through the differential and appropriate use of two lexigrams glossed as *food* and *tool* in a tightly controlled test situation. In other words, when shown their lexigram for banana, they called it a *food;* when shown their lexigram for wrench, they called it a *tool;* and so on. Their ability to do so indicated that for Sherman and Austen these lexigrams functioned representationally—that the lexigrams were meaningful. Consequently, the performance of these chimpanzees displayed a basic, though nonetheless elegant, capacity for *semantics*—the meaning of words (Savage-Rumbaugh 1986).

A Perspective of Language

We view language as being comprised of a number of interacting, nonspeech building blocks, the main one being semantics—the use of arbitrary symbols to represent things not necessarily present in space or time. These symbols can be used referentially with partners to plan and to coordinate behavior to an advantage. A main power of language as a form of communication is that it enhances the ability to predict and, hence, to coordinate with the behaviors and activities of others within a social group.

The use of arbitrary symbols for social communication and for the coordination of behavior is quite different from a highly predictable, species-specific pattern of behavior brought forth in response to well-defined stimuli and signals, for example fixed-action patterns in response to sign stimuli and releasers (Alcock 1984). Such species-characteristic behaviors—as, for example, the newborn gull's readiness to peck at the red spot on its parent's beak to obtain food—is predominantly under genetic control. In contrast, behavior involved in the calling of free-ranging monkeys evidences a representationally based communication system (Cheney and Seyfarth 1990). The transformation of an initially arbitrary stimulus into a communicative symbol is based on a representationally based communication system and requires complex learning and cognition.

The Primary Role of Comprehension in Language

In our view, the essence of *language* is the ability to comprehend symbols—be they spoken words, geometric symbols, or signs (Bellugi et al. 1991). That the symbols of a language system might be word-lexigrams does not preclude their acquiring basic semantic value and functioning as spoken words. Lexigrams and other graphic media can be language, albeit unspoken and silent (Sevcik et al. 1991).

From our perspective, it is the interpretation of symbols, then, and not speech itself that is the *sine qua non* of language. Without question, the mechanisms that provide for speech and hearing also provide very efficient systems for linguistic communication; but speech produces only sounds, and it is only those sounds—not meanings—that are propagated to the listener. It is in the assignment of meaning by the listener that language is accomplished and, in our view, whether the meaning is assigned to speech sounds or to geometric symbols is not critical.

Roots of Ape Language Skills in Early Experience

Early in this chapter, this question was posed: are the behavioral requisites to language unique to humans? Because *Pan* has evinced competence for acquiring significant dimensions of language, one wonders whether its capacity for language with humans takes other forms in the wild or whether its language ability is uniquely cultivated by ape-human interactions during early life.

An evaluation of these possibilities can be achieved in part by considering Matata, an adult female bonobo who had a history of feral living until the age of about five years. Not until she was a young adult did she enter language studies at our laboratory. Matata has always impressed us as being a very bright bonobo, particularly outdoors in the 55-acre forest that surrounds the laboratory. She was especially clever, for example, in finding insect-laden reeds. Matata is highly sociable and interacts comfortably with her human companions during cleaning, feeding, playing, and even while giving birth. In contrast, she has never excelled in learning the meanings of lexigrams or in any of the other tasks that were readily mastered by other apes in our programs. At best, she partially mastered the use of seven symbols, and those she used exclusively to request specific foods.

Matata may have failed at these tasks because she was introduced to them at an age well beyond that which provides for the structuring of cognitive requisites for language. This view would imply that the cognitive structures laid down for adaptation in the forest simply were not applicable to the electronics of a modern-day laboratory.

A contrasting perspective is obtained from her adopted son Kanzi. While we tried to teach lexigrams to Matata for more than two years, Kanzi played about the room. When Kanzi was 30 months of age, Matata was sent to the Yerkes Field Station to be bred. Even on the first day of her absence, we could immediately see that Kanzi had already mastered lexigrams. By observation alone, he had learned the lexigrams that Matata had been unable to learn. Moreover, he learned them *spontaneously,* that is, without any specific, food-reinforced, discrete-trial training regimen. Kanzi was able to use lexigrams to request items, to indicate a correspondence between the item requested and selected, and to comprehend the meanings of symbols used by others. He, not Matata, was the one who learned and he learned well. This learning occurred with no specific language training given to him—either before or after this discovery.

Within the next two years Kanzi also appeared to be comprehending spoken, English words. In fact, controlled tests revealed his comprehension of both individual words (see figure 1b) and novel requests made of him, again without specific training (Savage-Rumbaugh 1988; Savage-Rumbaugh et al. 1986, 1993).

Confirmation of the bonobo's ability to learn observationally and without the benefit of specific language training was provided by Mulika, born to Matata in 1983 (Sevcik 1989). Because her eye became infected, Mulika was separated from Matata four months after birth and only then entered our continuing project. When we applied the stringent criterion designed for her older sibling Kanzi, we found that Mulika, at 22 months of age, employed only six symbols *productively,* while her *receptive* skills with both lexigrams and spoken, English words included 42 symbols (Savage-Rumbaugh et al. 1986). In addition, Mulika had used 30 of these symbols at the keyboard only once—and two of them not at all. Observational learning was once again implicated as a powerful, profoundly unappreciated form of learning in the great apes.

Pan paniscus (bonobos) and Pan troglodytes (chimpanzees) Compared

Is the ability to acquire language by observation unique to the bonobo? To answer this question, a study was undertaken in which the bonobo Panbanisha and the chimpanzee Panzee were reared together from the age of about six weeks.

By two years of age, Panbanisha had demonstrated spontaneous, symbol learning and comprehension of speech. To our surprise, Panzee then also came to do so, although at a later age and not as efficiently (Savage-Rumbaugh et al. 1992). Panbanisha has remained superior to Panzee in language across the years, though Panzee has her own avenues of excellence—notably using tools, drawing, and mastering mechanical and spatial relations.

Learning Language via Comprehension, then Production

For both Kanzi and Mulika, competent use of word-lexigrams was facilitated by their prior comprehension of lexigrams used by others. Although the history of research on learning in this century emphasized a process involving the overt (external) response of an organism with the ensuing consequences (such as reinforcement), this process appears not to be the case in the learning of language during infancy by bonobos and chimpanzee. Rather, learning by these apes appears to depend more upon the extraction of predictive relationships (such as between the use of lexigrams by another person/ape and the ensuing results) than upon the actual use of keys embossed with lexigrams.

Language learning by the bonobos Kanzi, Mulika, and Panbanisha and by the chimpanzee Panzee parallels that of the human child in that *comprehension,* or the ability to grasp meaning, preceded productive use of language where *production* means speech in the child and use of lexigrams by the ape). What these apes had in common was rearing from shortly after birth in a language-saturated environment, one in which speech and the lexigram-embossed keyboard were used to attempt communication at every point. Their pattern of acquiring language competence stands in sharp contrast to learning done by the chimpanzees Lana, Sherman, and Austin who were between one-and-one-half and two years old when they entered language research.

Lana, Sherman, and Austin received specially designed training regimens to cultivate their use of lexigrams for requesting food and activities, for labeling items, and so on. The assumption was that, if they used their symbols, they would surely come to understand the meanings of these symbols. That assumption was, however, totally incorrect. That chimpanzees can use lexigrams, signs, or plastic tokens does not mean that the symbols function representationally. As an analogy, a digitized telephone information service that talks does not necessarily understand what it says.

Comprehension was cultivated in Sherman and Austin by having them label things. In this training, Sherman and Austen were asked to use the correct lexigram for each item as that item was presented. When the chimpanzees performed correctly, they did not receive the labeled item: rather, when they labeled correctly, the chimpanzees got some other incentive, such as a favorite food or drink. As a consequence, Sherman and Austin learned that the symbols stood for things regardless of whether those things were immediately obtainable. As a further consequence, Sherman and Austin developed the ability to give items to one another or to an experimenter upon specific request. This ability enabled Sherman and Austin to use communication to solve problems jointly. Then, they came to announce what they were about to do. For example, they would say wrench at the keyboards and, then, select a wrench from a tool kit and give it to the experimenter. This skill extended to a situation in which they went to an adjoining area with no one else present, saw a novel array of foods and drinks, returned to the keyboard, announced to others the food or drink that they would return with and, then, did so almost without error (Savage-Rumbaugh et al. 1983).

Grammar: Comprehension of Novel Sentences of Request

A formal, controlled test of the bonobo Kanzi's ability to understand the syntax of human speech was begun when he was eight years old; a similar test was given to the young girl Alia when she was about two years old (Savage-Rumbaugh et al. 1993). Alia's mother was a member of the research team that worked with Kanzi and, starting shortly after Alia's birth and continuing throughout testing, the mother worked with Alia half days at the Language Research Center. As a result, both Alia and Kanzi had extensive experience in this small social group (the research team) in which the members negotiated and coordinated their behaviors regarding what they would do, where they would go, what things were called, and so on, through the use of keyboards embossed with lexigrams. Although neither Kanzi nor Alia were required to use a keyboard to obtain an incentive or to engage in an activity, they were always encouraged to observe the use of the keyboard and to use it themselves. In addition, whenever they used the keyboard, their messages were taken at face value; that is, a researcher responded as though the subjects had intended to "say" whatever they did. We believe that this procedure encouraged Kanzi and Alia to attend to what they and others said with the lexigrams.

The test of comprehension included more than 400 sentences and was controlled in that the experimenter who spoke the sentence was hidden from view by sitting behind a one-way mirror. Additional persons who were part of the test procedures and in the room with the subject wore headphones and listened to loud music so that they could not hear what the subject was being asked to do.

A variety of objects were present on each trial, both in front of the subject and in other areas where the subject might need to go to fulfill a request. This assortment of objects was present to ensure that Kanzi and Alia fully understood the request and to diminish the possibility of the appropriate behavior occurring by chance (see figures 1c and 1d).

Several types of requests were posed. In some cases, the subject was asked to do something to an object or to use one object in relation to another, for example, "Kanzi, can you knife (cut) the sweet potato?" "Alia, can you put a rubber band on your ball?" "Kanzi, put some water on the carrots." In others cases, Kanzi or Alia were asked to do something with an object designated by its location or to engage in some activity, for example, "Kanzi, get the telephone that's outdoors," "Can you give Rose a carrot?" "Kanzi, can you take the gorilla (a stuffed toy) to the bedroom?"

The sentences of request presented to Kanzi and Alia were not only novel, but quite unusual ("Kanzi, can you get the lettuce that's in the microwave?"—where other objects as well as lettuce were placed for the test). The intent was to present requests that the subject had not previously been asked to carry out. Researchers posed such questions as these: "Kanzi/

Alia, can you make the doggie bite the snake?" "Can you tickle Linda with the bunny (a puppet)?" and so on—where the doggie, the snake, the bunny, and similar objects were frequently new toys that had not been used for such purposes in the past.

Analyses revealed that Kanzi and Alia were much more similar than they were different. In blind-test conditions, Kanzi was 74% correct overall on the sentences and Alia was 65% percent correct, where the probability of being correct by chance approximated zero. On the remaining trials, Alia and Kanzi were either partly correct or required additional assistance in order to be correct. A detailed scoring system distinguished such trials from those on which the subjects fulfilled the requests without hesitation or error. On the "error" sentences, however, the experimenters worked with the subjects so that the subjects fulfilled the request and never viewed themselves as having failed.

Alia and Kanzi also tended to make the same kinds of errors. Alia, however, was better than Kanzi at retrieving *two* objects in a single request (for instance, "Give Kelly the peas and the sweet potatoes"). On the other hand, Kanzi was the better at dealing with embedded phrases such as "Get the melon that's in the potty,"—where another melon lay directly in front of him.

Never before has a being other than a human demonstrated the capacity to comprehend novel sentences of request as has Kanzi. His ability to comprehend must certainly exceed the boundaries of the tests given him and, in all likelihood, is conservatively at the level of a normal human child of about two-and-one-half years of age.

With Kanzi's language comprehension abilities emerging before his language production abilities, for instance, understanding before talking, we have observed for the first time the pattern of language acquisition in an ape that characterizes the pattern of language acquisition of the human child. There seems little doubt that, if Kanzi could "speak," he would have much more to say than what is permitted by use of his keyboards. (An improved keyboard that is designed to facilitate his use of words, both individually and in combination, is under development.)

Grammar: Production

Kanzi has learned simple grammatical rules, modeled by humans, and has invented his own symbol-ordering rules as well (Greenfield and Savage-Rumbaugh 1991). The capacity to learn grammar, the rules through which symbols may be combined in a potentially infinite number of ways, is generally recognized as an essential requisite for competence in language.

In accordance with accepted methods for studying spoken and sign language in children, we classified Kanzi's lexigram and lexigram-gesture combinations. Each combination was classified by its semantic relationships;

for example, "MATATA BITE" was classified as agent-action, and "KEEP AWAY BALLOON" was classified as action-object. Kanzi's behavior that followed production of a combination, rather than an observer's interpretation of Kanzi's intent, determined the classification. Findings indicated that, although Kanzi employed a wide variety of semantic relationships, he tended to use specific orders.

Kanzi's use of the action-object order, "SLAP BALL" for instance, was significantly more frequent than his use of the alternative order, "SURPRISE HIDE". In this regard, he follows the rule of grammar used by his caretakers. Analyses further indicated that Kanzi made up his own rule for combining an agent gesture with an action lexigram, as in the example "CHASE YOU," with "chase" expressed with a lexigram and "you" with a gesture. His ordering rule for lexigram-gesture combinations, "Place lexigrams first," contrasted with the ordering strategy expressed by his caregivers' English-based rule. Furthermore, Kanzi's rule had considerable generality as well as originality.

Kanzi frequently combined two action lexigrams such as "BITE TICKLE", CHASE HIDE", or "CHASE BITE". From the perspective of human language, these combinations are merely unstructured lists in that they lack the minimum requirements of a statement (for instance, instead of providing one predicate and one subject, action-action combinations are a chain of two predicates). However, Kanzi's action-action combinations reflected both *natural* action categories and *preferred* action orders in social play.

In playing "chase," Kanzi would frequently use the lexigram for *chase* on the keyboard, then he would lead a person by the hand and place that person's hand on the person he apparently wanted chased. However, Kanzi's ability to signal a difference in meaning was limited: on other occasions, Kanzi would nominate himself as the one to be chased by scampering off after using the lexigram for chase and then gesturing to the person nominated to chase after him. Contrary to the claim that apes cannot make statements but are limited to demands or requests, Kanzi frequently made statements of what he was about to do. In sum, Kanzi has both learned and invented rules for ordering relationships between two categories of symbols and he does so with a competence that reflects the grammar of one-and-a-half year old children.

Development, Rearing, Adult Competencies, and Cultural Evolution

It is our view that our apes' language skills appear spontaneously and generally in accord with the pattern of language acquisition of the human child because our apes have been reared from birth in language-saturated environments where communication is emphasized. From shortly after birth, language (both speech and lexigrams) was used consistently to the end that the apes learned that their world was somewhat predictable. Through processes of observational learning, the apes came to discern a relationship between the actions of others (such as their use of lexigrams and speech) and the consequences that ensued (such as traveling to a place, getting some specific food or drink, engaging in play, and so on). We believe that, as this learning occurs, the ape's psychological competencies become structured in new and novel ways—ways that might be very foreign to their species in the field or in zoos, ways that approximate those of the developing human infant. Thus, our view as to why our apes came to comprehend speech and lexigrams goes well beyond the suggestion that they became facile learners (Krech et al. 1962; Davenport et al. 1973) as a result of generally stimulating environs during infancy.

Our apes give reason for us to believe that, during infancy, they learned more from observing the consequences of their social partners' use of the keyboard than they did from using it. Such learning extends to the infant ape's informational opportunities, which far exceed its motor ability to interact effectively with its physical and social environs. Such learning provides the infant ape a panoply of information from which it abstracts and integrates the basic framework of its language skills, skills that will not be made manifest until a later age. The ape's brain does this not because it is "pellet-driven" (or shaped by specific reinforcers), but because that is what its advanced primate brain was selected to do (Rumbaugh et al. 1991).

It seems likely to us that as the ape learns language it develops neurological networks that resemble those that were basic to human neuroevolutionary trends and to the evolution of language and its attendant branches (Riesen 1982; Greenough et al. 1990). A definitive answer to this question awaits future research.

This perspective holds an interesting implication for understanding the emergence of human culture. The implication is that the first steps taken toward culture through the inventions of adults had their maximum impact upon the developing cognitive structures of young children who grew up observing each invention/innovation and its merits. These cognitive structures may have served to channel the attention, creativity, and inventiveness of these children through the course of their development. Their own creations, in turn, would have provided still better, more focused contexts whereby the intellect of their own infants would be both patterned and directed. Cultural gains might have been made indirectly and quietly as each gain, in turn, served to direct cognitive development of the ever-observant child to specific topics for reflection and refinement as he or she matured. Thus, the bedrock for geometric gains in building culture and technology across eons was laid in the minds of babes.

Summary

Language, in its basic dimensions, may no longer rationally be held as the characteristic that separates humans from animals (Bates et al. 1991). Although as a species humans excel over all other species both in the mastery and the expression of language through speech, language is more than simply speech. Language is a reflection of capacities that are differentially structured depending upon the environment in which the being—human or ape—is reared.

Acknowledgments

This research was supported by Grant HD–06016 from the National Institute of Child Health and Human Development and the College of Arts and Sciences, Georgia State University.

References

Alcock, J. 1984. *Animal Behavior: An Evolutionary Approach.* Sunderland, Mass.: Sinauer Associates.

Bates, E., D. Thal, and V. Marchman. 1991. Symbols and syntax: A Darwinian approach to language development. In N.A. Krasnegor, D.M. Rumbaugh, R.L. Schiefelbusch, and M. Studdert-Kennedy, eds., *Biological and Behavioral Determinants of Language Development,* pp. 29–65. Hillsdale, N.J.: Lawrence Erlbaum Associates.

Bellugi, U., A. Bihrle, and D. Corina. 1991. Linguistic and spatial development: Dissociations between cognitive domains. In N.A. Krasnegor, D.M. Rumbaugh, R.L. Schiefelbusch, and M. Studdert-Kennedy, eds., *Biological and Behavioral Determinants of Language Development,* pp. 363–93 Hillsdale, N.J.: Lawrence Erlbaum Associates.

Cheney, D.L., and R.M. Seyfarth. 1990. *How Monkeys See the World.* Chicago: Univ. of Chicago Press.

Darwin, C. 1859. *The Origin of Species.* New York: Hurst and Company.

Davenport, R.K., C.W. Rogers, and D.M. Rumbaugh. 1973. Long-term cognitive deficits in chimpanzees associated with early impoverished rearing. *Developmental Psychol.* 9:343–7.

Descartes, R. 1956. *Discourse on Method.* New York: Liberal Arts Press. (Original work published 1637.)

Essock, S.M. 1977. Color perception and color classification. In D.M. Rumbaugh, ed., *Language Learning by a Chimpanzee: The LANA Project,* pp. 207–24. New York: Academic Press.

Gardner, R.A., and B.T. Gardner. 1989. Cross-fostered chimpanzees. In P.G. Heltne and L.A. Marquardt, eds., *Understanding Chimpanzees,* pp. 220–33. Cambridge, Mass.: Harvard Univ. Press.

Goodall, J. 1989. Gombe: Highlights and current research. In P.G. Heltne and L.A. Marquardt, eds., *Understanding Chimpanzees,* pp. 2–21. Cambridge, Mass.: Harvard Univ. Press.

Greenfield, P., and E.S. Savage-Rumbaugh. 1991. Imitation, grammatical development, and the invention of protogrammar by an ape. In N.A. Krasnegor, D.M. Rumbaugh, R.L. Schiefelbusch, and M. Studdert-Kennedy, eds., *Biological and Behavioral Determinants of Language Development,* pp. 235–58. Hillsdale, N.J.: Lawrence Erlbaum Associates.

Greenough, W.T., G.S. Withers, and C.S. Wallace. 1990. Morphological changes in the nervous system arising from behavioral experience: What is the evidence that they are involved in learning and memory? In L.R. Squire and E. Lindenlaub, eds., *The Biology of Memory, Symposia Medical Hoechst* 23:159–84. Stuttgart and New York: F.K. Schattauder Verlag.

Heltne, P.G., and L.A. Marquardt, eds. 1989. *Understanding Chimpanzees.* Cambridge, Mass.: Harvard Univ. Press.

Hewes, G.W. 1973. Primate communication and the gestural origin of language. *Current Anthro.* 14:5–32.

———. 1977. Language origin theories. In D.M. Rumbaugh, ed., *Language Learning by a Chimpanzee: The LANA Project,* pp. 3–53. New York: Academic Press.

Hopkins, W.D., and E.S. Savage-Rumbaugh. 1991. Vocal communication as a function of differential rearing experiences in *Pan paniscus:* A preliminary report. *Internat. J. Primatol.* 12:559–84.

Krech, D., M.R. Rosenzweig, and E.L. Bennett. 1962. Relations between brain chemistry and problem solving among rats raised in enriched and impoverished environments. *J. Comp. and Physiological Psychol.* 55:801–8.

Matsuzawa, T. 1990. *The Perceptual World of a Chimpanzee.* Tokyo: Ministry of Education Project No. 63510057.

Miles, L. 1990. The cognitive foundations for reference in a signing orangutan. In S.T. Parker and K.R. Gibson, eds., *Language and Intelligence in Monkeys and Apes,* pp. 511-39. New York: Columbia Univ. Press.

Pate, J.L., and D.M. Rumbaugh. 1983. The language-like behavior of Lana chimpanzee: Is it merely discrimination and paired-associate learning? *Anim. Learn. Behav.* 11:134–8.

Premack, D. 1975. Language in chimpanzees? *Science.* 172:808–22.

Riesen, A.H. 1982. Effects of environments on development in sensory systems. In W.D. Neff, ed., *Contributions to Sensory Physiology.* Vol. 6, pp. 45–77. New York: Academic Press.

Romski, M.A. 1989. Two decades of language research with great apes. *Am. Speech-Language-Hearing Association.* 31:81–3.

Romski, M.A., and R.A. Sevcik. 1991. Patterns of language learning by instruction: Evidence from nonspeaking persons with mental retardation. In N. Krasnegor, D.M. Rumbaugh, R. L. Schiefelbusch, and M. Studdert-Kennedy, eds., *Biological and Behavioral Determinants of Language Development,* pp. 429–45. Hillsdale, N.J.: Lawrence Erlbaum Associates.

Rumbaugh, D.M., ed. 1977. *Language Learning by a Chimpanzee: The LANA Project.* New York: Academic Press.

Rumbaugh, D.M., W.D. Hopkins, D.A. Washburn, and E.S. Savage-Rumbaugh. 1991. Comparative perspectives of brain, cognition, and language. In N. A. Krasnegor, D.M. Rumbaugh, R.L. Schiefelbusch, and M. Studdert-Kennedy, eds., *Biological and Behavioral Determinants of Language Development,* pp. 145–64. Hillsdale, N.J.: Lawrence Erlbaum Associates.

Rumbaugh, D.M., and J.L. Pate. 1984. The evolution of primate cognition: A comparative perspective. In H.L. Roitblat, T.G. Bever, and H.S. Terrace, eds., *Animal Cognition,* pp. 569–87. Hillsdale, N.J.: Lawrence Erlbaum Associates.

Savage-Rumbaugh, E.S. 1986. *Ape Language: From Conditioned Responses to Symbols.* New York: Columbia Univ. Press.

———. 1988. A new look at ape language: Comprehension of vocal speech and syntax. In D. Leger, ed., *The Nebraska Symposium on Motivation.* 35:201–55. Lincoln, Nebr.: The Univ. of Nebraska.

Savage-Rumbaugh, E.S., K.E. Brakke, and S.S. Hutchins. 1992. Linguistic development: Contrasts between co-reared *Pan troglodytes* and *Pan paniscus.* In T. Nishida, W.C. McGrew, P. Marler, M. Pickford, and F.B.M. de Waal, eds., *Topics in Primatology.* Vol. 1, *Human Origins,* pp. 51–66. Tokyo: Univ. of Tokyo Press.

Savage-Rumbaugh, E.S., K. McDonald, R.A. Sevcik, W.D. Hopkins, and E. Rubert. 1986. Spontaneous symbol acquisition and communicative use by two pygmy chimpanzees. *J. Exp Psychol.: General* 115:211–35.

Savage-Rumbaugh, E.S., J. Murphy, R.A. Sevcik, S. Williams, K. Brakke, and D.M. Rumbaugh. 1993. Language comprehension in ape and child. *Monographs of the Society for Research in Child Development,* Vol. 58, pp. 3–4.

Savage-Rumbaugh, E.S., J.L. Pate, J. Lawson, S.T. Smith, and S. Rosenbaum. 1983. Can a chimpanzee make a statement? *J. Exp. Psychol.: General.* 112:457–92.

Sevcik, R.A. 1989. A comprehensive analysis of graphic symbol acquisition and use: Evidence from an infant bonobo *(Pan paniscus).* Ph.D. diss., Georgia State University, Atlanta.

Sevcik, R.A., M.A. Romski, and K.M. Wilkinson. 1991. Roles of graphic symbols in the language acquisition process for persons with severe cognitive disabilities. *Augmentative and Alternative Communication.* 7:161-70.

Terrace, H. 1979. *Nim: A Chimpanzee Who Learned Sign Language.* New York: Knopf.

Tuttle, R.H. 1986. *Apes of the World.* Park Ridge, N.J.: Noyes Publications.

Individual Differences in the Cognitive Abilities of Chimpanzees

Sarah T. Boysen

After more than three decades of study of the chimpanzee, led by the landmark fieldwork of Goodall and her associates (Goodall 1968, 1986; McGrew 1992) and by the equally significant work by pioneers intent on characterizing the behavioral and cognitive characteristics of this remarkable species in captivity (Yerkes and Yerkes 1929; Hayes and Hayes 1951; Kellogg and Kellogg 1933), the chimpanzee has emerged as an animal whose capabilities cannot be easily compartmentalized. The chimpanzee has demonstrated its diverse capacities and traditions in use of tools, patterns of grooming, use of food resources, and capacity for attention as well as diversity in personality and temperament (Boysen 1992b; de Waal 1982, this volume; McGrew 1992, this volume). Indeed, the phrase that best describes the range of chimpanzee features and the behavior represented across chimpanzee populations in the wild and in numerous captive environments is *remarkable variability.*

Remarkable variability, including behavioral plasticity and flexibility, characterizes the range of cognitive abilities and cognitive potential that we studied in an effort to better define the learning capabilities and information-processing capabilities of chimpanzees. The Primate Cognition Project was established in recognition of our common primate heritage to bring potential shared cognitive similarities and differences of chimpanzees and humans more sharply into focus.

The past decade in the field of comparative psychology has witnessed a resurgence of interest in cognition and the comparative study of cognitive processes (Boysen and Capaldi 1993; Honig and Fetterman 1992; Ristau 1991; Roitblat et al. 1984). Moreover, chimpanzees have been reported to show cognitive abilities that are not in evidence for other animals, with

the possible exception of other apes. Those cognitive abilities include cross-modal matching of stimuli (Davenport and Rogers 1970), self-recognition (Gallup 1970, 1991; Lin et al. 1992; Parker 1991), tool use and construction (McGrew 1992), and social attribution processes not shown in monkeys (Povinelli et al. 1990; Premack 1986; Premack and Woodruff 1978). Additional highly sophisticated cognitive capabilities have been suggested for the chimpanzee (Premack 1976; Premack and Woodruff 1978) and, thus, chimpanzees may afford a unique model with which to study the comparative development of cognition.

The studies undertaken in our laboratory have explored a range of cognitive abilities and capacities in the chimpanzee, including recognition of individual humans and conspecifics, vigilance and attention; and number-related skills (Boysen 1992a; Boysen et al. 1987, 1989a, 1989b, 1990, 1993; Boysen and Capaldi 1993). While the chimpanzees exhibited success with most of the tasks that we presented to them, it became apparent early in training that our three original chimpanzees represented a unique constellation of temperament, personality, and learning styles (Boysen 1992b). While we needed to adhere to the same experimental procedures with each animal on a given task, we readily perceived differences in the relationship of each chimpanzee with its human teacher, in the individual chimpanzee's abilities, and in the animal's approach to learning. To ignore the animals' individuality would have been to deny the real contribution of individual differences to the experimental context. These individual differences were significant parameters for precisely the cognitive potential we hoped to study (Boysen 1992b; Oden and Thompson 1992).

Physiological Correlates of Social Recognition.

The relationships between our animals were dramatically different and readily discernible, even with only three chimpanzees. The two males Kermit and Darrell are from the collection of the Yerkes Regional Primate Research Center, Emory University. They are six months apart in age and have been together since early infancy at the Yerkes Center. Upon arrival at Ohio State at ages 3 and 3½, they had been housed together continuously. The young female Sheba was originally from the Columbus Zoo. At the time she joined the project at age 2½, she was physically too small to be housed with the boisterous and much larger males. However, all three animals spent a significant portion of each day interacting with one other and their human teacher.

From the first introduction of Sheba to the males, Darrell was aggressive; he maintained an antagonistic attitude toward her for the next nine years. On the other hand, although physically smaller than and subordinate to Darrell at the time, Kermit was immediately protective of Sheba, literally shielding her body with his from Darrell's blows and bites. These reactions by both males were exhibited within the very first seconds of their introduction to Sheba. And, like the enduring reaction of Darrell toward Sheba, Kermit has remained protective and highly prosocial toward her for the same nine-year period. Similarly these chimpanzees clearly exhibited distinct likes and dislikes toward humans who visited or worked in the laboratory.

With these observations in mind, two studies were undertaken to explore underlying physiological reactivity to individual social relationships via cardiac measures. To explore a potential objective index of recognition of individuals relative to prior social experience, we recorded Sheba's heart rate as she viewed visual stimuli in the form of color slides in three social categories: her human caregivers, familiar individuals who she was likely to recognize but who were not directly involved in the project, and strangers; blank slides were used as a control (Berntson and Boysen 1986). Heart rate responses were recorded by disposable adhesive sensors attached to Sheba's chest and secured with a wide elastic bandage. Sheba required no pretraining for this procedure, and she readily cooperated in the preparation for recording her heart rate. We imposed no restraint (which we would not have done, despite our interest in the questions). The ease of testing and her apparent lack of concern for the recording apparatus and the attached sensors were likely supported by the strong bond between Sheba and her primary caregiver (Sarah Boysen), coupled with a gamelike approach to the task. For example, Sheba was provided with a full frontal view of herself in a large mirror, and she sought the opportunity to see herself.

We recorded Sheba's heart-rate responses to color slides of six human female faces in the three categories noted earlier and to blank control slides. Each slide was presented for eight seconds, with baseline heart rate recorded four seconds before, and twelve seconds after each slide was presented. Sheba simply sat and viewed the slides with an experimenter who was both unaware of the sequence of the slides and unable to see the projection screen. The results of the study revealed that heart rate responses were deceleratory to most slides (see figure 1) and that Sheba exhibited significant differences in her heart rate responses across the three categories. The most striking finding was Sheba's consistently larger deceleratory response to color slides of her caregivers.

The larger deceleratory heart rate response to the caregiver category appeared consistently over sessions, and to each of the individual color slides in this category, for an average of minus 9.4 beats per minute. The differential heart rate response was also apparent in measures of heart-period variability, reflected in a significantly longer deceleratory response to caregiver slides as compared with other slides. These changes in heart-rate response

Figure 1
The heart rate response patterns are shown during slide presentations of human, categorized as caregivers, familiar individuals, strangers, and blank (white) control slides.

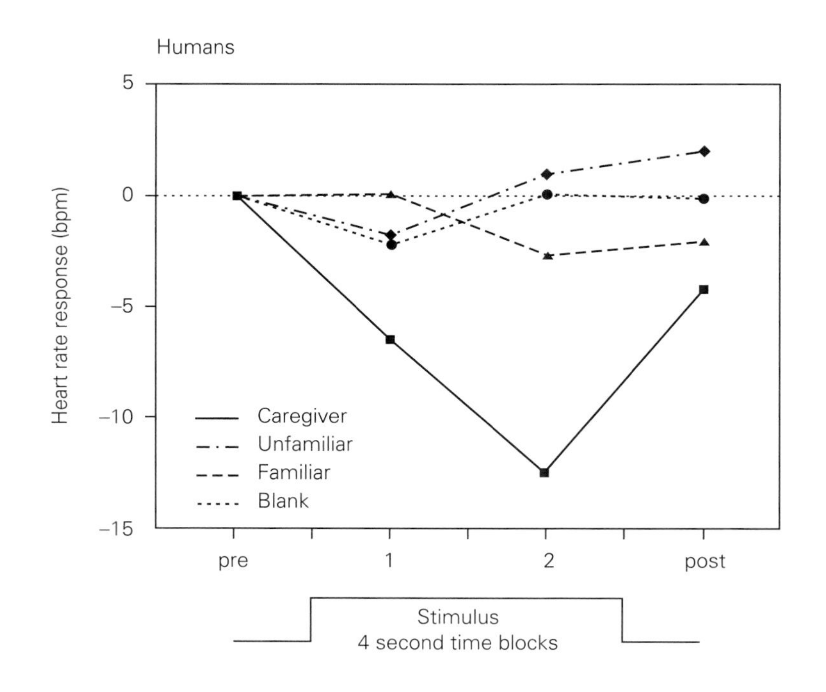

were not associated with any apparent pattern of overt behavioral change although, in one interesting instance, Sheba extended her hand toward the projected image of a caregiver.

These findings suggest that Sheba could recognize individuals from visual representations and that her heart rate might provide an objective measure of her social relationship with the individuals represented (Boysen and Berntson 1986). Most notably, these differences in heart rate response to the three social categories were exhibited in the absence of explicit training, task demands, or prior experience with slides of humans. The slower heart rate was likely part of an *orienting response* (Graham 1979), a response that is typically found associated with cardiac deceleration in numerous species including humans, apes, and monkeys (Graham 1979; Weisbard and Graham 1971; Boysen and Berntson 1984). Sheba's heart-rate responses in this study appeared to be dependent upon the inherent significance of the slides, and those responses reflected her strong social bond with her caregivers. The results raised an obvious question. Would Sheba's heart rate patterns similarly reflect her dramatically different social relationships with the other chimpanzees in her group?

To answer the question, Sheba was given the opportunity to view color slides of chimpanzees in three social categories: an aggressor category represented by Darrell; Kermit, with whom Sheba shared a gregarious social relationship and who therefore represented a companion category; and an unfamiliar chimpanzee in the stranger category. Blank slides served as a control (Berntson and Boysen 1989a). The testing procedures were identical to those outlined for the human-recognition study. Similar to the findings from that experiment, Sheba exhibited consistent heart rate patterns to one category of slides. However, in contrast to the deceleratory responses exhibited when viewing the human caregiver slides, the significant heart rate response to a chimpanzee category was acceleratory (see figure 2). Moreover, this response was made only to slides of the aggressor Darrell, with whom Sheba had a long history of agonistic interactions, or conflicts.

The consistent acceleratory pattern likely represented a *defensive response,* a response that is exhibited under aversive or threatening circumstances (Graham 1979). We concluded that the defensive response to Darrell's slides reflected the aggressive and negative social relationship that Sheba and Darrell maintained and that the heart rate response was an objective, physiological reflection of their acrimonious history (Berntson and Boysen 1989a). Thus, both visual-recognition studies revealed that social history is a potent source of individual differences in chimpanzees. Social learning can powerfully shape the behavior, preferences, and dispositions of chimpanzees and humans alike. Even in the absence of genetic variation, differential social experiences would serve to carve individual features into the chimpanzee, given its exceptional behavioral plasticity.

Figure 2
The heart rate responses are shown during slide presentations of chimpanzees, including categories of Aggressor, Companion, Stranger, and blank control slides.

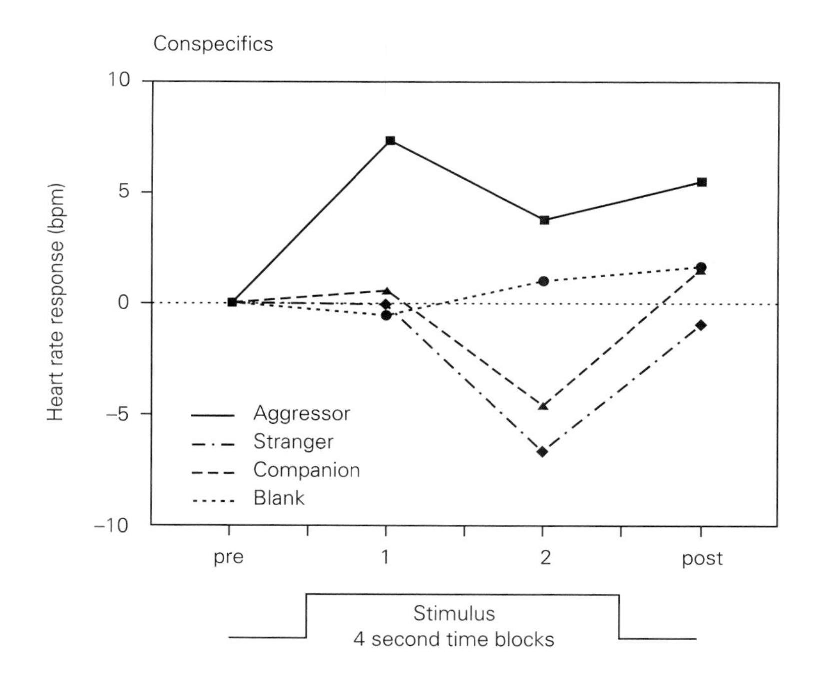

Cross-modal Recognition

Vocalizations in addition to physical features provide a rich source of information about individuals and play an important role in recognition. A number of species have been studied to determine if recognition of conspecifics through vocal input alone is possible (Cheney and Seyfarth 1980; Marler and Hobbett 1975; Snowdon and Cleveland 1980; Waser and Waser 1977). For example, several studies have demonstrated that human infants and juveniles respond selectively to vocalizations of their fathers (DeCasper and Prescott 1984) and that primate mothers respond differentially to vocalizations of their offspring (Kaplan et al. 1978; Newman 1985; Symmes and Biben 1985). Recognition of individual conspecifics may be particularly important among species such as chimpanzees who live in relatively large social communities. The ability to distinguish individuals in such a community would aid the establishment and maintenance of critical relationships and alliances and would contribute to the overall stability of the group (de Waal 1982).

As demonstrated by Sheba, many primate species have shown the ability to recognize visual representations of other primates and objects (Bruce 1982; Davenport and Rogers 1970, 1971; Fujita and Matsuzawa 1986; Rosenfield and van Hoesen 1979; Sands and Wright 1980a, 1980b, 1982; Swartz and Rosenblum 1980). Moreover, Bauer and Phillip (1983) reported that chimpanzees learned to match an individual's facial portraits with that individual's corresponding pant-hoot vocalizations. However, no testing of transfer using novel stimuli was conducted and, thus, the generality of performance was unclear.

To further explore the capacity of the chimpanzee for cross-modal recognition, we used auditory and visual stimuli as follows: vocal greetings of humans were played during presentation of pictures of individual humans and chimpanzee food-bark vocalizations were played during presentation of pictures of individual chimpanzees (Nelson 1989). In addition, novel auditory and novel visual stimuli, different from the familiar stimuli employed in training, were used for testing.

Four chimpanzees participated in the study including Kermit who was 9 years old at the time of testing; Darrell, 9½ years old; Sarah, 29 years old; and Sheba, 8 years old. Since their arrival at Ohio State, all the animals had participated in a variety of cognitive studies, and all had previous experience with mirrors, which they used for self-directed activities and for viewing one another from adjacent cages. Darrell and Kermit were housed in the same cage regularly, as were Sarah and Sheba, but on occasion each animal had been with the other chimpanzees for limited play sessions. In addition, they had all spent repeated, brief periods of time (one to four hours) housed together in the same cage.

Front and lateral views of the faces and heads of three familiar persons were used as visual stimuli for training and testing. Those six visual images were used in two forms; as separate color photographs, which could be presented one-by-one by an experimenter, and as slides. The slide images were presented three at one time with one blank slide as a control in a two-by-two matrix of four quadrants on a color television monitor. All possible combinations of the six images were presented in counterbalanced order and position on the color monitor. The response apparatus consisted of a touch-sensitive frame (Carroll Touch-Screen) that was placed over the color monitor and linked to a microcomputer that recorded and stored the responses.

Initial training involved two sets of simple matching tasks. The animals were required to choose the slide image presented on the color monitor that matched a color photograph presented by the experimenter. Subsequently, cross-modal matching of auditory and visual stimuli was required. A 15-second recording of a vocal greeting by one of the familiar persons was presented; the animal's task was to select the photograph that represented the person giving the recorded greeting. Throughout the phases of the study, photographs of three humans were presented on the monitor during each trial, with the fourth quadrant of the screen blank, and order and position of the visual stimuli counterbalanced over trials.

The second phase of this experiment employed the slide images instead of the color photographs that had been used previously as visual stimuli. Prerecorded vocalizations by each of the three individuals were now presented on a given trial. These auditory stimuli were presented immediately prior to presentation of the slide matrix, with vocal segments consisting of a brief greeting that had been matched for sound level and duration. The chimpanzees received 20 trials in each daily session until they were performing consistently at levels greater than chance.

The third phase of this experiment evaluated the chimpanzees' capabilities for transferring cross-modal matching skills to novel stimuli. The slide matrices were modified to include three novel stimuli, which were visual images of different, but familiar, persons. These novel stimuli were randomly placed in five of the 20 trials. The testing procedure using novel stimuli was the same as the training phases, with the exception that the experimenters gave no guidance or feedback.

Results

In reviewing the results of these experimental phases, the results of the match-to-sample training, during which the animals were required to match the slide image to a photograph, revealed considerable variability in the day-to-day performance of each chimpanzee. Although each animal exhibited a clear trend toward continued improvement across sessions, each varied considerably in their rate of acquisition. For example, Kermit and Sarah required only 20 daily sessions to reach a criterion of 85% correct responses over two successive sessions, while Darrell did not reach the same criterion until day 52, and Sheba needed 72 days.

These differences in performances were intriguing, particularly given Kermit's history of measurable attentional difficulties (Boysen 1992b). When Kermit was tested on a vigilance task developed for very young children with Attention Deficit Disorder, his performance was compromised if the stimulus items appeared in degraded form or in very rapid succession. These changes in the task produce performance deficits in children with demonstrated learning disabilities (O'Dougherty et al. 1988). Thus, Kermit's attention deficits impacted on his ability to perform tasks that required sustained attention. However, in the case of the match-to-sample training with photographs, Kermit's performance proved superior to that of Darrell and Sheba, who typically outperformed the other animals on most tasks.

In the match-to-sample cross-modal training phase, the sample photographs were replaced by audio stimuli, and the animals were required to match vocalizations to an individual depicted on the color television monitor. Again, the animals varied considerably in the number of trials required to reach criterion. For example, both Kermit and Sarah required 60 daily sessions, while Darrell reached criterion after only 28 sessions, and Sheba, who had performed the most poorly on the initial match-to-sample task, met criterion in only six sessions.

In the third phase, the match-to-sample transfer test entailed the addition of random trials containing novel visual stimuli and corresponding novel auditory stimuli embedded within the videotaped matrices that had been used during training. Of the four animals, only Darrell exhibited a statistically significant performance with the novel stimuli within the first five trials, and he was correct on each novel trial. For all animals, performance in the first two sessions with the novel stimuli approximated the performance with the familiar stimuli (55% for novel tests, 59% for familiar stimuli). (See table 1). Analysis of all test trials across sessions revealed that three animals (Sheba, Kermit, and Darrell) were performing well above

Table 1
Transfer test with novel human photographs and vocalizations of humans.

Subjects	Sessions 1–5 (%)	Sessions 6–10 (%)	Overall performance (%)
Darrell	76*	64*	70*
Kermit	52**	64*	58*
Sarah	48	48	48
Sheba	64*	72*	68*

Notes: * $p < .001$; ** $p < .05$.

chance levels, while one animal (Sarah) was performing only marginally above chance. Performance on training stimuli throughout the ten sessions was maintained at significant levels by all animals, although Kermit's performance declined over the last several sessions.

The second experiment of the study consisted of a transfer test in which the novel stimuli depicted conspecifics. In this study, images of the four chimpanzee subjects were presented on a color monitor in a matrix of four quadrants, and recordings of food-bark vocalizations of Darrell and Sheba were played. The animals were initially tested using two different vocal segments from Darrell and two different vocal segments from Sheba for 10 sessions of 20 trials per session, a total of 200 trials per animal. The animals then completed an additional block of five sessions of 20 trials per session for a total of 100 trials, during which four novel vocalizations (two different vocalizations from Kermit and two different vocalizations from Sarah) were presented as test stimuli. In this transfer test with conspecifics, Sheba and Kermit both performed significantly above chance in the first session, and Darrell's performance reached significance after two sessions. Evaluation of overall performance across the first 10 sessions revealed that each animal was performing above chance (see table 2). Darrell, Sheba and Sarah demonstrated significant improvement in performance from the first five-session block to the second five-session block, while Kermit maintained his initial accuracy throughout all 10 sessions. Considerable day-to-day variability was apparent in the performance of each animal.

An analyses of errors revealed that Darrell was the only subject to demonstrate a significant bias relating to position, making significantly more errors in the lower-left quadrant and significantly fewer errors on the upper-right quadrant. A bias relating to stimulus was also observed for Kermit and Sheba: Kermit made significantly more errors by selecting the image of Sarah and significantly fewer errors by selecting the image of Darrell. Sheba, in contrast, made a significantly more errors selecting the image of Darrell.

Table 2
Novel transfer test for cross-modal recognition of chimpanzees.

Subjects	Sessions 1–5 (%)	Sessions 6–10 (%)	Overall performance (%)
Darrell	49*	59*	54*
Kermit	41**	46*	44*
Sarah	35**	45**	40**
Sheba	48*	61*	55*

Notes: * $p < .001$; ** $p < .05$.

Transfer to the remaining novel stimuli in the last five-session block proved to be relatively rapid for each animal (see table 3). In the first session, Darrell, Kermit, and Sarah each responded with significant accuracy. Sheba's performance reached statistical significance within the first two sessions. Performance across all five sessions of this transfer test was maintained at greater than chance levels by each subject.

In light of these experiments, evaluation of the ability to recognize both humans and other chimpanzees cross-modally through auditory and visual representations revealed that these chimpanzees were capable of matching visual images of humans and conspecifics to the appropriate vocalizations and that they could do so with considerable accuracy. Moreover, after initial training, three of the four chimpanzees demonstrated efficient transfer to novel auditory and visual stimuli that represented humans. Generalization to cross-modal matching of visual images and vocalizations of conspecifics was similarly rapid. In a final transfer phase, three of the four chimpanzees responded accurately to novel auditory and visual stimuli representing conspecifics during the first session, and the fourth animal reached significance after two sessions.

While no consistent pattern of errors was apparent, errors were not randomly distributed. One source of errors was likely related to attentional variables, particularly for those animals with a prior history of attentional difficulties (Kermit and, in some cases, Sarah). Each animal demonstrated lengthy runs of consecutive correct responses throughout testing, with errors also tending to occur in consecutive trials. In addition, considerable session-to-session variability was observed, with incorrect choices often associated with signs of distraction or lack of attention to the task, such as moving away from the testing area or failure to orient to the test display.

Considerable differences existed in the prior social interactions between the animals, although this did not appear to be a primary determinant of response biases. Darrell and Kermit were generally housed together, as were Sarah and Sheba. Nevertheless, response biases were not clearly related to these housing conditions or to prior social history. Thus, while Kermit made significantly fewer errors on stimuli depicting Darrell, both Sarah and Sheba

Table 3
Novel generalization test for conspecific visual/auditory recognition.

Subjects	Correct response (%)
Darrell	54
Kermit	59
Sarah	53
Sheba	66

Note: Total trials = 100.

demonstrated a similar pattern of performance. Moreover, during the final test phase, Sheba made a significant number of errors by choosing her own image. Each apparent stimulus bias in the final transfer concerned stimuli that had been reinforced, indicating that interference from a previous phase may have contributed to these incorrect choices. However, because the transfer tests in the cross-modal study employed novel auditory and visual stimuli, the chimpanzees could not have responded on the basis of previously formed associations.

Thus, the present findings appear to reflect a robust capacity for visual-auditory, cross-modal recognition in the chimpanzee. These data suggest that chimpanzees possess the capacity to form sufficiently rich auditory and visual representations of individual humans and other chimpanzees, as well as of themselves, to permit such cross-modal recognition. The fact that each subject was capable of recognizing individuals depicted in color slides and sound recordings, both of which are inherently impoverished representations, indicates that the recognition of individuals may be a relatively fundamental capacity of chimpanzees. Moreover, each animal's capability of correctly matching novel visual stimuli and novel auditory stimuli in transfer tests suggests that each animal may have previously established cross-modal identifications.

Concluding remarks

A variety of studies using our chimpanzees have revealed various individual strengths and weaknesses with respect to specific cognitive capabilities in some testing contexts among these subjects. Had Kermit been our only subject, our conclusions regarding the cognitive dynamics of the chimpanzee may have been quite different. In view of such considerations, an appreciation of the capacity for attention, developmental and experiential histories, and the temperament style of individual animals appears to be important in the interpretation of behavioral studies and, consequently, in the comparative evaluation of cognitive processes of apes.

For example, over the past year, following significant renovation of the chimpanzee housing area that has permitted our chimpanzees to interact as a group for the first time, Kermit's performance on some tasks has been dramatically enhanced. As noted earlier, Kermit and Darrell had been housed together continuously since infancy. A group of three other chimpanzees formed over the years as additional animals arrived. Thus, at 28 years of age following a lifetime of species isolation at another laboratory where other chimpanzees could be seen but not interacted with, Sarah was introduced to 6-year-old Sheba. Two years later, 3-year-old Bobby joined this pair. All five chimpanzees were introduced to one another following the laboratory renovation, and they now spend every day together.

Darrell had been dominant to Kermit for the past decade prior to the renovations. We predicted a change in their rank when the females began interacting directly with the two males, when Kermit's massive increase in size after adolescence size became apparent. Prior to the renovation, the two groups of chimpanzees had very limited visual access and no physical access to each other. Following renovation, when the two groups were introduced, the females' became increasingly solicitous, particularly to Kermit, and he did emerge as the dominant male.

Other changes in Kermit ensued after the renovation. Most interesting was Kermit's remarkable new ability to acquire number concepts, despite five to six years of continuous training that had yielded essentially no stable and definitive understanding of number and quantity associations. Variables other than changes caused by the renovation also likely impacted on Kermit's recent acquisition of number relationships. Among these variables was the addition of a computer-interfaced, touch-frame testing system that very likely provided Kermit a structure for greater attention to the task. However, his concomitant rapid change in social rank, power over the group, and enhanced control over his environment have also very likely contributed in immeasurable ways to his new abilities to acquire number concepts. Given the flexible social structure of chimpanzees in the wild and in captivity, each individual makes a potentially significant contribution toward cooperative social living. Such contributions are readily revealed within even the small social group studied in our laboratory.

In many respects, the range of conceptual tasks offered in our laboratory and the chimpanzees' resulting opportunities to demonstrate differing abilities have provided a rich picture of the range of capabilities in individual chimpanzees as well as new insights into the potential cognitive capacities of chimpanzees in general. Thus, focusing on only a singular chimpanzee, a singular approach, or a singular task, while perhaps providing evidence for a particular skill or related skills, may limit the questions that can be explored in attempting to characterize the cognitive domain of the chimpanzee.

Acknowledgments

Support for these studies was provided by National Science Foundation grants BNS-9022355 and IBN-9222637 to Sarah T. Boysen. The Ohio State University Laboratory Animal Center and the Yerkes Regional Primate

Research Center are fully accredited by AAALAC. The Yerkes Center is supported by Base Grant RR-00165 from the Division of Research Resources. The contributions of Michelle Hannan, Traci Shreyer, and Kurt Nelson to the care and teaching of the chimpanzees are gratefully acknowledged. The continued cooperation of the Columbus Zoological Gardens staff, including Jerry Borin, Director; Don Winstel, Assistant Director; Dr. Lynn Kramer, D.V.M., and Director of Research, and Diana Frisch, Great Apes Department, is greatly appreciated.

References

Bauer, H.R., and M.M. Phillip. 1983. Facial and vocal individual recognition in the common chimpanzee. *Psychol. Record* 33:161–70.

Berntson, G.G., and S.T. Boysen. 1990. Cardiac correlates of cognition in infants, children and chimpanzees. In L.P. Lipsett and C. Rovee-Collier, eds., *Advances in Infancy Research*, pp. 187–220. New York: Ablex Publishing.

Boysen, S.T. 1992a. Counting as the chimpanzee views it. In W.K. Honig and J.G. Fetterman eds., *Cognitive Aspects of Stimulus Control*, pp. 367–83. Hillsdale, N.J.: Lawrence Erlbaum Associates.

———. 1992b. Pongid pedagogy: The contribution of human/chimpanzee interaction to the study of ape cognition. In H. Davis and D. Balfour, eds., *The Inevitable Bond: Scientist-Animal Interactions.* Cambridge, England: Cambridge Univ. Press.

Boysen, S.T., and G.G. Berntson. 1984. Cardiac startle and orienting responses in the great apes. *Behav. Neurosci.* 98:914–18.

———. 1986. Cardiac correlates of individual recognition in the chimpanzee *(Pan troglodytes). J. Comp. Psychol.* 100:321–4.

Boysen, S.T., G.G. Berntson and J. Prentice. 1987. Simian scribbles: A reappraisal of drawing in the chimpanzee *(Pan troglodytes). J. Comp. Psychol.* 101:82–9.

———. 1989a. Conspecific recognition in the chimpanzee *(Pan troglodytes):* Cardiac responses to significant others. *J. Comp. Psychol.* 103:215–20.

———. 1989b. Numerical competence in a chimpanzee *(Pan troglodytes). J. Comp. Psychol.* 103:23–31.

———. 1990. The emergence of numerical competence in the chimpanzee *(Pan troglodytes).* In S.T. Parker and K.R. Gibson, eds., *Language and Intelligence in Animals: Developmental Perspectives,* pp. 435–50. Cambridge, England: Cambridge Univ. Press.

Boysen, S.T., G.G. Berntson, T.A. Shreyer, and K. Quigley. 1993. Processing of ordinality and transitivity by chimpanzees *(Pan troglodytes). J. Comp. Psychol.* 107:208–15.

Boysen, S.T. and E.J. Capaldi. 1993. Counting in chimpanzees: Nonhuman principles and emergent properties of number. In S.T. Boysen and E.J. Capaldi, eds., *The Development of Numerical Competence: Animal and Human Models,* pp. 39–59. Hillsdale, N.J.: Lawrence Erlbaum Associates.

Bruce, C. 1982. Face recognition by monkeys: Absence of an inversion effect. *Neuropsychologia* 20:515–22.

Cheney, D., and R. Seyfarth. 1980. Vocal recognition in free-ranging vervet monkeys. *Anim. Behav. Monogr.* 28:362–7.

Davenport, R.K. and C.M. Rogers. 1970. Inter-modal equivalence of stimuli in apes. *Science* 168:279–80.

———. 1971. Perception of photographs by apes. *Behavior* 39:318–20.

DeCasper, A.J., and P.A. Prescott. 1984. Human newborns' perception of male voices: Preference, discrimination, and reinforcing value. *Developmental Psychobiol.* 17:481–91.

Fujita, K. and T. Matsuzawa. 1986. A new procedure to study the perceptual world of animals with sensory reinforcement: Recognition of humans by a chimpanzee. *Primates* 27:283–91.

Gallup, G.G., Jr. 1970. Self-recognition in chimpanzees. *Science* 167:86–7.

———. 1991. Towards a comparative psychology of self-awareness: Species limitations and cognitive consequences. In G.R. Goethals and J. Strauss, eds., *The Self: An Interdisciplinary Approach,* pp. 121–35. New York: Springer-Verlag.

Goodall, J. 1968. The behavior of free-living chimpanzees in the Gombe Stream Reserve. *Anim. Behav. Monogr.* 1:161–311.

———. 1986. *The Chimpanzees of Gombe: Patterns of Behavior.* Cambridge, Mass.: Belknap Press.

Graham, F.K. 1979. Distinguishing among orienting, defensive and startle reflexes. In H.D. Kimmel, E.H. van Olst, and J.F. Orlebeke, eds., *The Orienting Reflex in Humans,* pp. 137–67. Hillsdale, N.J.: Lawrence Erlbaum Associates.

Hayes, K.J. and C. Hayes. 1951. The intellectual development of a home-reared chimpanzee. *Proc. of the Am. Philosophical Society* 95:105–9.

Honig, W.K., and J.G. Fetterman. 1992. *Cognitive Aspects of Complex Stimuli.* Hillsdale, N.J.: Lawrence Erlbaum Associates.

Kaplan, J.N., A. Winship-Ball, and L. Sim. 1978. Maternal discrimination of infant vocalizations in squirrel monkeys. *Primates* 19:187–93.

Kellogg, W.N. and L.A. Kellogg. 1933. *The Ape and the Child.* New York: McGraw-Hill.

Lin, A.C., K.A. Bard and J.R. Anderson. 1992. Development of self-recognition in chimpanzees *(Pan troglodytes). J. Comp. Psychol.* 106:120–7.

Marler, P., and L. Hobbett. 1975. Individuality in long-range vocalizations of wild chimpanzees. *Z. Tierpsychol.* 38:97–109.

McGrew, W.C. 1992. *Chimpanzee Material Culture.* Cambridge, England: Cambridge Univ. Press.

Nelson, K.E. 1989. *Cross-modal Recognition of Conspecifics and Humans.* Columbus Ohio: University Microfilm.

Newman, J.D. 1985. The infant cry of primates: An evolutionary perspective. In B.M. Lester and C.F.Z. Boukydis, eds., *Infant Crying: Theoretical and Research Perspectives,* pp. 307–23. New York: Plenum Press.

Oden, D. and R.K. Thompson. 1992. In H. Davis and D. Balfour, eds., *The Inevitable Bond: Scientist-Animal Interactions.* Cambridge, England: Cambridge Univ. Press.

O'Dougherty, M., G.G. Berntson, S.T. Boysen, F.S. Wright, and D. Teske. 1988. Psychophysiological predictors of attentional dysfunction in children with congenital heart defects. *Psychophys.* 25:305–15.

Parker, S.T. 1991. A developmental approach to the origins of self-recognition in great apes. *J. Human Evol.* 6:435–49.

Povinelli, D.J., K.E. Nelson, and S.T. Boysen. 1990. Inferences about guessing and knowing by chimpanzees *(Pan troglodytes). J. Comp. Psychol.* 104:203–10.

Premack, D. 1976. *Intelligence in Ape and Man.* Hillsdale, N.J.: Lawrence Erlbaum Associates.

———. 1986. *Gavagai.* Cambridge, Mass.: MIT Press.

Premack, D., and G. Woodruff. 1978. Does the chimpanzee have a theory of mind? *Behav. and Brain Sci.* 1:515–26.

Ristau, C. 1991. *Cognitive Ethology: The Minds of Other Animals (Essays in Honor of Donald R. Griffin).* Hillsdale, N.J.: Lawrence Erlbaum Associates.

Roitblat, H.L., T.G. Bever, and H.S. Terrace. 1984. *Animal Cognition.* Hillsdale, N.J.: Lawrence Erlbaum Associates.

Rosenfield, S.A., and G.W. van Hoesen. 1979. Face recognition in the rhesus monkey. *Neuropsychologia* 17:503–9.

Sands, S.F., and A.A. Wright. 1980a. Primate memory: Retention of serial list items by a rhesus monkey. *Science* 209:938–48.

———. 1980b. Serial probe recognition performance by a rhesus monkey and a human with 10- and 20-item lists. *J. Exper. Psychol.: Anim. Behav. Processes* 6:386—96.

———. 1982. Monkey and human pictorial memory scanning. *Science* 216:133–4.

Snowdon, C.T., and J. Cleveland. 1980. Individual recognition of contact calls by pygmy marmosets. *Anim. Behav. Monogr.* 28:717–27.

Swartz, K.B., and L.A. Rosenblum. 1980. Operant responding by bonnet macaques for color videotaped recordings of social stimuli. *Anim. Learn. Behav.* 8:311–21.

Symmes, D., and M. Biben. 1985. Maternal recognition of individual infant squirrel monkeys from isolation call playbacks. *Am. J. Primatol.* 9:39–46.

de Waal, F.B.M. 1982. *Chimpanzee Politics.* New York: Harper and Row.

Waser, P.M., and M.S. Waser. 1977. Experimental studies of primate vocalizations: Specializations for long-distance propagation. *Z. Tierpsychol.* 45:239–63.

Weisbard, C. and F.K. Graham. 1971. Heart-rate change as a component of the orienting response in monkeys. *J. Comp. and Physiol. Psychol.* 76:74–83.

Yerkes, R.M., and A.W. Yerkes. 1929. *The Great Apes: A Study of Anthropoid Life.* New Haven, Conn.: Yale Univ. Press.

Field Experiments on Use of Stone Tools by Chimpanzees in the Wild

Tetsuro Matsuzawa

Introduction

Chimpanzees *(Pan troglodytes verus)* at Bossou, Guinea, West Africa, have been observed to use stones as a hammer and anvil to open the nuts of the oil palm (Sugiyama and Koman 1979b). Nut cracking with stones is a strong example of the diversity of material culture among chimpanzees (McGrew 1992). This tool-use behavior has been reported only among chimpanzees of West Africa (Boesch and Boesch 1984; Boesch 1991a; Hannah and McGrew 1987; Kortlandt 1986); no chimpanzees in East Africa have been observed to crack nuts with stones although stones and nuts are available (Goodall 1986; Nishida 1990).

The use of stone tools by chimpanzees in the wild was analyzed using field experiments in an outdoor laboratory at the site. This chapter describes the recent research highlighting new findings in the cognitive abilities of chimpanzees in the wild including the use of a *metatool* (a tool that serves as a tool for another tool), the possibility of a critical period for learning, and cultural transmission between groups and across generations of chimpanzees.

Wolfgang Köhler demonstrated the abilities of chimpanzees to use sticks, to stack boxes, or to obtain a banana that was out of reach (Köhler 1927). Since then, many researchers have studied the cognitive skills of chimpanzees, as, for example, linguistic studies (Gardner and Gardner 1969; Premack 1971; Fouts 1973; Rumbaugh 1977; Terrace 1979; Matsuzawa 1985a; Savage-Rumbaugh 1986).

Focusing on issues of comparative cognitive science, I have compared the perceptual and the cognitive skills of humans and chimpanzees using the same apparatus and following the same experimental procedure. My

experiments have dealt with color perception (Matsuzawa 1985b), form perception (Matsuzawa 1990a; Tomonaga and Matsuzawa 1992), face perception (Tomonaga et al. 1993), number conception (Matsuzawa 1985a; Matsuzawa et al. 1991), and linguistic abilities (Matsuzawa 1989, 1990b; Fujita and Matsuzawa 1990; Itakura and Matsuzawa 1993).

These laboratory studies illuminated similarities between the chimpanzees and humans as well as a few differences. This work also raised a question: is the intelligence demonstrated by chimpanzees in the laboratory paralleled in their natural habitat? To answer this question, it was necessary to go to West Africa to explore the corresponding cognitive skills of chimpanzees in the wild. My colleagues and I studied the use of stone tools by chimpanzees at Bossou, Guinea, West Africa (Sakura and Matsuzawa 1991; Matsuzawa 1991b; Fushimi et al. 1991; Sugiyama et al. 1993; Yamakoshi and Matsuzawa 1993b).

In the field it is difficult to observe nut cracking directly. The bush beneath the palm trees is thick, and the chimpanzees are too timid when they are in the secondary forest near the village. Therefore, to study the use of stone tools by chimpanzees at Bossou, an outdoor laboratory (Kortlandt 1967) was established in the home range of the chimpanzees. This chapter presents findings revealed by these field experiments on nut cracking of chimpanzees.

Method

Subjects

The behavior and ecology of Bossou chimpanzees, with no supplemental feeding, have been investigated since 1976 by Sugiyama and his colleagues (Sugiyama 1984; Sugiyama and Koman 1979a; Matsuzawa et al. 1991) and each member of this chimpanzee group has been individually identified since 1976.

I started field experiments on nut cracking in 1987 (Sakura and Matsuzawa 1991). Since then, my colleagues and I have accumulated longitudinal observations on the development of skills in the use of stone tools by the Bossou chimpanzees. I collected the experimental observations for this study during three dry seasons: from December 1990 to February 1991, from December 1991 to February 1992, and from December 1992 to February 1993.

The Bossou group consisted of 18 chimpanzees in the first period, 18 chimpanzees in the second period (three disappeared and three were born after the end of the first period), and 16 chimpanzees in the third period (one died and one was born after the end of the second period).

General Procedure

The outdoor laboratory was established at the top of a small hill named Gban, which is the core area of the chimpanzees' home range. The research site was located at 7° 38' 38" north latitude and 8° 30' 28" west longitude measured by Global Positioning System satellites. The altitude is 700 meters above sea level. I brought stones (57 stones at maximum and 5 at minimum) and nuts of the oil palm *(Elaeis guineensis)* to the top of this hill. I controlled the number of stones available, the quantity and freshness of the nuts, and the location and arrangement of stones and nuts according to the purpose of each experiment.

The nut of the oil palm weighs 7.2 grams on average (SD = 2.2 grams) and has a hard shell. The nuts are oblong and round, shaped like a rugby ball. A chimpanzee places a nut on an anvil stone with one hand and then cracks the nut using a hammer stone with the other hand. The kernel inside the shell of the nut weighs 2.0 grams on average (SD = 0.4 grams) and is edible.

The observer hid behind a screen made of grass about four meters long and two meters high. The distance between the observer and the cracking site was about 20 meters. The observer watched continuously from 7 am to 6 pm. All episodes of nut cracking were directly observed and videotaped in this experimental setting within the natural habitat.

Results

In the first study period, December 1990 to February 1991, the chimpanzees came to the outdoor laboratory on 20 days out of 26 days of the field experiment (44 parties and 1,208 minutes in total observation time). In the second study period, December 1991 to February 1992, they came on 37 out of 44 days of the field experiment (111 parties and 2,516 minutes in total). In the third study period, December 1992 to February 1993, they came on 57 out of 63 days of the field experiment (185 parties and 2,840 minutes in total).

The field experiment revealed some interesting characteristics of nut cracking by Bossou chimpanzees. Table 1 shows general information about nut cracking for each subject and specific topics are discussed in this chapter.

Handedness

Nut cracking consists of a series of complex actions in which each hand has a very different role, as shown in figure 1. Suppose that an adult chimpanzee holds a hammer stone in the left hand. Nut cracking proceeds in the following three steps: picking up a nut with the right hand, usually

Table 1

Summary of stone-tool use by chimpanzees in the wild at Bossou, Guinea. The hand used for the hammer stone in the nut-cracking behavior was recorded since the field experiment project began in 1988. In general, the data were collected from January to March in the dry season of each year. The present study is based on data from the three most recent study periods.

				Year Observed				
Name	Sex	Age in 1992	Mother	'88	'90	'91	'92	'93
Tua	m	adult	unknown	?	L	L	L	L
Kai	f	adult	unknown	?	R	R	R	R
Nina	f	adult	unknown	?	X	X	X	X
Fana	f	adult	unknown	?	L	L	L	L
Jire	f	adult	unknown	?	L	L	L	L
Velu	f	adult	unknown	?	R	R	R	R
Yo	f	adult	unknown	?	L	L	L	L
Pama	f	adult	unknown	?	X	X	X	X
Kie*	f	16	Kai	?	R	R	—	—
Foaf	m	11.5	Fana	?	R	R	R	R
Puru*	m	11.5	Pama	R	R	R	—	—
Vube*	f	10	Velu	?	L	—	—	—
Ja	f	9	Jire	?	R	R	R	R
Yunro	f	8	Yo	?	X	X	X	X
Na	m	7.5	Nina	?	R	R	R	R
Kakuru*	f	6.5	Kie	?	A	R	—	—
Vui	m	6.5	Velu	?	X	X	L	L
Pili	f	6	Pama	—	X	R	R	R
Jokro	f	4.0	Jire	—	X	X	X	—
Yela*	f	3.5	Yo	—	X	—	—	—
Fotayu	f	1.5	Fana	—	—	—	X	X
Vuavua	f	1.5	Velu	—	—	—	X	X
Yoro	m	1.0	Yo	—	—	—	X	X
Poni	f	0.0	Pama	—	—	—	—	X

Notes: * = The individual had disappeared or died by February, 1993. L = always used left hand for hammer. R = always used right hand for hammer. A = ambidextrous use for hammer. X = no successful hammer use but eating nuts cracked by others. ? = data available because no observation at the cracking site. — = data unavailable because the subject was not yet born, disappeared, or died in the research period. Age represents the estimate in February, 1993.

Source: Matsuzawa 1988, 1991, 1992 and 1993 data; Fushimi et al. 1990 data.

with the second and the third fingers, and placing it on an anvil stone; hitting the nut repeatedly with a hammer stone held by the left hand until the shell is open; and using the right hand to take the kernel from the cracked shell and bringing the kernal to the mouth. Sometimes, the chimpanzee swept the broken shell from the surface of the anvil stone with the back of the fingers of the right hand before placing the next nut.

Nut cracking should be characterized as the collaboration of the two hands rather than as the dominance of one hand over the other. However, to conveniently describe the behavior, the hand holding the stone hammer

Figure 1
A chimpanzee at Bossou cracks open a nut of the oil palm with a pair of stones used as a hammer and anvil.

is referred to as the *preferred hand.* The term *right-hander* simply refers to the fact that the chimpanzee used the hammer with the right hand when cracking nuts.

The preference for using one hand or the other to hold the hammer was very clear among the Bossou chimpanzees. They showed consistent preference at the individual level; each adult chimpanzees more than 12 years of age always used one particular hand exclusively for the hammer. Even the younger chimpanzees showed strong preference, with a few exceptions. Observations at Bossou provided the first report of a consistent handedness (laterality) among chimpanzees in the wild.

Table 1 shows that at the population level, right-handedness was not significantly more frequent than left handedness. This result is congruent with other reports from other research sites on hand preference in tool use (Boesch 1991b; McGrew and Marchant 1992). The observed hand preference in nut cracking of the Bossou chimpanzees is comparable in terms of its exclusiveness with hand preference of humans. However, the hand preference at the population level in chimpanzees was totally different from that in humans, where right-handedness is much more frequent.

Table 1 provides evidence regarding the hereditability of handedness in chimpanzees. The comparison of the preferred hand within mother-offspring pairs and brother-sister pairs showed both congruent and incongruent hand preference. Thus, these data show no significant evidence for inheritance of hand preference. Hand preference in nut cracking is analyzed in more detail (Sugiyama et al. 1993).

Taking these finding together, it may be concluded that hand preference in chimpanzees is somehow different from human handedness with its relationship to hemispheric dominance.

Development of Nut-Cracking Skill and the Critical Period

Table 1 shows that no chimpanzee less than three years old succeeded in cracking a nut with a pair of stones. In general, the Bossou chimpanzees succeeded in using a pair of stones as cracking tools between the age of three and five years (see figure 2). However, the beginner's performance was not as efficient as the performance of mature chimpanzees. Chimpanzees did not reach an adult level of skill in nut cracking until they were nine to ten years old.

Longitudinal observation of an individual's behavior starting at infancy illuminated the development of skill in tool use among chimpanzees in the wild. There are at least four recognizable stages in that development. The first stage is characterized by single-object manipulation. For example, an infant chimpanzee raked through the nuts with its hands, picked up a nut and, perhaps, mouthed it. Also, infants touched, hit, rolled, or held a stone. This typical action can be seen at around one year of age. Infants also

Figure 2
Development of the skill of using a pair of stones as a hammer and anvil. The percentage of individuals who succeeded in the use of these tools is plotted as a function of age.

acquired kernels from their mothers by directly reaching for a kernel in the mother's hand or in the mother's mouth or by reaching for the cracked nut on the anvil stone (see figure 3).

The second stage is characterized by object-association manipulation, an action simultaneously involving at least two objects. For example, an infant chimpanzee pushed a nut against a stone, put a nut on a stone, held a stone and then pushed it against a nut, and so on. This typical action can be seen around two years of age. In this stage, there was also a tendency to repeat the same action. For example, an infant chimpanzee put nuts on the anvil stone again and again. In addition to simply relating two objects such as pushing one against the other, the chimpanzees started to perform actions that eventually would be involved in nut cracking: hitting a nut with a hand, pounding a nut with a foot, and so on.

The third stage is characterized by the relationship of object-association sets. In the previous stage, the chimpanzee had already started to show the actions of putting things together. In this stage, the chimpanzees performed multiple actions in a collaborative way. For example, an infant chimpanzee placed a nut on the anvil stone and hit it with a hand. This behavior can be seen at around three years of age.

In the fourth and final stage, older than three years of age, the chimpanzees succeeded in coordinating multiple actions to actually crack open a nut with a pair of stones. Once their skill level is reached, it still takes a long time to attain the refinement displayed by adult chimpanzees. For example, young chimpanzees who hold a hammer in one hand have a tendency to use the same hand for placing a nut on the anvil and for picking up the kernel from the cracked shell. This means that left-right coordination is not as complete as that seen in adult nut cracking. Young chimpanzees

Figure 3
An infant chimpanzee, 1½ years old, tries to acquire the kernel of an oil-palm nut that was cracked by her mother.

in this stage do not show perfect laterality of hand use like adult chimpanzees. It means that they change the hand for the hammer stone and use the hand also for manipulating the nuts.

Critical Period of Learning

There were three chimpanzees—the juvenile female Yunro and the full-grown Nina and Pama—who did not use stones to crack nuts. Nina and Pama ate broken pieces of kernel produced by other members. They did not attempt to use stones. However, all of their offspring, Na, Pru, and Pili, had no difficulty in cracking oil-palm nuts with stones. The two mothers foraged for the kernels when their offspring cracked nuts with stones.

Yunro, who was 8 years old in 1993, failed to crack nuts. She placed a nut on an anvil stone and tried to crack it using her left knuckle or her right foot, but never used a stone hammer (see figure 4). This behavior of hitting a nut on the anvil by hand or foot is typical of infants around three years of age just before acquiring the skills of stone-tool use in nut

Figure 4
An eight-year-old female named Yunro tries to crack the nut of an oil palm by hitting the nut on an anvil stone with her left hand or her right foot rather than a hammer stone.

cracking. Yunro was observed to forage for the broken pieces of kernel. She also continued to acquire the kernel from her mother up to seven years of age but not thereafter. As she became older, she simply foraged broken pieces, as did Nina and Pama.

Why had Yunro not developed the same skills as her peers? Her mother Yo was one of the most skillful nut crackers at Bossou. The real reason might be that when Yunro was about four years old, her left ankle was caught by a wire trap set by poachers. She could not move around freely for months until the wire was finally taken off. Her opportunity to manipulate objects was severely limited at a critical period of learning. She did not learn to use a pair of stones as tools within that time.

We don't know the reasons why the two adult females did not acquire the nut-cracking skill. One possibility is that they were immigrants from a community in which there was no culture involving the use of stone tools. Another possibility is that they were suffering from a physical disability similar to that of Yunro when they were young. Actually, Nina is malformed in her right hand and Pama is abnormal in her left eye, although we don't know the onset of these physical disabilities.

Aside from speculation, the important point is that the offspring of the nonadept mothers acquired the skill of nut cracking and the offspring of the skillful mother did not. These results clearly indicate that the skill of nut cracking is not genetically transmitted to the next generation. As a result of these findings, I postulate that the skill of nut cracking can be acquired within a critical period (several years) followed by an even longer learning process in which refinement of the skill occurs.

Flexible Use of Stone Tools

When the Bossou chimpanzees use tools to crack open an oil-palm nut, the hammer and anvil set were always stones found in their natural habitat. The average weight of a hammer was 700 grams and that of an anvil was 2,200 grams. However, there can be substitution of tools. In a previous study, chimpanzees used the hardest spot on a fallen tree trunk as an anvil when no stone anvil was available (Sakura and Matsuzawa 1991).

In the present study, there were three examples in one experiment of chimpanzees using wood as a tool instead of stones when the number of stones available was reduced to 10. One young male chimpanzee (Vui, 5½ years old) came to the site with 10 other members. He used a fallen stick 20 centimeters long as a hammer. He also dug a wood block from the ground and used it as an anvil. This wooden hammer and anvil set did not work for cracking nuts. Two young females, Ja, eight years old and Pili, five years old, used a stone hammer and a wooden anvil. They succeeded in cracking the nuts with the wooden anvil although it took more than 20 hits to crack a nut open.

In the first period, I observed that seven stone tools, hammers or anvils, were broken into two or more pieces when used for nut cracking. Then, four out of seven stones were used again as stone hammers. The chimpanzees broke a stone anvil with a hammer and the broken piece became a better hammer. This may actually be the first step in the making of stone tools among chimpanzees. These findings clearly demonstrate that the chimpanzees used the tools not in a stereotyped fashion but in a flexible, insightful way. In other words, they can recognize the functions of the tools and which stone or combination of stones function best.

Invention of a Metatool

One of the most interesting findings is how the chimpanzees make a metatool, that is a tool that serves as a tool for another tool (see figure 5). For example, a metatool is a third stone placed beneath the anvil stone as a wedge to keep the surface of the anvil stone flat and stable. So far, I have observed three instances in which three separate individuals have made such complex anvils (see table 2).

In all three cases, the chimpanzees used the same anvil stone (ID# 401) while the hammer stones and the wedge stones were all different. This indicates that the slant surface of that specific anvil stone was the real cause for invention of a metatool. The sequential actions in making a hierarchical structure of stone tools have features in common with the actions by captive chimpanzees involved in solving cognitive tasks (Matsuzawa 1989). The hierarchical organization of the metatool was described by Matsuzawa (1991a).

Human Children and Stone Tools

Human children in the village Bossou (N=28, Ages 1–13) were tested in the same situation as the chimpanzees. Nuts and stones were provided by the experimenter. Children were instructed to crack nuts with stones. Children younger than 2½ years old did not succeed in cracking the nuts.

Table 2
Episodes of using the hammer-anvil-wedge stone tool by chimpanzees in the wild. The chimpanzees used a third stone as a wedge to keep the surface of the anvil stone flat and stable.

				ID number of stones used (weight in grams)		
Name	Sex	Age	Date	Hammer	Anvil	Wedge
Kai	female	adult	01/16/91	36 (765)	401 (4,100)	73 (1,950)
Na	male	6.5	01/13/92	34 (785)	401 (4,100)	81 (745)
Foaf	male	10.5	01/20/92	87 (660)	401 (4,100)	8 (480)

Figure 5
The chimpanzees at Bossou used a metatool, a tool for another tool. A third stone used as a wedge was placed beneath the anvil stone to keep the surface flat and stable.

Their actions were similar to those of chimpanzees who were less than three years old. For example, they used a hammer stone to hit a nut on the ground, hit a nut on a stone anvil by hand, and so on. The children older than three years used a pair of stones and cracked open the nuts. Also, taking efficiency into consideration, human children reached a refined level of skill at around ten years of age, again similar to the chimpanzees.

The making of a metatool in humans was also tested in a special situation called the three-stones test. The children were given a set of three stones: large, medium, and small. The large stone was the anvil stone (ID# 401) that was used with a wedge by the chimpanzees. The medium stone was useful for a hammer and the small stone was useful for a wedge. The three stones were placed randomly in front of the children with a pile of oil-palm nuts. The same test was given to the chimpanzees at the outdoor laboratory where, in that case, the set of three stones and the nuts were placed a little farther away from the central area of the outdoor laboratory.

There were a variety of solutions when using the slanted anvil stone. I identified three types of solutions: the first type was common to both species; the second type was unique to chimpanzees; and the third type was unique to humans. The first-type solution, common to both species, involved a human or a chimpanzee holding a nut on the anvil with one hand and trying to crack it with the hammer in the other hand. This did not work well. Subjects moved or rotated the anvil stone to make the slant surface better. Subjects turned the anvil stone upside down. This solution was a little better but the anvil stone was still unstable. Use of the metatool was also a solution common to humans and chimpanzees and was among the best of solutions. In human subjects, the youngest metatool user was a young boy, 6 years and 9 months old (see figure 6).

Figure 6
A boy of 6 years and 9 months successfully responded to the three-stones test. The boy spontaneously found the solution of using the third stone as a wedge to keep the surface of the anvil stone flat and stable, in a manner similar to the solution of the chimpanzees.

An example of the second-type solution, unique to chimpanzees, was when young chimpanzees picked up a handful of nuts and threw them onto an anvil stone. Adult chimpanzees held and supported the anvil stone with one foot to keep the surface flat. This type of solution was never observed in human subjects.

The third-type solution, unique to humans, occurred when children older than five years showed a special solution to the three-stones test. They adjusted the position of the nut placed on the anvil to prevent the nut from rolling off. Another solution was to place the nut on the tiny spot on the tip of the anvil stone. Both solutions involved fine control and balance of the nuts.

Although there was a slight difference between the two species, the course of skill development for nut cracking by stones was in general similar in humans and chimpanzees.

Cultural Transmission

In West Africa, chimpanzees use stone tools to crack nuts, but the nuts they crack differ from one community to the other. For example, the chimpanzees at Bossou crack oil-palm nuts, but the chimpanzees at Taï in the Ivory Coast crack five different species of nuts (Boesch and Boesch 1982). There is a neighboring community about 10 kilometers away from Bossou at Mt. Nimba within the borders of Guinea, Ivory Coast, and Liberia,. The chimpanzees at Mt. Nimba crack the coula nut *(Coula eduris)* with stones (Yamakoshi and Matsuzawa 1993a).

In January 1993, we provided the 18 chimpanzees at Bossou with the unfamiliar coula nuts. In addition to the pile of oil-palm nuts, three coula nuts at a time were provided in the outdoor laboratory. In the first encounter with the coula nuts, some of the chimpanzees (8 out of 14 members excluding the infants less than three years old) sniffed the nuts, picked them up, tried to bite them, but did not attempt to crack them. Some of the chimpanzees (5 out of 14 members) simply neglected the coula nuts. But the adult female Yo (estimated to be 31 years old) immediately placed the coula nuts on her stone anvil, cracked them, and ate the kernels (see figure 7); then she started cracking the oil-palm nuts. The coula nuts are encased within a thick shell, so Yo must have known that the nuts were inside before cracking.

When Yo cracked the coula nuts, a group of juveniles gathered around and peered at her while she was cracking and eating the strange nuts. The next day, an unrelated 6½-year-old male named Vui cracked open a coula nut without any practice. Four days later, a six-year-old female named Pili did the same. The two juvenile chimpanzees cracked the nuts and sniffed the kernel and chewed and spat it out. Though coula nuts were continuously provided for another two weeks, only these two juveniles cracked the nuts (Yamakoshi and Matsuzawa 1993b).

Figure 7
The adult female Yo knew how to crack open the coula nuts. Yo may be responsible for cultural transmission between communities if she is an immigrant from the Mt. Nimba region.

These findings suggest the hypothesis that the coula-eating female Yo was born in the neighboring community and traveled the 10 kilometers to join the community at Bossou. Perhaps she had already learned to crack coula nuts at her community of origin, which had that cultural tradition. Nut cracking was transmitted from the immigrant female Yo to the young members of the Bossou group while other adult members remained unskilled.

General Discussion

This study outlines development of skill in the use of stone tools for cracking nuts among the Bossou chimpanzees. The significance of these results can be interpreted from the viewpoint of cognitive development.

The list of tools used by wild chimpanzees has become long (Goodall 1986; Nishida 1990; McGrew 1992). However, "one tool for each target object"—such as in termite fishing using a twig, catching ants using a wand, drinking water using a leaf sponge, and throwing a stone at an enemy—is the generalized rule of tool use by chimpanzees in the wild. The exception to this rule is nut cracking using a pair of stones. Nut cracking involves a set of two tools for a target object. Additionally, the examples involving a metatool show that chimpanzees are able to use three tools to make a set: a hammer, an anvil, and a wedge. This is the most complex tool use ever found among chimpanzees in the wild.

Studies of the use of symbols in chimpanzees in captivity show that chimpanzees can relate one thing to another. For example, they can relate a visual symbol called a *lexigram* to its referent. The chimpanzees in captivity can be trained to combine lexigrams and to construct a lexigram from its elements, a series of geometric figures (Matsuzawa 1989). Similarly, the use of tools in the wild also show a one-to-one, or the isomorphic, relationship between tool and target object. Construction of a combined hammer-anvil-wedge tool is a good example of the hierarchical nature of relating different objects. It must be noted that the invention of the combined tool, or metatool, is not simply a series of repetitive actions such as repeatedly stacking boxes. With the metatool there is a necessary order of manipulating each piece. Therefore, the chimpanzees need to learn the series of actions to combine the pieces in the proper order.

The present study also reveals that it takes a long time to learn the series of actions and develop the motor coordination needed to develop the skill of stone-tool use. Development proceeds in stages to reach a high level of skill in the use of complex tools. The first stage was the action of manipulating a single piece: holding a stone, mouthing a nut, and so on. The second stage was the action of relating one piece to another, the direction of the action, such as learning to hit a nut with a stone and not to hit a stone with a nut. The third stage was coordinating multiple actions. Nut cracking with stones consists of a series of actions that follow an "action grammar", organized sequential behavior (Greenfield 1991; Matsuzawa 1991a; 1991b).

It must be noted, as well, that there are social aspects involved in the process of learning to crack nuts. In the course of development, the infant chimpanzees looked carefully at the actions of other individuals including their mothers. Not only did they observe tool use but also some of the objects were actually transferred from the tool user to the observer.

There are at least three important findings related to social aspects. First, all the infant chimpanzees younger than four years old were observed to acquire the kernels that were cracked by the mothers.

Second, the infant female Kakru, four years old, interfered with her mother's nut cracking and got the nut on the anvil. Kakru put the nut on her own anvil and then cracked the nut to get the kernel. In another episode, the infant female Jokro, 1½ years old, stole a nut held by her elder sister Ja, seven years old, and mouthed it. Although she had not yet succeeded in cracking a nut with stones, she was very interested in the nuts to be cracked.

Third, the young chimpanzees had a strong tendency to use the set of stones that had been used by the adults (who were not always the mothers) after the adults left the nut cracking place. These facts show that the infants observed the nut cracking and also obtained objects involved in the action: kernels, nuts, and stones. In a sense, the infants learned complex tool use through their interaction with other individuals in the community.

In addition, the experiment relating to cultural transmission suggests a possible mechanism of propagating a new behavior, in this case, cracking a new nut. One female was the only individual apparently familiar with coula nut cracking. Through observation, young chimpanzees quickly learned to crack open coula nuts. They eventually acquired a taste for the coula nuts and began to eat them. However, the adults in the community remained uninterested in the new fruit and thus did not expand their food repertoires.

The use of stone tools established for oil-palm nuts was easily transferred to a new kind of nut but only among juveniles who were still developing their nut cracking skills. Although further research is necessary, the present study illuminates a possible mechanism for transmission of cultural behavior between different communities and across generations of chimpanzees.

Summary

The use of stone tools by chimpanzees in the wild was analyzed by field experiments in an outdoor laboratory set up at Bossou, Guinea. The chimpanzees at Bossou crack open oil-palm nuts with a pair of stones used as a hammer and anvil. Longitudinal study of the nut-cracking behavior revealed the cognitive development involved in this behavior. In general, the chimpanzees succeeded in using the stone tool between three and five years of age, the critical period for learning this skill. It takes almost ten years to acquire the refined level of skill shown by adults. The chimpanzees at Bossou showed the ability to use a metatool. This metatool is the most complex tool known to be used by chimpanzees in the wild. Analysis of the action grammar reveals the hierarchical nature of the multiple actions involved in the use of stone tools. This complexity of the use of stone tools corresponds to the complexity of cognitive skills shown by chimpanzees in captivity. Social interaction in the community also has an important role in learning the use of tools. The test with a new nut, *Coula eduris,* revealed the possible learning mechanism of cultural transmission between communities and across generations.

Acknowledgments

The present study was carried out with the collaboration of DRST (Direction de la Recherche Scientifique et Technologique) of Guinea and the villagers at Bossou. Field research at Mount Nimba was supported by the government of Côte d'Ivoire and the villagers at Yeale. I wish to thank Dr. Yukimaru Sugiyama who has conducted field research at Bossou since 1976 and who gave me the opportunity to do this research. Special thanks are due to the following colleagues during the research period: Jeremy Koman, Guano Gumi, Tino Camara, Osamu Sakura, Takao Fushimi, Hisato Ohno, Miho Nakamura, Kiyonori Kumazaki, Norikatsu Miwa, Naoto Yokota, Rikako Tonooka, Noriko Inoue, and Gen Yamakoshi. Helpful comments on the early draft were made by Iver Iversen and Joseph Soltis. Financial support was given by a grant (No. 01041058) from the International Scientific Research Program of Ministry of Education, Science, and Culture, Japan.

References

Boesch, C. 1991a. Teaching among wild chimpanzees. *Anim. Behav.* 41 :530–2.

———. 1991 b. Handedness in wild chimpanzees. *Internat. J. Primatol.* 12:541.

Boesch, C., and H. Boesch. 1982. Optimization of nut-cracking with natural hammers by wild chimpanzees. *Behav.* 3:265–86.

———. 1984. Possible causes of sex differences in the use of natural hammers by wild chimpanzees. *J. Human Evol.* 13:415–40.

Fujita, K., and T. Matsuzawa. 1990. Delayed figure reconstruction in a chimpanzee *(Pan troglodytes)* and humans *(Homo sapiens). J. of Comp. Psych.* 104:345–51.

Fouts, R. S. 1973. Acquisition and testing of gestural signs in four young chimpanzees. *Science* 180:978–80.

Fushimi, T., O. Sakura, T. Matsuzawa, H. Ohno, and Y. Sugiyama. 1991. Nutcracking behavior of wild chimpanzees *(Pan troglodytes)* in Bossou, Guinea (West Africa). In A. Ehara, T. Kimura, O. Takenaka, and M. Iwamoto, eds., *Primatology Today,* pp. 695–6. Amsterdam: Elsevier

Gardner, R. A., and B. T. Gardner. 1969. Teaching sign language to a chimpanzee. *Science* 165: 664–72.

Goodall, J. 1986. *The Chimpanzees of Gombe: Patterns of Behavior.* Cambridge, Mass.: Belknap Press.

Greenfield, P. M. 1991. Language, tools and brain: The ontogeny and phylogeny of hierarchically organized sequential behavior. *Behav. and Brain Sci.* 14:531–51.

Hannnah, A. C., and W. C. McGrew. 1987. Chimpanzees using stones to crack open oil palm nuts in Liberia. *Primates* 28:31–46.

Itakura, S., and T. Matsuzawa. 1993. Acquisition of personal pronouns by a chimpanzee. In H. L. Roitblat, L. M. Herman, and P. E. Nachtigall, eds.,

Language and Communication: Comparative Perspectives, pp. 347–63. Hillsdale, NJ: Lawrence Erlbaum.

Köhler, W. 1927. *The Mentality of Apes,* 2d ed. London: Kegan Paul, Trench, Trubner.

Kortlandt, A. 1967. Experimentation with chimpanzees in the wild. In D. Starck, R. Schneider, and H. J. Kuhn, eds., *Neue Ergebnisse der Primatologie-Progress in Primatology,* pp. 208–24. Stuttgart: Fischer.

———. 1986. The use of stone tools by wild-living chimpanzees and earliest hominids. *J. Human Evol.* 15:77–132.

McGrew, W. C. 1992. *Chimpanzee Material Culture: Implications for Human Evolution.* Cambridge, England: Cambridge Univ. Press.

McGrew, W. C., and L. F. Marchant. 1992. Chimpanzees, tools, and termites: Hand preference or handedness? *Current Anthro.* 32: 114–9.

Matsuzawa, T. 1985a. Use of numbers by a chimpanzee. *Nature* 315:57–9.

———. 1985b. Color naming and classification in a chimpanzee *(Pan troglodytes). J. Human Evol.* 14:283–291.

———. 1989. Spontaneous pattern construction in a chimpanzee. In P. Heltne and L. Marquardt, eds., *Understanding Chimpanzees,* pp. 252–65. Cambridge, Mass: Harvard Univ. Press.

———. 1990a. Form perception and visual acuity in a chimpanzee. *Folia Primatol.* 55:24–32.

———. 1990b. Spontaneous sorting in human and chimpanzee. In S. Parker and J. Gibson, eds., *Language and Intelligence in Monkeys and Apes: Comparative Developmental Perspectives.* pp. 451–68. Cambridge, England: Cambridge Univ. Press.

———. 1991a. Nesting cups and metatools in chimpanzees. *Behav. and Brain Sci.* 14:570–1.

———. 1991 b. *Chimpanzee Mind.* Tokyo: Iwanami-Shoten Publishers (in Japanese).

Matsuzawa, T., S. Itakura, and M. Tomonaga. 1991. Use of numbers by a chimpanzee: A further study. In A. Ehara, T. Kimura, O. Takenaka, and M. Iwamoto, eds., *Primatology Today.* pp. 317–20. Amsterdam: Elsevier.

Matsuzawa, T., O. Sakura, T. Kimura, Y. Hamada, and Y. Sugiyama. 1991. Case report on the death of a wild chimpanzee *(Pan troglodytes verus). Primates* 31:635–41.

Nishida, T. 1990. *The Chimpanzees of the Mahale Mountains: Sexual and Life History Strategies.* Tokyo: Univ. of Tokyo Press.

Premack, D. 1971. Language in chimpanzees? *Science* 172:808–22.

Rumbaugh, D. 1977. *Language Learning by a Chimpanzee.* New York: Academic Press.

Sakura, O., and T. Matsuzawa. 1991. Flexibility of wild chimpanzee nutcracking behavior using stone hammers and anvils: an experimental analysis. *Ethology* 87:237–48.

Savage-Rumbaugh, E. S. 1986. *Ape Language: From Conditioned Response to Symbol.* New York: Columbia Univ. Press.

Sugiyama, Y. 1984. Population dynamics of wild chimpanzees at Bossou, Guinea, between 1976 and 1983. *Primates* 25:391–400.

Sugiyama, Y., T. Fushimi, O. Sakura, and T. Matsuzawa. 1993. Hand preference and tool use in wild chimpanzees. *Primates* 34: 151–9.

Sugiyama, Y., and J. Koman. 1979a. Social structure and dynamics of wild chimpanzees at Bossou, Guinea. *Primates* 20:323–39.

———. 1979b. Tool-using and making behavior in wild chimpanzees at Bossou, Guinea. *Primates* 20:513–24.

Terrace, H. 1979. *Nim: A Chimpanzee Who Learned Sign Language.* New York: Alfred A. Knopf.

Tomonaga, M., and T. Matsuzawa. 1992. Perception of complex geometric figures in chimpanzees and humans: Analysis based on choice reaction time. *J. Comp. Psych.* 106:43–52.

Tomonaga, M., Itakura, S., and Matsuzawa, T. 1993. Superiority of conspecfic faces and reduced inversion effect in face perception by a chimpanzee. *Folia Primatol.* 61:110-4

Yamakoshi, G., and T. Matsuzawa. 1993a. Preliminary surveys of the chimpanzees in the Nimba Reserve, Côte d' Ivoire. *Primate Research* 9:13–7. (in Japanese with English summary)

———. 1993b. A field experiment in cultural transmission between groups of wild chimpanzees at Bossou, Guinea. Paper presented at the 23rd International Ethological Conference, Torremolino, Spain.

Section 4

Afterword and Postscript

Fifi

Review of Recent Findings on Mahale Chimpanzees

Implications and Future Research Directions

Toshisada Nishida

An increasing number of reports from West and Central Africa continue to substantiate that chimpanzees have remarkably varied behavioral repertoires. Despite the fact that I have spent several thousand hours with familiar chimpanzees, I still witness patterns of behaviors totally new to me whenever I go to Mahale.

The purpose of this afterword is to explore little studied aspects, new trends of research, and unanswered questions about the behavior of chimpanzees from the perspectives gained through my research at Mahale and from the previous chapters of this volume. These aspects, trends, and questions include the following topics: predation; scavenging; medicinal plants; night and bedmaking behaviors; long- and short-distance communication; behavioral cues; innovative behavior and culture; female relationships; lone males, ostracism and xenophobia; DNA typing; reciprocal altruism; coalition; deception and morality; and chimpanzee/bonobo comparisons.

Predation on Chimpanzees

Recently, new evidence of predation on chimpanzees by big cats has appeared from both East and West Africa. Three to six chimpanzees were eaten by lions at Mahale (Tsukahara 1993), and at least four were killed by leopards at Taï (Boesch 1991).

The lack of evidence of lion predation from sites other than Mahale may be explained simply by the absence of these predators. Few studies on chimpanzees have been conducted where lions are regularly seen. The past study sites of Kasakati, Filabanga, Mount Assirik, and Ugala harbored lions, but the periods of study conducted at these sites were relatively short. A less likely explanation may be that the lions that ate the chimpanzees were exceptional, that is, they were "chimp-eaters" in the same sense that some lions are "man-eaters" (Patterson 1979).

In those years when chimpanzees were victimized at Mahale, two local people were also killed and eaten by a lion at the park boundary, and this man-eater was shot in 1991. Since that time, we have not heard roaring or seen footprints of lions within the study area, and attacks on people and chimpanzees have ceased. No other evidence indicates that the man-eater was also the chimp-eater, but this remains a possibility. The lack of evidence of predation by leopards at Mahale and Gombe is more puzzling, given that leopards kill even adult chimpanzees at Taï.

I suggest that feces of lions and leopards should be collected systematically at every study site. It is easy to examine whether any of the hairs found are from chimpanzees (Inagaki and Tsukahara 1993). We should compare across study sites how often leopards or lions are heard (McGrew 1983). Such information is needed in savanna woodlands where lions live and, in particular, where chimpanzees in large parties regularly cross from one riverine forest to another to establish whether predation plays a role in shaping the social organization of chimpanzees and protohominids. This information is important because the prevailing theory of the ape social structure (Wrangham 1979) was formulated without taking the risk of predation by big cats into consideration.

Scavenging

Scavenging may be defined as feeding on a dead animal that either was killed by another species or died of disease or an accident. It is a matter of controversy and speculation to what extent early hominids (*Australopithecus* species) scavenged from carcasses (Lewin 1989).

The first reports of chimpanzees feeding on fresh carcasses of animals killed by baboons while chimpanzees were present came from Gombe (Morris and Goodall 1977). However, reports of chimpanzees feeding on animals that were killed earlier by other species while chimpanzees were absent have come only from Mahale. In addition to the cases already reported (Hasegawa et al. 1983), three more cases of scavenging have been observed at Mahale. These cases involved the following carcasses: three blue duikers, two bushbucks, one red-tailed monkey, and one red colobus monkey. All the scavenged individuals were adult or near adult, similar in size to or larger than those typically killed by chimpanzees.

By contrast, at Mahale, about 10 chimpanzees including adult males were found sitting about 10 meters from the carcass of a large bushpig that, from the evidence of injuries on the throat and chest of the victim, had probably been killed by a leopard. It was likely that the leopard fled upon hearing the approach of a large party of chimpanzees. The body was fresh and still infested with many ticks. The chimpanzees only gazed at the pig for a long time and departed without eating it. (We, however, did partake of the meat, which was quite palatable!). Perhaps the chimpanzees' reluctance to eat was a result of either the unusually large size of the carcass (greater than 60 kilograms) or anxiety about the leopard's return. In addition, Goodall (1968) described responses to large dead mammals (adult baboons in two cases, adult bushpig in one case) by chimpanzees. Kortlandt and van Zon (1967) stated that dead mammals were very frightening to the forest-dwelling chimpanzees of Zaire.

It will be useful to collect data at other sites or in situations where chimpanzees do not scavenge. Why, for example, has no case of scavenging on carcasses been recorded in other study sites, such as Taï? In trying to model human evolution, it is important to know the kind of environmental conditions that favor the development of scavenging in chimpanzees.

Medicinal Plants

Two lines of evidence on the use of medicinal plants have been reported. Observations of the slow swallowing of leaves without chewing by apparently healthy individuals and of the eating of unusually bitter piths by sick individuals have opened a new avenue of research called *zoopharmacognosy* (Huffman and Wrangham this volume; Rodriguez and Wrangham 1993). To date, however, only one species of *Vernonia* and one species of *Aspilia* have been chemically analyzed in detail. While recent analyses of *Vernonia* demonstrate which compounds are likely to be effective when ingested (Huffman and Wrangham, this volume), a debate is going on as to what kind of physiologically active agents are being ingested from *Aspilia* (Page et al. 1992).

According to my observations of leaf swallowing, when chimpanzees swallow the leaves of such plants as *Aspilia, Commelina, Ficus,* or *Trema* in the early morning, they never touch the leaf blade itself. Instead, they hold the stalk and bring the leaf to the mouth, or bring the mouth directly to the leaf (see figure 1). Later in the afternoon, they do not swallow, but

Figure 1
In the early morning, an adult male swallows a leaf of *Commelina benghalensis* without chewing it.

quickly chew a lot of the leaves of the plants *Aspilia* and *Ficus* using the usual "pull-through" technique of picking with the hand. These observations suggest that the physiologically active agents are on the surface of the leaves, and that these agents are concentrated only in the early morning. Therefore, biochemical analyses should compare leaves collected in the early morning (especially the washings from the surfaces of these leaves) with the leaves collected in the afternoon. Without such procedures, no decisive conclusions will emerge, given that so many bioactive chemical substances are found in plants. The development of this type of research is especially significant because many people are interested in the conservation of nature only when human utility is clearly indicated; chimpanzees may point the way to new medicinal knowledge!

Night Behavior and Bedmaking Behavior

Most observation of chimpanzees and bonobos has been limited to daytime, neglecting the nighttime activities that occupy one-half of their lives. Except for Goodall's earliest observations (1968), descriptions of nighttime activities have been anecdotal in chimpanzees and virtually nil in bonobos (Fruth and Hohmann, this volume).

We often hear the calls of chimpanzees in the middle of the night, and we occasionally have evidence of a party of chimpanzees moving from one place to another, vacating the beds that they made in the evening. Observations of night combat include the following: Goodall (1986) described Figan attacking his rival who was lying in an evening bed. On another occasion, male chimpanzees born in the wild attacked a human-raised female chimpanzee at a rehabilitation camp in the middle of the night (Brewer 1978). During my observation of a power struggle continuing on and off from daytime to nighttime, the rivals had to make evening beds at least three times on one occasion (Nishida 1981). A fatal attack on an alpha male by two males occurred in the Arnhem Zoo at night (de Waal 1986). Finally, I have often noticed fresh wounds on adult male chimpanzees in the early morning; the fresh wounds suggest that aggressive confrontations occur at night far more often than we would imagine.

Chimpanzees sometimes mate at night. Once at dawn, one of my focal females in estrus visited and mated with adult males one after another, while they were still lying in their beds. It was so dark that I could hardly identify these males. Therefore, it seems necessary to study mating behavior at night.

A study of the nightlife of chimpanzees and bonobos will surely reward researchers with interesting observations and create a new base of knowledge.

Long-distance Communication

Vocal communication is the least understood behavioral pattern of chimpanzees. Even the function of the *pant-hoot,* the most conspicuous call, has just begun to be the object of detailed study at Mahale and Gombe (Mitani, this volume).

The chimpanzees of M Group at Mahale usually move in a large integrated party consisting of 50 to 80 individuals, particularly from July to December. I have been curious about who leads the nomadic movements and who is most responsible for attracting other members. We have conducted several multiple, simultaneous focal observations in which five or six field-workers each followed a target male and recorded every pant-hoot call he emitted, responses to the target male's pant-hoots from others, every audible pant-hoot from others, and the target males response to pant-hoots from others. The preliminary analysis suggests that an alpha male and a few of the oldest males are most influential. Such study should be repeated at Mahale as well as at other sites.

Goodall (1986) claims the existence of a variety of context-specific pant-hoots. Mitani (this volume) did not find evidence of this supposed diversity. This issue could be resolved by using experimental playbacks of vocalizations but, to date, this method has been virtually untried in the wild with great apes. Although field playback is difficult under natural conditions, it may be the only method for bringing about a breakthrough in explaining the long-distance, acoustic communication of chimpanzees in the wild.

Short-distance Communication

No detailed, systematic study of the short-distance, vocal communications of wild chimpanzees has occurred. In fact, chimpanzees emit a wide range of short-distance sounds: calm grunts, pant-grunts, pant-screams, food grunts, food screams, and so on (see figure 2). A female chimpanzee in estrus does not pant-grunt as often as when she is not in estrus. Moreover, even when an estrous female pant-grunts, she often abbreviates the vocalization by emitting short, inconspicuous grunts, although these sounds still appear to signal her identity. So, for example, it appears as if an adult female greets a male by saying "Good morning!" when not in estrus and by saying

Figure 2
A juvenile male pant-grunts to an adult male.

only "Hi!" when in estrus. This variation may reflect both an emotional change on the part of the female and her expectation of a change in the attitude of the male as a result of her desirable state.

The functions of food grunts, food screams, and food aha calls (Goodall 1986) remain puzzling. It is unlikely that these calls advertise the presence of food to other chimpanzees, for the calls are too soft to attract chimpanzees from a distance. Second, two or more individuals who have already found a lot of food usually emit the food aha simultaneously. Sometimes, many chimpanzees travel while emitting the food aha, long before they arrive at a food patch known to have lots of fruit. Such sounds may function to coordinate the feeding of several chimpanzees at the same food patch, or they may simply express the joy or excitement of eating.

Detailed studies of gestural communication are needed. Leaf grooming, grooming handclasp, peering, and ground grooming are among the behavioral patterns for which function has not been clearly elucidated.

The ability of captive bonobos and, probably, captive chimpanzees to understand human speech (Rumbaugh, Savage-Rumbaugh and Sevcik this volume; Savage-Rumbaugh 1993) should have an immense impact on the field study of communication in bonobos and chimpanzees in the wild. Specifically, is the animals' ability to understand human speech somehow of use in the natural environment, where no speech occurs? Diamond (1992) has pointed out the possibility of a crude language in chimpanzees, but I do not imply that chimpanzees in the wild use a system of communication like human speech. However, we should still investigate sequences of their short-distance communications, both vocal and gestural. Savage-Rumbaugh's study, which suggests that logical thinking preceded speech in human evolution, is a challenge to field primatologists, who may find unexpected capacities of chimpanzees in the wild.

Behavioral Cues to Identity

The human capacity to exploit widely varied environments comes partly from our ability to synthesize information from as many sources. Chimpanzees also have numerous information sources in the natural environment but their use of these sources is not yet understood.

Like other nonhuman primates, chimpanzees respond differently to the vocalizations of different individuals. For example, a chimpanzee mother immediately runs to her offspring when the latter screams from somewhere out of her sight. An adult male excitedly mounts his associate and repeats pelvic thrusts when he hears the distant pant-hoot calls of his rival. Moreover, pant-grunt calls are as idiosyncratic as are pant-hoot calls; I myself can discriminate about 10 females from their pant-grunt calls.

It is a common observation that you can often use sounds to recognize people even though they are not in sight. I have the impression that chimpanzees can make the same educated guesses. The adult male Jilba was once lying alone on the ground after feeding. Then, the sound of footfalls were heard. The passing chimpanzee could not be seen because of dense underbush. Jilba sat and listened intently to the footsteps, then set off after the passing chimpanzee, who proved to be Bembe, his close associate of the same age.

When following target males, I have noticed that they respond markedly to some stamping or slapping sounds (see figure 3) but not to others, just as they respond to some pant-hoot calls but not to others. Even the sound of a rock falling into water may elicit different responses. These observations, admittedly anecdotal, suggest that chimpanzees often identify individuals by the sounds those individuals produce. Thus, it is likely that chimpanzees use not only their own vocal sounds but also their own nonvocal sounds for communication and social coordination. Nonvocal sounds vary from one individual to another, just as the pattern of agonistic displays vary from one individual to another.

Figure 3
(A) Below
The adult male Kalunde always slaps at the wall of this metal house with both hands simultaneously.
(B) Right
Another adult male, Ntologi, kicks more often than slaps at the wall in his spectacular display.

Male chimpanzees differ in the frequency of sound-producing, nonvocal displays; the display elements that each individual prefers to use are variable. For example, the alpha male Ntologi performed nonvocal displays nine times more often that did his two age-mates. Rock-throwing displays by M Group chimpanzees, both male and female, old and young, were observed 107 times during three months of 1992; 66% of these displays were done by two adult males, Ntologi and Kalunde. Even more interesting is that both of the males—and Ntologi, in particular—like to throw rocks into water. So, if you hear a rock landing in water, the probability is 43% that Ntologi threw it and 23% that Kalunde threw it. Moreover, both of them prefer to throw heavy rocks. In these cases, they use both hands, and the rocks often weigh 10 kilograms or more. So, if you hear a heavy rock falling into water, there is a 67% chance that Ntologi threw it, and a 27% chance that Kalunde threw it. This means that if an M Group chimpanzee is alone with Kalunde and hears a heavy rock falling into water, that chimpanzee will be able to imagine correctly that Ntologi is now displaying.

How chimpanzees exploit information in the natural environment and use idiosyncratic display to coordinate social relationships is one of the most interesting areas of future study. Rigorous testing can be done by playbacks.

Innovative Behaviors and Culture

McGrew (1992, this volume) neatly summarizes the differences in habitual tool-use behavior between local populations. New and probably habitual tool-use patterns have already been observed since the 1991 "Understanding Chimpanzees" conference in Chicago. Thus far, for example, evidence of ant-dipping behavior has been reported for the first time from Sierra Leone (Alp 1993). Furthermore, the ant-dipping technique at Bossou (Matsuzawa 1993) and Taï (Boesch and Boesch 1990) in West Africa has proved to be different from that reported at Gombe: the West African tools were much shorter, and no pull-through action was involved (McGrew 1974). Finally, a chimpanzee at Ndoki dislodged a hornbill from its nest with a stick (Kuroda and Suzuki in press), and a female at Mahale used a modified branch to capture a squirrel hiding in a hole (Huffman and Kalunde 1993). Undoubtedly, more and more tool-use patterns will be reported in the future from, in particular, Central and West Africa.

Here I want to point out the importance of habitual but idiosyncratic new behavior patterns. At Mahale, new types of tool use have been observed as idiosyncratic habits: The adult male Kalunde was seen at four separate times inserting a small probe into one of his nostrils to stimulate sneezing, so that a deeply placed mucous plug was discharged from his blocked nasal passage (Nishida and Nakamura in press). The adult male Musa grasped and pulled a blade of bamboo grass through his hand with an upward motion of one arm. This process produced a conspicuous sound that attracted the attention of an estrous female. This may be a brand-new behavior related to the leaf clipping used in courtship.

Idiosyncrasy is not limited to the use of tools. For example, the adult male Musa was once seen to "wash" a colobus skin in a riverbed by dunking and shaking it violently under water and by even stamping on the skin with one foot while the skin was placed on the rock (see figure 4). After that he chewed on the skin continuously. The adolescent male Alofu began habitually pressing his fingers on his nipples when pant-grunting to adult males, apparently showing them his extreme respect. In 1991, an adult male Jilba groomed dry leaves and a moth on the ground or he groomed rocks and the ground itself while he was being groomed on the back or even while he was engaged in handclasp grooming by the alpha male (see figure 5). This apparent displacement activity, which I call *ground grooming,* was later shown in 1992 by Toshibo, another young adult male, who was one of Jilba's age-mates. All the cases mentioned above occurred spontaneously without any human intervention.

The recording of such innovative, idiosyncratic behavior is important. First, it gives us an idea of the frequency of innovation in a conservative society. Second, follow-up studies will clarify what types of behavior will

Figure 4
The old male Musa dunked and shook the colobus skin in the river and occasionally stamped on it.

Figure 5
The young adult male Jilba just grooms the ground despite the fact that the alpha male Kalunde is grooming Jilba in a typical handclasp posture, in which the participants normally groom each other simultaneously.

survive in a particular local environment and become a feature of local culture instead of being ignored by the other members of the local group. Such clarification is an important potential of a long-term study.

Goodall's (1973) pioneering work showed that most innovative behaviors that she observed disappeared after a few years of being in fashion. At the same time, Tomasello (this volume) asserts that the social learning process in chimpanzees differs from that in human beings. Is the fact that many innovations never progress beyond idiosyncratic behaviors related to this claimed difference between humans and chimpanzees? The answer remains to be elucidated. However, it is important to remember that early hominid technological culture did not change rapidly and that conservatism was the rule rather than the exception. Oldowan and Acheulean hand-ax technologies lasted without significant changes for over one million years each, which suggests that most innovations of presapiens hominids vanished without being socially transmitted over generations.

Female Relationships

The investigation of social relationships among adult females continues to be the most neglected area in field studies of social organization. Reasons for this include the difficulty involved in following adult females, the infrequency and subtlety of interactions among them, and the lack of interest shown to females by researchers in comparison with the interest shown to the conspicuous adult males. Thus, many important questions have remained incompletely answered.

Moreover, findings are inconsistent between different wild populations, between captive and wild groups, and even within the same population. For example, females are said to be less aggressive than males at Gombe (Goodall 1986) but to be on a similar level of aggression at Kibale (Wrangham et al. 1992). While all but one or two females born into the K and M Group at Mahale emigrated as adolescents (Nishida et al. 1990), many females of the Kasakela community at Gombe remained with their natal group and gave birth there (Goodall 1986). At Kibale, one study shows that females associate more with adult males than with other females (Wrangham et al. 1992) but another study (Ghiglieri 1984) shows that females associated more with other females than with males. While adult females had a great influence on the outcome of a power takeover reported for a captive group (de Waal 1982), female involvement in a similar episode was practically nil in the wild (Nishida 1981).

The reasons for the inconsistencies appear multifold. First, we must consider observational biases. Comparisons should be made only between equivalently habituated populations. For instance, vigils of unhabituated individuals at fig trees may sample some subjects disproportionately and miss others entirely.

Second, we must look to our methods of analysis. For example, we lack information on the frequency with which unprovisioned females encounter one another and on the frequency of competitive interactions in such encounters (Baker and Smuts this volume).

Third, local environments differ, especially with regard to human encroachment. This factor probably explains the difference in the degree of female transfer between Gombe and Mahale and probably explains differences in many aspects of social relationships between Bossou and elsewhere (Muroyama and Sugiyama this volume). Sugiyama (1989) reports that females do not transfer at Bossou, but recently female transfer was indirectly inferred there. Only one old female of Bossou immediately cracked the *Coula* nuts that had been collected from the Nimba Mountains and presented by a researcher at the feeding place where chimpanzees regularly crack open oil-palm nuts with stone tools. This strongly suggests that the female transferred from Nimba to Bossou (Yamakoshi 1993) when she was young, perhaps 20 years ago. At that time, Bossou may have been less isolated from other groups geographically and florally.

Fourth, behavioral variations may arise from different degrees of social stability, which has been proposed to explain the contrasts in female aggression seen between the Detroit and Arnhem Zoos (Baker and Smuts this volume). This explanation is consistent with the observation that frequent aggression occurs between newly immigrated and resident females in the wild (Nishida 1989). Finally, de Waal (this volume) integrates beautifully the differences in female behavior seen between populations as instances of the varied adaptive potentials of chimpanzees.

Lone Males, Ostracism, and Xenophobia

Ostracism, which is temporary banishment, and *xenophobia,* which is the fear of strangers, appear to be deeply rooted in human nature. In this respect, the existence of lone male chimpanzees and the processes of creating such males are interesting.

To date, lone chimpanzees have been recorded only as ex-alpha males who had been defeated in power takeovers. This process has been observed three times at Mahale (Uehara et al. 1993) and once at Gombe (Goodall 1992). Now, two other sequences resulting in lone males have been seen at Mahale.

In 1991, the young adult male Jilba, who had had several confrontations with the alpha male Kalunde, was severely attacked by a group of eight: five adult males including Kalunde, two adult females, and one adolescent male. Jilba then spent three months alone to recover before returning to the M Group. Jilba had been an unusual young adult male in that he had rarely pant-grunted to higher ranking males, the alpha male Kalunde in particular. He had bullied adult females for no apparent reason. The communal aggression to Jilba may, therefore, have been punishment or revenge for his impertinence. The episode showed the strong tendency of chimpanzees to follow an alpha male (or support the winner), because no one tried to attack Jilba until Kalunde began to chase him.

Another case of a lone male involves the discovery of a surviving member of K Group (Uehara et al. 1993). After all the K Group adult males disappeared and all the cycling females moved to M Group by 1983 (Nishida et al. 1985), an 11-year-old male named Limongo, an adult female named Wakiluhya, and her four-year-old daughter named Ashura remained in K Group's range. Limongo's old, shy mother was suspected to have died in 1982. Since Wakiluhya's family moved to M Group in 1987, Limongo has been the sole surviving member of K Group. He has been intermittently observed roaming alone in the original range of K Group ever since.

Why are lone males produced in chimpanzee society? Limongo may be a special case. We understand why he is living alone: he cannot emigrate as an adult because of the risk of being killed. However, ex-alpha males and Jilba were originally members of the group. It appears that chimpanzees, both male and female, tend to adopt the strategy of following the male who is currently strongest. Even if an ex-alpha male used to be a favorite companion, other chimpanzees shun him once he is ostracized and, instead, support the new alpha male. For example, the ostracized Ntologi incidentally met the new alpha male Kalunde and several adult males, females, and immatures in 1991. When Kalunde began to chase Ntologi, all of Ntologi's former favorite friends, allies, consorts, and even probable offspring followed suit and aggressed against him. Why such extreme opportunism has evolved in chimpanzee society awaits explanation.

Recently, much attention has been paid to the fact that at Gombe, the Kasakela community exterminated the Kahama community (Manson and Wrangham 1991). However, it is sometimes forgotten that the two communities were one and the same at least 10 years before. From the anthropological perspective, information about the splitting of one originally compatible kin group into two hostile neighboring groups is probably even more important than the striking observation that a warlike phenomenon may exist in chimpanzees. Human communities continuously split into branch communities that gradually cease to communicate peacefully with each other. Hence, a huge number of small ethnic groups may exist in a relatively small area as, for example, the 1,000 tribes in New Guinea (Diamond 1992).

DNA Typing

The application of DNA typing techniques to chimpanzees is underway at Gombe (Morin et al. 1993), at Bossou (Sugiyama et al. 1993), and at Mahale (Takasaki unpubl. data). Furthermore, a Mahale study is planned to investigate the reproductive success of M Group males, focusing on the alpha male Ntologi. However, the DNA typing technique should be used in a more basic way, too.

Recent observations at Mahale teach us of the danger of overlooking close kinship between two individuals because we do not see close associations or social relationships between them. For example, the 10-year-old female Tula has an 18-year-old brother Nsaba. They are known to be siblings, as each has been followed since birth. However, they have neither associated closely nor socialized with each other for at least the past three years after the death of their mother in 1986. Although I followed Nsaba for more than 70 hours, I never saw Nsaba and Tula sitting closely or grooming each other. On the other hand, Tula adopted the three-year-old orphan Maggie, whose mother was suspected to have fallen victim to a lion; Tula has provided Maggie with all maternal care except lactation for almost two years now. Newly arrived observers would never know that Nsaba and Tula are siblings and would infer that Tula and Maggie are siblings. As a matter of fact, Maggie has a biological 15-year-old brother, Masudi, who has not shown affectionate behaviors to her except on the first day after their mother disappeared when Masudi carried her on his back.

Currently there are three older-brother younger-sister pairs whose sibling relations could not be inferred from observation. The presence or absence of the mother and differences in age may influence sibling relationships. For all the above pairs, the mother is dead, and ages of the siblings differ by at least eight years.

These observations remind us that we cannot be too careful in inferring or negating kin relations from the observational data alone. I strongly recommend that any suspected or unknown kin relations should be corroborated by DNA analysis.

Reciprocal Altruism

One of the most unique social characteristics of humankind is the occurrence of reciprocal altruism and the developments that arise from it, from silent trade, potlatch, and kula in preliterate societies to money, lotteries, insurance, banks, taxation, and many other economic and political institutions in modern societies. Reciprocal altruism has not yet been well documented among nonhuman primates, although some remarkable progress has been noted recently (Hemelrijk and Ek 1991; Muroyama 1991; de Waal 1989).

I once tried to investigate reciprocated grooming in chimpanzees in the wild, but one problem of such study is determining the currency of trade. For instance, adult males may reciprocate grooming by aid in combat or by meat sharing as well as by grooming. On the other hand, if a female with her infant is groomed by a nulliparous female, she may reciprocate by lending her infant rather than grooming. However, it is unlikely that newly immigrated females have meat to share, infants to lend, or power to give aid in combat for reciprocation of being groomed. My study regime was to select newly immigrated nulliparous females as study subjects in order to limit the currency for reciprocation of grooming only to grooming. Preliminary results suggest that such young females tend to reciprocate grooming with the same amount of grooming, although the sample size was too small to assert this decisively.

Such a line of study would be appropriate for a captive colony as at the Arnhem or the Detroit Zoo, where continuous, detailed observations are possible.

Reciprocal meat sharing is difficult to study in the wild. However, by concentrating on one particular individual, such as an alpha male (Nishida et al. 1992), one may accrue a sufficient body of data in the long run. Boesch (this volume) claims that the chimpanzees of Taï collaborate in hunting; this has not been confirmed at Mahale. If chimpanzees that play a real role as beaters, rather than onlookers, are given meat by the hunter for their cooperation, or if roles are changed in subsequent hunting (as in reciprocal altruism), then a cooperative system could be at work. It will be interesting to see if either is going on at Taï.

Coalition Strategies

Coalition strategies by adult males have been my major research topic for the last two years. The most impressive episode illustrating such a strategy has been the reacceptance of the ex-alpha male Ntologi (see figure 6). He had been the alpha male for almost 12 years, from July 1979 until May 1991, when he was expelled from M Group by a coalition of two or more adult males, although he was never seen to pant-grunt to any other male. The details of the power takeover were not seen. According to my student Miya Hamai, it was clear that the second-ranking Kalunde became alpha. Thereafter, throughout 1991, Ntologi was occasionally seen roaming alone or with one other male, keeping some distance from M Group. Kalunde seemed to be not as powerful an alpha male as Ntologi had been, as Kalunde was sometimes attacked by two adult males, Jilba and Nsaba.

Figure 6
Ntologi and Kalunde are walking in tandem, followed by a young adolescent male.

There was a curious triangular relationship, to coin a phrase, in which Kalunde was very dominant to the second-ranking Shike, but not so to the third-ranking Nsaba; Shike was very dominant to Nsaba, but fearful of Kalunde; and Nsaba was not afraid of Kalunde, but fearful of Shike. Although both males occasionally pant-grunted to Kalunde, he did not appear very confident in his status, often mounting one of the nearby adult males for reassurance when he was confronted by Nsaba or Jilba.

The reacceptance of Ntologi probably stemmed from this triangle. Near the end of 1991, Shike became seriously ill. As a result, his rank descended to the bottom of the adult male hierarchy, and then he disappeared. Presumably he died. Nsaba, now second ranking, became much more confident because a troublesome obstacle to his rise had been removed. Consequently, no researcher knew whether Nsaba or Kalunde was alpha.

Just at this time, Ntologi began to approach the male cluster of M Group. According to my student Kazuhiko Hosaka, Ntologi took advantage of the discord between Kalunde and Nsaba. Kalunde, who needed aid from others against Nsaba's challenge, tended to form coalitions with Ntologi. At the same time, Ntologi jeopardized Kalunde's alpha status, allowing Ntologi's reacceptance. So, medicine was cure and poison at the same time.

Kalunde and Nsaba once united forces against Ntologi, who ran away screaming. However, Kalunde made it a rule to ally with Ntologi rather than with Nsaba and finally collapsed before Ntologi. Thus, Ntologi again became the alpha male, while Kalunde became subdominant even to Nsaba. The reacceptance of Ntologi became possible only with the disappearance of Shike.

This episode, which awaits Hosaka's detailed analysis, shows how complicated the political confrontations and coordinations can be among wild chimpanzees. The coincidence of Ntologi's return to M Group and Shike's disappearance from it may not have been at all accidental. It is likely that Ntologi collected information about changes in the composition and power dynamics of M Group from hearing pant-hoot calls and other displays even when he was roaming alone. What constitutes the sources of information that an individual chimpanzee collects may be another interesting topic to study.

Deception and Morality

Deceptive behavior provides the material with which field-workers can approach the chimpanzee mind. Concealment of intention and information appears to be considered antisocial or, at least, unfriendly behavior in chimpanzee, as well as in human, society. So, if deception is detected, moralistic aggression may well emerge on the part of the deceived. Following such a line of reasoning (de Waal 1991), I have explored deceptive behavior of chimpanzees in the natural environment.

One day in 1991, the postprime male Bakali was grooming a young adult male on the ground. Meanwhile, another postprime but slightly more dominant male, Musa, approached them. I thought because Musa was approaching in such a relaxed manner, that Musa was about to groom Bakali's back. To my surprise, he suddenly kicked Bakali on the back and ran away. Bakali chased after Musa, screaming. The chase continued on and off for more than three minutes, while Bakali screamed incessantly. I call Musa's surprise attack *deceptive,* because Musa appeared to conceal his intention when approaching. Balaki's unusually long bout of screaming suggests that he did not expect to be kicked. What struck me about this episode is that Musa appeared to feel guilty for his surprise attack: Musa never attacked Bakali again but, instead, only avoided Bakali's counterattack. On the other hand, Bakali continued to chase Musa, as if Bakali were confident that, because he was on the side of justice, there would be no retaliation.

In the same year, Kalunde, who was the alpha male and Musa's intimate ally, intervened by charging at Musa when Musa was chasing a female to get her colobus meat. This is an example of *partisan intervention* (Boehm this volume). As a result, the female escaped from Musa's chase. Immediately, Musa changed his target to Kalunde, chasing fiercely after him. Kalunde, the alpha male, ran away screaming and climbed a tree, where he continued to scream for a long time. Kalunde even mounted his rival Nsaba, while Musa sat under the tree. Why was Kalunde screaming when he could easily have intimidated Musa? My interpretation again is that Kalunde felt guilty for attacking his ally in favor of the female, that is, he felt guilty for betraying his friend. This morally weak position of Kalunde may explain why Musa was more assured in his counterattack, while Kalunde could not afford to make a second attack. A similar case was reported by de Waal (1982). Here, the female Puist attacked her friend Luit, who had not complied with Puist's request for aid in combat despite the fact that Puist had helped Luit in a confrontation with his rival.

I suggest that we should record instances in which less dominant chimpanzees assuredly chase more dominant chimpanzees. Unusually fierce, prolonged retaliation on the part of a subordinate party and the corresponding reluctance to escalate the fight on the part of the dominant party may be one of the factors discriminating moralistic aggression from a conventional counterattack by a subdominant.

Psychologists may offer dozens of alternative hypotheses for these observations. However, faithfully recording episodes is the only way for a field-worker to gain materials to study the psychology of wild primates. Field-workers produce anthropomorphic hypotheses from their observations, as anthropomorphism is a human (and, perhaps, chimpanzee) way of interpreting events (Humphrey 1976). This enables us to formulate testable hypotheses. I hope that psychologists will corroborate or refute the hypothesis about moralistic aggression by ingenious experimental approaches (Boysen this volume; Povinelli this volume; van Hooff this volume).

Comparison of Chimpanzees and Bonobos

It has become increasingly evident that the behaviors of African apes differ between and within species (Doran and Hunt this volume; Hashimoto and Furuichi this volume; Tutin this volume). One of the interesting questions is how can the differences in social organization of the apes be explained by ecological factors, such as the availability of food or the presence of predators. The relationship between party size and food supply has been extensively discussed (Chapman et al. this volume; Malenky et al. this volume). However, as the concept of party size is notoriously ellusive, between-site comparison by different researchers is often problematic. My proposal is that within-site comparisons of party size reflecting daily, seasonal, and annual change should be established first. For within-site comparison, the most objective definition of *party size* would be the largest party that includes at least one adult male (preferably the alpha male) in a day. This definition would be similar to that for the *acoustic party size* of Mitani and Nishida (1993). Although we have only one sample a day, we can avoid observer bias to a great extent.

If the size of bonobo parties are larger than those of chimpanzees, why is that the case? A large food supply is a necessary condition for a large party but not a sufficient explanation. Abundant food and a large food patch *permit* many individuals to aggregate, but do not *force* them to do so unless the food patches are few and localized. How, then, do bonobos benefit from forming large parties? Predation pressure and between-group aggression would be the likely causes. However, there are virtually no leopards at Wamba (Ihobe pers. com.), and between-group aggression is usually so mild that some individuals from different unit-groups mingle with each other peacefully (Idani, 1990). Nevertheless, a violent fight between groups resulting in serious injuries to several individuals has been recorded once (Kitamura unpubl. data; Kano and Mulavwa 1984). The relative peacefulness between bonobo groups contrasts with the fact that aggression between chimpanzee groups is severe (Goodall 1986; Nishida et al. 1985). Predation is not a negligible factor of chimpanzee mortality. Clearly, the size and extent of localization of food patches should be compared between the two species.

Conclusion

Future study should put more emphasis on the differences rather than on the similarities between chimpanzees and human beings. People tend to think that the behaviors of chimpanzees and bonobos are interesting because they resemble the behaviors of human beings. This view has been overemphasized and, because of anthropocentrism, can do an injustice to these creatures. They are charming in their own right. They are inferior to us in some aspects, but superior in others. For example, Matsuzawa (1991) found that the chimpanzee Ai can identify her caretakers or cage mates from photographs with upside-down images as easily as from right-side-up images, while human beings find it difficult to recognize their friends from the upside-down images. This difference probably reflects the different environments in which the two species have evolved separately for about five million years; two-dimensional terrestrial versus three-dimensional arboreal.

When observing in the natural environment, we often are surprised by the sudden arousal and excitement of chimpanzees who have been quietly resting on the ground. This is apparently a reaction to the very distant group-mate calls that we, as human beings, never notice. Moreover, it appears likely that chimpanzees can detect the emotion and thoughts of their companions from facial expressions and postures alone. How do chimpanzees develop this perceptivity? Tomasello (this volume) proposes possible differences in social learning between chimpanzees and humans. Tomasello shows that chimpanzees may learn by emulation rather than by imitation. My fieldwork supports his hypothesis by suggesting that chimpanzees do not imitate each other as much as humans imitate each other.

The ultimate purpose of our studies may be the elucidation of a worldview of chimpanzees in the wild. We cannot accomplish this by field studies alone, but only through interdisciplinary research. Psychological study of chimpanzees in captivity should follow and corroborate or correct the hypotheses generated by field research. Field experiments using playback will be useful in the proof or disproof of some questions. DNA fingerprinting techniques may clarify unknown kin relationships and father-offspring relationships and may drastically change our views of the social relationships of chimpanzees. I do not doubt that the next decade will provide still more exciting new discoveries for a worldview of chimpanzees in the wild.

Acknowledgments

I thank P. Heltne and L.A. Marquardt for an invitation to the "Understanding Chimpanzees: Diversity and Survival" symposium. I am grateful to the Tanzanian Government for permission to conduct my fieldwork. I am also indebted to W.C. McGrew, M.A. Huffman, P. Heltne and E. Altman for their useful comments and editing. For sharing unpublished information cited above, I am indebted to my colleagues M. Hamai, H. Hayaki, K. Hosaka, M.A. Huffman, K. Kawanaka, J.C. Mitani, H. Takasaki, and S. Uehara in particular. Most episodes recorded here come from my last two visits to Mahale funded by a grant under the Monbusho International Scientific Research Program (#03041046).

References

Alp, R. 1993. Meat eating and ant dipping by wild chimpanzees in Sierra Leone. *Primates* 34:463–68.

Boesch, C. 1991. The effects of leopard predation on grouping patterns in forest chimpanzees. *Behav.* 117:220–41.

Boesch, C., and H. Boesch. 1990. Tool use and tool making in wild chimpanzees. *Folia Primatol.* 54:86–99.

Brewer, S. 1978. *The Forest Dwellers.* London: Collins.

Diamond, J. 1992. *The Third Chimpanzee.* New York: Harper Perennial.

Ghiglieri, M.P. 1984. *The Chimpanzees of Kibale Forest.* New York: Columbia Univ. Press.

Goodall, J. 1968. The behaviour of free-living chimpanzees in the Gombe Stream Reserve. *Anim. Behav. Monogr.* 1:161–311.

———. 1973. Cultural elements in a chimpanzee community. In E. Menzel, ed., *Precultural Primate Behavior,* pp. 144–84. Basel: S. Karger.

———. 1986. *The Chimpanzees of Gombe.* Cambridge, Mass.: Belknap Press.

———. 1992. Unusual violence in the overthrow of an alpha male chimpanzee at Gombe. In T. Nishida, W.C. McGrew, P. Marler, M. Pickford, and F.B.M. de Waal, eds., *Topics in Primatology.* Vol. 1, *Human Origins,* pp. 131–42. Tokyo: Univ. of Tokyo Press.

Hasegawa, T., M. Hiraiwa-Hasegawa, T. Nishida, and H. Takasaki. 1983. New evidence of scavenging behavior in wild chimpanzees. *Current Anthro.* 24:231–32.

Hemelrijk, C.K., and A. Ek. 1991. Reciprocity and interchange of grooming and support in captive chimpanzees. *Anim. Behav.* 41:923–35.

Huffman, M.A., and M.S. Kalunde. 1993. Tool-assisted predation on a squirrel by a female chimpanzee in the Mahale Mountains, Tanzania. *Primates* 34:93–8.

Humphrey, N. 1976. The social function of intellect. In P.P.G. Bateson and R.H. Hinde, eds., *Growing Points in Ethology,* pp. 303–17 Cambridge, England: Cambridge Univ. Press.

Idani, G. 1990. Relations between unit-groups of bonobos at Wamba, Zaire: Encounters and temporary fusions. *African Study Monogr.* 11:153–86.

Inagaki, H., and T. Tsukahara. 1993. A method of identifying chimpanzee hairs in lion feces. *Primates* 34:109–12.

Kano, T., and M. Mulavwa. 1984. Feeding ecology of the pygmy chimpanzee *(Pan paniscus)* of Wamba. In R.L. Susman, ed., *The Pygmy Chimpanzee: Evolutionary Biology and Behavior,* pp. 233–74 New York: Plenum Press.

Kortlandt, A., and J.C.J. van Zon. 1967. Experimentation with chimpanzees in the wild. Lecture at the 10th International Ethological Conference, Stockholm, Sweden.

Kuroda, S., and S. Suzuki. In Press. Preliminary report on predatory behavior and interactions over meat in tschego chimpanzees *(Pan troglodytes troglodytes)* in the Ndoki Forest, Congo. *Primates* 35:2.

Lewin, R. 1989. *Human Evolution.* 2d ed. Oxford, England: Blackwell.

Manson, J.H., and R. W. Wrangham. 1991. Intergroup aggression in chimpanzees and humans. *Current Anthro.* 32:369–90.

Matsuzawa, T. 1991. *The World Viewed by a Chimpanzee.* Tokyo: Univ. of Tokyo Press.

———. 1993. Lecture at the Symposium on the Great Apes, Kyoto Univ. Primates Research Institute, February 28.
McGrew, W.C. 1974. Tool-use by wild chimpanzees in feeding upon driver ants. *J. Human Evol.* 3:501–8.
———. 1983. Animal foods in the diets of wild chimpanzees *(Pan troglodytes):* Why cross-cultural comparison? *J. Ethol.* 1:46–61.
———. 1992. *Chimpanzee Material Culture.* Cambridge, England: Cambridge Univ. Press.
Mitani, J.C., and T. Nishida. 1993. Contexts and social correlates of long-distance calling by male chimpanzees. *Anim. Behav.* 45:735–46.
Morin, P.A., J. Wallis, J. Moore, R. Chakraborty, and D. S. Woodruff. 1993. Non-invasive sampling and DNA amplification for paternity exclusion, community structure, and phylogeography in wild chimpanzees. *Primates* 34:347–56.
Morris, K., and J. Goodall. 1977. Competition for meat between chimpanzees and baboons of the Gombe National Park. *Folia Primatol.* 28:109–21.
Muroyama, Y. 1991. Mutual reciprocity of grooming in female Japanese macaques *(Macaca fuscata). Behav.* 119:161–70.
Nishida, T. 1981. *The World of Wild Chimpanzees,* Tokyo: Chuokoronsha.
———. 1989. Social interactions between resident and immigrant female chimpanzees. In P. Heltne and L.G. Marquardt, eds., *Understanding Chimpanzees,* pp. 68–89. Cambridge, Mass.: Harvard Univ. Press.
Nishida, T. and M. Nakamura. In Press. Chimpanzee tool-use to clear a blocked nasal passage. *Folia Primatol.*
Nishida, T., T. Hasegawa, H. Hayaki, Y. Takahata, and S. Uehara. 1992. Meat-sharing as a coalition strategy by an alpha male chimpanzee? In T. Nishida, W.C. McGrew, P. Marler, M. Pickford, and F.B.M. de Waal, eds., *Topics in Primatology.* Vol. 1, *Human Origins,* pp. 159–74. Tokyo: Univ. of Tokyo Press.
Nishida, T., M. Hiraiwa-Hasegawa, T. Hasegawa, and Y. Takahata. 1985. Group extinction and female transfer in wild chimpanzees in the Mahale National Park, Tanzania. *Z. Tierpsychol.* 67:284–301.
Nishida, T., H. Takasaki, and Y. Takahata. 1990. Demography and reproductive profiles. In T. Nishida, ed., *The Chimpanzees of the Mahale Mountains,* pp. 63–97. Tokyo: Univ. of Tokyo Press.
Page, J.E., F. Balza, T. Nishida, and G.H.N. Towers. 1992. Biologically active diterpens from *Aspilia mossambicensis,* a chimpanzee medicinal plant. *Phytochemistry* 31:343–9.
Patterson, J.H. 1979. *The Man-Eaters of Tsavo.* London: MacMillan Publishers.
Rodriguez, E., and R.W. Wrangham. 1993. Zoopharmacognosy: The use of medicinal plants by animals. In K.R. Downum et al., eds., *Phytochemical Potential of Tropical Plants,* pp. 89–105. New York: Plenum Press.
Savage-Rumbaugh, S. 1993. *Kanzi.* Tokyo: NHK Books.
Sugiyama, Y. 1989. Population dynamics of chimpanzees at Bossou, Guinea. In P. Heltne and L.G. Marquardt, eds., *Understanding Chimpanzees,* pp. 134–45. Cambridge, Mass.: Harvard Univ. Press.
Sugiyama, Y., S. Kawamoto, O. Takenaka, K. Kumazaki, and N. Miwa. 1993. Paternity discrimination and inter-group relationships of chimpanzees at Bossou. *Primates* 34:545–52.
Tsukahara, T. 1993. Lions eat chimpanzees: The first evidence of predation by lions on wild chimpanzees. *Am. J. Primatol.* 29:1–11.

Uehara, S., M. Hiraiwa-Hasegawa, K. Hosaka, and M. Hamai. 1994. The fate of defeated alpha male chimpanzees in relation to their social networks. *Primates* 34:49–55.

Uehara, S., T. Nishida, H. Takasaki, R. Kitopeni, M., Kasagula, K., Norikoshi, T. Tsukahara, R. Nyundo, and M. Hamai. A lone male chimpanzee in the wild: The survivor of a disintegrated group. *Primates* 35:(3).

de Waal, F.B.M. 1982. *Chimpanzee Politics.* London: Jonathan Cape.

———. 1986. The brutal elimination of a rival among captive male chimpanzees. *Ethol. and Sociobiol.* 7:237–51.

———. 1989. Food sharing and reciprocal obligations among chimpanzees. *J. Human Evol.* 18:433–59.

———. 1991. The chimpanzee's sense of social regularity and its relation to the human sense of justice. *Am. Behav. Scientist* 34:335–49.

Wrangham, R.W. 1979. On the evolution of ape social systems. *Social Sci. Information* 18:335–68.

Wrangham, R.W., A.P. Clark, and G. Isabirye-Basuta. 1992. Female social relationships and social organization of Kibale Forest chimpanzees. In T. Nishida, W.C. McGrew, P. Marler, M. Pickford, and F.B.M. de Waal, eds., *Topics in Primatology.* Vol. 1, *Human Origins,* pp. 81–98. Tokyo: Univ. of Tokyo Press.

Yamakoshi, G. 1993. A female chimpanzee carrying "culture." Abstracts of Spoken Papers, p. 15, 30th Congress of Japanese Association of Africanists, Hirosaki University, May 30.

Postscript—Conservation and the Future of Chimpanzee and Bonobo Research in Africa

Jane Goodall

At this point we should face a sobering fact: unless there is radical change in conservation policies across Africa, the opportunities for further research into behavioral diversity in populations of wild chimpanzees will become increasingly scarce—because, one after the other, those populations are vanishing. At the time of writing (March 1994), chimpanzees are being studied, or habituated for study, at 12 sites in 10 countries. We know that different behavioral traditions exist in each of these areas, and we can expect to find further variations as new communities are observed. But we can never know the true extent of cultural diversity of chimpanzees because so many communities, along with their cultures, are already gone. It is vital that we should study as many groups of chimpanzees in as many parts of Africa as possible, before it is too late. It is desperately important that scientists join conservationists in trying to protect chimpanzee habitats in as many countries as possible.

Chimpanzees were once present in 25 African nations. They are still found in 21 of those countries, but only in four—Congo, Zaire, Cameroon, and Gabon—in significant numbers. Elsewhere, remaining populations are small and often fragmented. Everywhere, including those four countries in the heart of the chimpanzees' range, numbers are continually decreasing as a result of habitat destruction and hunting. The story is similar for bonobos, who were probably never found north of the Zaire River (formerly the Congo River). In their known range, in Zaire, the size of the remaining bonobo populations continues to decrease.

One of the most important sessions at the 1986 conference was that on conservation. It was profoundly disturbing. One speaker after another

described the habitat destruction and the hunting and/or trade in live infant chimpanzees that was going on in his or her area. The overall picture was grim; it led us to form the Committee for Conservation and Care of Chimpanzees, known as the Four C's, that was chaired by Geza Teleki. Five years later, during the 1991 conference, the sessions on conservation showed that little had improved. However, the sense of urgency generated by the conference did lead to a number of surveys, undertaken with a view to establishing an overall chimpanzee/bonobo action plan, to be used for the coordination of conservation efforts across Africa. There were other projects carried out independently that provided us with additional information. Paul and Nathalie Marchesi carried out a comprehensive survey in Ivory Coast; Bob Dowsett and Francoise le Maire conducted a general survey in southern Congo and included information on chimpanzees; Mike Fay and Richard Carroll worked in Central African Republic and northern Congo; Peter Jenkins and Liza Gadsby gathered some information about chimpanzees during their mandrill survey in Nigeria; and Jim Moore carried out two surveys, one on the small remaining population in Mali, the other in the very arid Ugalla region of western Tanzania.

Those surveys, along with field studies started after 1986—such as those in Endoke, Niokolo Koba, and Kigali National Parks—certainly increased our knowledge of the distribution and size of chimpanzee populations in certain areas. Despite heavy hunting there are still, as mentioned, significant numbers of chimpanzees in the heartland of their range in Gabon, Cameroon, Congo and Zaire, but the rate of decline is alarming. The greater number of chimpanzees are found these countries because that is where we find the last of Africa's great rain forests–but how long can the rain forests last? Like forests and woodlands everywhere, rain forests are disappearing. Commercial timber companies are given huge concessions. Although environmentally-friendly, selective logging is possible, it is seldom practiced. Also, the ever-increasing growth of human populations leads to clear-cutting for agriculture, grazing, and building. Once the forest has gone, the chimpanzees who lived there will soon be gone too. They may move away from devastated areas and try to exist elsewhere, but it is unlikely that they can become integrated into another social group. Chimpanzees are aggressively territorial; refugees are more likely to be attacked and killed than accepted. All across Africa there are pathetic remnants of chimpanzee communities, hanging on for as long as they can, in patches of forest in areas so steep and rugged that even desperate human populations can see no way of clearing these areas for their own purposes. The healthy populations that remain are becoming increasingly fragmented, forced to exist in smaller and smaller areas and with ever-decreasing opportunities for gene exchange between neighboring social groups. That is the case at Gombe.

By and large, during the five years between the two conferences, nothing much had been done to protect the beleaguered chimpanzee populations and, in some cases, the situation had deteriorated. In Zaire, for example, an effort by the Jane Goodall Institute (JGI) in conjunction with a local group of Zairian, European and American conservationists initiated an educational program to stimulate interest in chimpanzees and bonobos. This effort got off to a good start in 1990. The government was persuaded to enforce an existing law and confiscate infant chimpanzees and bonobos that were offered for sale in the markets of Kinshasa. After seven such confiscations, these pathetic infants were seen far less often in the markets. But that almost certainly did not mean that chimpanzee hunting had been reduced–merely that the infants were no longer brought to the market. The infants offered for sale in Kinshasa in the past had been sent down the great Zaire River from the north. If hunters from the north received word that such sales were no longer possible, they probably either kept the infants in their villages as play things for their children–or cooked them, along with their mothers, for the little extra meat their tiny carcasses would provide. We had plans for attempting to educate the hunters, but it would have been a slow and thankless task; and the 1991 rioting, when most foreigners left the country, brought all such efforts to an end. Both a JGI plan, in association with the University in Luanda, for chimpanzee conservation in the Cabinda area of northern Angola, and an attempt to conduct a survey of chimpanzees in Liberia had to be abandoned as a result of the renewed outbreak of civil war in those countries.

In the People's Republic of Congo, the government has been persuaded to confiscate infant chimpanzees sold illegally by hunters or middlemen. Two organizations receive these confiscated youngsters: JGI and H.E.L.P. JGI cares for chimpanzees in Brazzaville (in the zoo) and in the Pointe Noire area (in the 'Mpili Sanctuary built by Conoc, Congo). HELP has placed many young chimpanzees on a cluster of islands near the border of Gabon. By 1992 more than 50 infants were received by these two organizations. It has been estimated that 10 chimpanzees die for every infant that survives the trauma of capture and transport–mothers, infants who die either when their mothers are shot or during the journey to the marketplace, and other adults who are shot as they try to defend victims. If we accept this estimate, the capture of those 50 infants means that 500 others died in four years. Even that number is but a percentage of the total number of chimpanzees killed by hunting, mostly for bushmeat.

Recent reports, particularly that of Swiss photographer Karl Ammann, suggest that the hunting of chimpanzees for bushmeat poses a much more serious threat to the remaining chimpanzee populations throughout west and central Africa than was previously supposed, and that the number of orphan chimpanzees, by-products of this cruel trade, is far greater than I had supposed. Many of these youngsters are bought as pets, often to rescue them from market or roadside; others are purchased to attract customers to hotels, bars and other places of business. In either case, the end result is the same: once they become too strong to handle and potentially dangerous, they are killed or banished—tied on chains or confined in tiny cages or crates, often in appalling conditions.

Some young chimpanzees are still sold to dealers for onward shipment to other parts of the world as there is still a demand for wild chimpanzees for entertainment and, in some countries, for biomedical research. However, since chimpanzees were reclassified as *endangered* under the U.S. Endangered Species Act, this trade is not as flourishing as it was in 1986. The actual number shipped out of African countries is not known, but there are reports of youngsters arriving in various European ports as well as Dubai, Israel, Mexico, and South American countries. Often it is only too easy for unscrupulous dealers to smuggle chimpanzees out of Africa–out of countries such as Congo with its miles of unpatrolled coastline in the southwest and Zaire, where further rioting took place in February 1993 and where the political situation is such that illegal trade in wildlife ranks low on the government's list of concerns. In fact, many of the young chimpanzees smuggled into Uganda, Burundi, Angola and Zambia, as well as Congo, originated in the forests of Zaire.

I have painted a grim picture. We cannot escape the stark reality that vast areas of rain forest and woodland are destroyed each year and, concurrently, thousands of living creatures die or are killed for food. Is there any hope? Can chimpanzees and bonobos survive in the wild? Can we save areas of natural habitat that are large enough to supply their needs and that are sufficiently well protected to ensure their safety? I have always been an optimist and I believe that, at least in some countries, conservation can work. For example, Uganda and Burundi were once flourishing centers for illegal traffic in chimpanzees and other wildlife. Then governments clamped down on this trade so that today it is very seldom that one hears of orphan chimpanzees for sale there. Moreover, Burundi is fighting to save the last of its unique montane rain forest in the north of the country. True, only some 350 chimpanzees live there, but the forest is contiguous with a larger area of similar habitat in Rwanda; the total number of chimpanzees in that particular forest, for the two countries combined, is probably around 500. The recent civil war brought our project in Burundi to an end temporarily, but we expect to return to the site. In Uganda, the government has designated its nine main forest blocks as reserves and is actively encouraging the development of chimpanzee tourism in place of commercial logging.

In Congo, the magnificent northern forests are being preserved in the newly designated Ndoke National Park. In Ivory Coast, Christophe Boesch is working with the government to safeguard the unique flora and fauna—including the chimpanzees—of the Taï forest. Similar conservation activities are occurring in many other parts of Africa.

Even more important, perhaps, is the growing awareness, in so many African countries, that there *is* a problem and that conservation of natural resources is desperately important for the people of the continent and not just a whim of wealthy capitalists from the developed world. African governments, by and large, admit that conservation measures are necessary to prevent deforestation, knowing that deforestation will lead to erosion, changing climate and desertification, and will leave an environment equally inhospitable to human and to nonhuman animal alike.

There is, of course, a major problem. Even when governments–and peasants–fully appreciate the effect of poor land management, the level of economic poverty in the country may prevent the implementation of sound conservation measures *unless population growth can be stabilized.* Until this happens, villagers must continue to cut down trees to grow extra food, to build more houses, to graze their cattle, and to cook their food. What else can they do? For those who have not been to Africa and who have not seen the conditions for themselves, the crippling poverty in which so many people live is hard to imagine. The subsistence farmer, even when aware that loss of trees leads to erosion and to loss of fertility in years to come, cannot concern himself with the future when he must struggle to feed himself and his family each day.

Nevertheless, although the solution is not always obvious, the fact that the nature of the problem is understood—that along with an appreciation of the dire consequences that will result if the problem is not solved–is an important first step. The political climate in many African countries has changed over the past ten years, so that it is easier now for governments and local authorities to implement new conservation strategies. These include family planning projects; programs to educate women and give them more control over their lives; rural development programs such as agro-forestry (growing trees along with other crops for building poles, firewood, charcoal, shade and so forth) and game farming; and the crafting of artifacts to sell. If village economies can be boosted, it will no longer be necessary for the inhabitants to grow all their food; some can be bought and forests can be saved. Then, valuable forest products such as medicinal plants can be harvested in a sustainable manner.

None of this may sound particularly relevant to the researchers and their students studying chimpanzees or bonobos (or any other species) in Africa. However, if we hope to continue studying wild populations in the future, conservation of the natural habitat is of utmost importance. The field researcher can contribute to conservation efforts in several ways. First, researchers can employ local people as field staff to help in the collection of data. This is an important part of our methodology at Gombe and at some other field sites. Second, researchers can help in the development of tourism. I am well aware that tourism is a two-edged sword and a highly controversial topic. It is absolutely true that too many visitors can damage, even destroy, the very places and species they travel so far to see. "Forest tourism" is particularly difficult to operate in an environmentally friendly manner; the numbers of people allowed into a forest at any given time must be strictly controlled. Yet even small numbers of visitors can make a difference to local economy—especially if villagers from around the area are employed as guides and in as many other ways as possible. The Gombe National Park was opened for tourism in the late seventies and more and more visitors have made the difficult journey to see for themselves the chimpanzees made famous by the National Geographic specials. Gradually the revenue generated by these visits has become increasingly beneficial to the Tanzanian government. This is good in some ways, but there are disadvantages: the chimpanzees and the visitors actually pose a danger to each other. Because the chimpanzees we have studied for so long have absolutely no fear of (or respect for) people, they allow visitors to get very close to them and, thus, run the risk of catching infectious human diseases that are brought unwittingly to the park by visitors. The visitors, for the most part, do not realize that they may constitute a health hazard to the chimpanzees. Nor do they usually know that chimpanzees are four times stronger than the average human male and can be dangerous—many visitors try to get closer than they should. So, for the past several years, we have been habituating a chimpanzee community in the north of the park, with a view to moving the tourist operation away from the over-habituated study community. We hope this will improve the situation for visitors and chimpanzees alike.

In eastern Zaire, the Tongo chimpanzees were habituated for the express purpose of starting chimpanzee tourism in an attempt to conserve the area as a national park. The exercise has been very successful and has helped, undoubtedly, to save those particular chimpanzees. Dean Anderson, working for JGI, has initiated a similar project in the Kigali forest in Burundi, and a number of Peace Corps volunteers are habituating chimpanzees in protected forests in Uganda. Of course, students can and should collect valuable data on the behavior and ecology of the chimpanzees while habituating communities for tourism and, subsequently, after the tourist visits have begun.

Field researchers can contribute to chimpanzee conservation by helping to educate the local people, teaching them about the behavior of chimpanzees and other animals and about the need for preserving the natural habitat. This task will be much easier if an economic benefit, in the form of tourism, has already been demonstrated. At Gombe, as mentioned, we employ local people not only as trail cutters and so forth, but also to help collect data. These men are proud of their jobs, they care about the chimpanzees as individuals, and they talk about them to family and friends in the villages. We have virtually no poaching at Gombe. In other words, our own staff act as educators in their own villages. Recently, we have arranged for some of these local people to visit the local schools and talk about conservation as this relates to village life. It is, of course, particularly important to develop programs for children, and we have started bringing small groups of youngsters to Gombe to show them something of the wonder of their cultural heritage.

We are also starting conservation education programs for the local people, particularly children, at our three sanctuaries—two in Congo and one in Burundi. Very few of the children (or adults either) who visit the sanctuaries have seen chimpanzees in the wild. The opportunity to see chimpanzees close up, and to observe how, in so many ways, they behave like us (as when they kiss, embrace, beg, share food, and stare intently into a visitor's face) is clearly a unique and rewarding experience. The chimpanzees thus serve as a focal point as we introduce the children, through pictures, to the wildlife that is their heritage.

A few other sanctuaries exist in Africa. Stella Brewer tried to release a group of chimpanzees into the wild in Senegal, but there was a free-ranging community already in residence, and they mounted a series of attacks on the new group. Brewer then moved the group to Baboon Island in the Gambia, and the project was taken over by Janice Carter. In Zambia, Dave and Shiela Siddle have built a large sanctuary on their farm and are caring for more than 40 chimpanzees, refugees from all over the world. Lonrho (Kenya) Ltd. has built another sanctuary in Kenya for refugees from other parts of Africa and plans are underway for similar refugees in Sierra Leone and Uganda.

Is it appropriate to spend relatively large amounts of money on a few captive individuals? Would it be more appropriate to use the funding for conservation projects? If the answers to those two questions are *no* and *yes* respectively, which is the view of many conservation organizations, how should we resolve the problem posed by the many orphan chimpanzees–those confiscated by government officials and those who were once pets? Should they be allowed to die for lack of proper care? Or be humanely killed?

There are three reasons why I believe that sanctuaries are justified. First, the sanctuaries can play an important role in education. Second, managed correctly, sanctuaries should bring in tourist money so that not only will they become self-supporting, but additionally they will help to boost the economy of the local people. Third, it is important to try to care for abandoned chimpanzees on ethical grounds. We are trying to teach local people the true nature of our closest relatives, that each has his or her own personality. If we permit euthanasia or tolerate mistreatment this would surely be seen as a contradiction of our message. How much better to care for the orphans, so that each one, named and with a known history, may become an ambassador for the species as a whole.

We have the same attitude towards the animals in the zoos where we are trying to improve conditions. In most African zoos the conditions are pretty grim. This is hardly surprising in view of the poverty of the *human* animals. In Congo, for example, few people living in a city can afford more than one meal a day, if that. It is inevitable that the budget for a zoo is meager and that the animals should often go hungry. But that doesn't mean we should condone this. When people ask how we can justify "wasting" food on "animals" when there are so many starving people, I always tell them that, as we (the human race) put the animals in the cages, these animals are our responsibility. Zoo animals cannot get out to find food for themselves. So we must either open the cage doors and set them free, kill them, or feed them. Slow death by starvation and neglect is simply not an appropriate solution. We are helping the Brazzaville Zoo administration by providing a little extra food and helping to care for the animals. Conditions have improved and, as more people visit, we hope to develop a conservation education center there. A similar center is planned for the Entebbe Zoo in Uganda. It would be better if there were no zoos at all but, as the animals are already captive, we should ensure that their living conditions are adequate and hope that they will stimulate interest in wildlife and conservation.

In summary, the second chimpanzee/bonobo conference organized and hosted by the Chicago Academy of Sciences, once again, served to increase greatly our knowledge of our closest living relatives and to highlight their plight worldwide: the destruction of their natural habitat, the decrease in their numbers throughout their range, and the often deplorable conditions in which they may languish in captivity. As the human primate continues to increase in numbers and to destroy, pollute and make ugly more and more places around the world, the outlook for slow-breeding species such as the great apes becomes correspondingly more bleak. Only if the scientific community allies itself with the conservation and environmental communities and seeks the cooperation of local citizens, governments, business and industry, is there hope for the genus *Pan* to continue through the 21st century.

Biographies of the Authors

Kate C. Baker, Ph.D., Biological Anthropology, University of Michigan; Researcher, White Sands Research Center, The Coulston Foundation; coauthor, "Social Relationships of Female Chimpanzees: Diversity Between Captive and Social Groups."

Gender differences as related to dominant and aggressive behavior and to play are her research interests. In the course of her work, she conducted research on chimpanzees in captivity in Texas and in Detroit, and she assisted in the development and implementation of the environmental enrichment and research program at the New Mexico Regional Primate Research Laboratory.

White Sands Research Center,
The Coulston Foundation
Building 1264
Holloman AFB, New Mexico 88330 USA

Christopher Boehm, Professor of Anthropology, University of Southern California; Director, Jane Goodall Research Center, University of Southern California; author, "Pacifying Interventions at Arnhem and Gombe."

Social relationships of primates and humans with a focus on conflict management are his primary concern. As an anthropologist, he is interested in communication and language both in humans and in apes. His articles apply his chimpanzee studies to the understanding of human behavior.

University of Southern California
Department of Anthropology
Los Angeles, California 90089 USA

Christophe Boesch, Assistant Professor, Institute of Zoology, University of Basel; Chimpanzee Researcher, Taï National Park, Côte d'Ivoire; author, "Hunting Strategies of Gombe and Taï Chimpanzees."

In the course of his long-term research, he and his wife Hedwige Boesch-Achermann focused primarily on nut-cracking behaviors and the behavioral ecology of chimpanzees in a comparative perspective. His recent articles focus on food sharing, the use of tools, the notion of culture in chimpanzees, and cooperation in hunting.

University of Basel
Institute of Zoology
4051 Basel, Switzerland

Sarah T. Boysen, Ph.D., Associate Professor, Ohio State University; Director, Comparative Cognition Project, Emory University; author, "Individual Differences in Cognitive Abilities in the Chimpanzee."

Cognitive processes and mechanisms underlying cognition as well as methods of evaluating cognitive processes both in great apes, monkeys, dogs, and pigs are her present research interests. In the course of her work, she made a significant contribution to our understanding of the evolution of cognition through study of cognitive processes of chimpanzees in captivity.

The Ohio State University
Primate Cognition Project
1885 Neil Avenue, Townshend Hall
Columbus, Ohio 43210 USA

Colin A. Chapman, Ecologist, Assistant Professor, Department of Zoology, University of Florida; coauthor, "Party Size in Chimpanzees and Bonobos: A Reevaluation of Theory Based on Two Similarly Forested Sites."

The ecological determinants of group size and social structure of primates are his primary research interests. In the course of his research he has studied spider monkeys in Costa Rica for six years and has been working in Kibale Forest since 1989.

University of Florida
Department of Zoology
Gainesville, Florida 32611 USA

Diane M. Doran, Professor, SUNY at Stony Brook; Past-Director, Karisoke Research Center, Rwanda; coauthor, "Comparative Locomotor Behavior of Chimpanzees and Bonobos: Species and Habitat Differences."

Locomotive and positional behavior of both chimpanzees and gorillas as that behavior relates to human origins are her primary research interests. In the course of her work, she studied bonobos in Zaire, chimpanzees at Taï Forest, Ivory Coast, and mountain gorillas at Karisoke, Rwanda.

SUNY at Stony Brook
Department of Anthropology
Stony Brook, New York 11794 USA

Barbara Fruth, Researcher, Department of Zoology, University of Munich coauthor, "Comparative Nest-Building Behavior in Bonobos and Chimpanzees."

Ecological and social aspects of primate behavior are her research interests. In the course of her work, she studied chimpanzees in Ivory Coast and in Zaire, and she observed the social behavior of bonobos as those behaviors relate to nests and to nest building at Lomako in Zaire.

Ludwig Maximilians Universität
Zoologisches Institut
Postfach 20 21 36
Munich 80021 Germany

Takeshi Furuichi, D. Sc., Department of Zoology, Kyoto University, Japan; Assistant Professor, Meiji-Gakuin University, Japan; coauthor, "Social Role and Development of Noncopulatory Sexual Behavior of Wild Bonobos."

Sexual behavior and the life history of the gender that transfers between groups are his research interests. In the course of his work, he observed bonobos and Japanese macaques in the wild.

Meiji-Gakuin University
Laboratory of Biology
1518 Kamikurata
Totsuka, Yokohama 244, Japan

Jane Goodall, Ph.D., Field Researcher, Gombe, Tanzania; Founder, Gombe Stream Research Center, dedicated to the study of chimpanzees; Founder in 1977, The Jane Goodall Institute, a nonprofit organization committed to conservation of chimpanzees, field research on chimpanzees in the wild, and studies of chimpanzees in captivity; author, numerous books on chimpanzees; recipient, numerous awards; author, "Foreword" and "Postscript—Conservation and the Future of Chimpanzee and Bonobo Research in Africa."

Treatment of chimpanzees in captivity and their conservation in the wild are her deepest concerns. She first traveled to Gombe, Tanzania, in 1960 to conduct field research for the late anthropologist Louis B. Leaky. She has studied chimpanzees in the wild for more than 30 years.

The Jane Goodall Institute
P.O. Box 599
Ridgefield, Connecticut 06877 USA

The Jane Goodall Institute, a nonprofit organization dedicated to wildlife research, education and conservation; contributed five illustrations of chimpanzees to *Chimpanzee Cultures.*

In addition to supporting the ongoing research of chimpanzee behavior in Gombe, the Institute operates two sanctuaries in Burundi and in The

Congo that are dedicated to caring for chimpanzees rescued from inhumane captive conditions. In 1991, JGI launched a major environmental education initiative called Roots and Shoots dedicated to teaching young people an environmental ethic based on respect for all living beings. JGI's activities are funded solely by donations and grants. Your contribution helps to continue this important work.

The Jane Goodall Institute
P.O. Box 599
Ridgefield, Connecticut 06877 USA

Chie Hashimoto, M. Sc. and Post-graduate Student, Primate Research Institute, Kyoto University, Japan; coauthor, "Social Role and Development of Noncopulatory Sexual Behavior of Wild Bonobos."

The development of social behavior in immature individuals is her research interest. In the course of her work, she studied bonobos, chimpanzees, and wild Japanese macaques.

Kyoto University
Primate Research Institute
Inuyama, Aichi 484, Japan

Paul G. Heltne, President, Chicago Academy of Sciences; trained as a primatologist, he has organized major symposia on diversity and conservation of chimpanzees; editor of *Neotropical Primates, The Lion-Trailed Macaque, Science Learning in the Informal Setting,* and *Understanding Chimpanzees.* He is an editor of *Chimpanzee Cultures* and author of the "Preface."

In his work as leader of a museum, he seeks to promote scientific literacy for all citizens. He has established professional programs for teachers, founded the International Center for the Advancements of Scientific Literacy with Jon Miller and served as a member of the Task Force on Education of the American Association of Museums.

The Chicago Academy of Sciences
2001 N. Clark Street
Chicago, Illinois 60614

Gottfried Hohmann, Researcher, Max Planck Institute for Behavioral Physiology, Andechs, Germany; coauthor, "Comparative Analysis of Nest-Building Behavior in Bonobos and Chimpanzees."

Bonobos at Lomako in Zaire and at Planekendael in Belgium are his research interest. In the course of his work, he conducted field research in India and Sri Lanka.

Max Planck Institute for Human Ethology
Von-der-Tann-Str. 3–5
Andechs, 82346, Germany

Jan A.R.A.M. van Hooff, Professor, Universiteit Utrecht, The Netherlands; Scientific Director, Arnhem Zoo, The Netherlands; Founder in 1971, Arnhem Chimpanzee Consortium; author, "Understanding Chimpanzee Understanding."

Communication, behavior, and social organization of primates are his research interests; at the Consortium, he continues to document the social processes by which individual chimpanzees regulate their position in the community. He also conducts research on tropical forest primates in Indonesia focusing on the effects that ecological conditions have on social phenomena.

Universiteit Utrecht
Ethologie and Socio-ecology
PB 80086
3508 TB Utrecht, The Netherlands

Michael A. Huffman, Researcher, Kyoto University, Japan; coauthor, "Diversity of Medicinal Plant Use by Wild Chimpanzees."

The use of medicinal plants by chimpanzees, social processes, aging in chimpanzees, and ethnomedicine are his research interests. In the course of his work, he has conducted long-term studies on chimpanzees in the wild and free-ranging Japanese macaques.

Kyoto University
Department of Zoology
Laboratory of Evolution Studies
Sakyo, Kyoto 606-01, Japan

Kevin D. Hunt, Professor of Anthropology, Indiana University; coauthor, "Comparative Locomotor Behavior of Chimpanzees and Bonobos: Species and Habitat Differences."

Functional morphology, the evolution of human bipedalism, and australopithecine and ape ecology are his current research interests. He has studied chimpanzees at Mahale and Gombe in Tanzania and at Kibale in Uganda. His most recent research has dealt with the feeding ecology and functional morphology of *Australopithecus afarensis.*

Indiana University
Department of Anthropology
Bloomington, Indiana 47405 USA

Suehisa Kuroda, Associate Professor, Department of Zoology, Kyoto University, Japan; coauthor, "The Significance of Terrestrial Herbaceous Food for Bonobos, Chimpanzees, and Gorillas."

All aspects including evolution, anatomy, conservation, and behavior patterns of African great apes are his research interest. In the course of his work, he studied interspecies relationships between chimpanzees and gorillas in northern Congo.

Kyoto University
Laboratory of Physical Anthropology
Faculty of Science
Sakyo, Kyoto, 606, Japan

Mark Maglio, Wildlife Illustrator; illustrator of the six chimpanzee and bonobo illustrations used throughout this volume.

Mark Maglio Designs
P.O. Box 872
Plainville, Connecticut 06062 USA

Richard K. Malenky, Researcher, Lomako Forest, Zaire; Recipient with his wife, Nancy Thompson-Handler, of the 1991 L.S.B. Leakey Great Ape Fellowship for Research and Conservation, conferred for expansion of their efforts at Lomako; coauthor "The Significance of Terrestrial Herbaceous Foods for Bonobos, Chimpanzees, and Gorillas."

The influence of food resources, diet, and nutrition on grouping and on social behavior of the great apes are his central research interests. In the course of his work, he conducted field studies of feeding ecology of great apes at Lomako.

SUNY at Stony Brook
Department of Anthropology
Stony Brook, New York 11794 USA

Tetsuro Matsuzawa, Professor, Section of Language and Intelligence, Department of Behavioral and Brian Sciences, Primate Research Institute, Kyoto University, Japan: author, "Field Experiments on Use of Stone Tools in the Wild."

Language, perception, and cognition are his research interests. In the course of his work, he conducted field studies in the use of stone tools by chimpanzees in the wild at Bossou, Guinea, and he studied the behavior of the captive chimpanzee Ai, one of the world's most intelligent chimpanzees. His work with Ai was highlighted in the acclaimed Japanese television film, "Ai: the Chimpanzee Mind."

Kyoto University
Primate Research Institute
Inuyama, Aichi, 484, Japan

W.C. McGrew, Professor of Anthropology and Zoology, Miami University (Ohio); author, "Tools Compared: The Material of Culture" and "Overview: Diversity in Social Relations;" coauthor of "Introduction—The Challenge of Behavioral Diversity."

Comparative socio-ecology of the African apes, but especially the use of tools in chimpanzees, is his research focus. Twenty years of fieldwork in East, Central and West Africa was presented in

Chimpanzee Material Culture: Implications for Human Evolution (Cambridge University Press 1992).

Miami University
Department of Sociology and Anthropology
Oxford, Ohio 45056 USA

John C. Mitani, Ph.D. Anthropology, University of California, Davis; Associate Professor, Department of Anthropology, University of Michigan; author "Ethological Studies of Chimpanzee Vocal Behavior."

Research interests include primate behavior and animal communication. He does field research on the vocal behavior of chimpanzees in Tanzania; on orangutans in the Gunung Palung Nature Reserve in Indonesia; and on the male swamp swallow in Dutchess County, New York.

University of Michigan
Department of Anthropology
1054 LSA Building
Ann Arbor, Michigan 48109 USA

Yasuyuki Muroyama, Researcher, Université Louis Pasteur, Strasbourg, France; coauthor, "Grooming Relationships in Two Species of Chimpanzee."

Social behavior in primates, primarily social interactions and cognition of social relationships are his current research interests. In the course of his work, he conducted field study of Japanese macaques in Japan as well as chimpanzees and patas monkeys in Cameroon.

Université Louis Pasteur
Laboratoire de Psychophysiologie
7 rue de l'Université
6700 Strasbourg, France

Toshisada Nishida, Professor of Zoology, Kyoto University, Japan; Director, Mahale Mountains Chimpanzee Research Project, Tanzania; Affiliate, numerous conservation and anthropological societies, author, "Afterword, A Review of Recent Findings on Mahale Chimpanzees: Implications and Future Research Directions."

Application of physical anthropology to the study of behavioral ecology and social dynamics of chimpanzees in the wild are his research interests. In the course of his work, he studied chimpanzees extensively in Tanzania.

Kyoto University
Department of Zoology, Faculty of Science
Kitashirakawa-Oiwakecho,
Sakyo, Kyoto, 606, Japan

Evelyn Ono Vineberg, editor, *Pan paniscus/Bonobo Newsletter;* coauthor of "The Significance of Terrestrial Herbaceous Foods for Bonobos, Chimpanzees and Gorillas."

Her research focus is the behavioral ecology of bonobos, especially the influence of food distribution. Her visits to study bonobos in Wamba have spanned more than 15 years since 1979.

Pan paniscus/Bonobo Newsletter
10603 Sunset Ridge Drive
San Diego, California 92131 USA

Daniel J. Povinelli, Director of the Division of Behavioral Biology, New Iberia Research Center, University of Southwestern Louisiana; author, "What Chimpanzees Might Know About the Mind."

His research interests are in cognitive development, utilizing approaches from the fields of evolutionary biology, physical anthropology, and comparative and developmental psychology. He has investigated the evolution of social intelligence and has compared the development of the mind in human children, chimpanzees, and rhesus monkeys, as well as in several other species.

University of Southwestern Louisiana-New Iberia Research Center
Laboratory of Comparative Behavioral Biology
100 Avenue D
New Iberia, Louisiana 70560 USA

Duane M. Rumbaugh, Ph.D in 1955, University of Colorado; Past-President, Physiological and Comparative Division 6, American Psychological Association; Project Director, the NICHD-supported LANA project and its subsequent elaborations; Professor, Georgia State University; coauthor, "Biobehavioral Roots of Language: A Comparative Perspective of Chimpanzee, Child, and Culture."

Comparative and developmental psychology, language requisites, and cognition are his research interests. Research accomplished defines a close relationship between brain complexity and complex learning skills in primates.

Language Research Center
3401 Panthersville Road
Decatur, Georgia 30034 USA

E. Sue Savage-Rumbaugh, Professor, Georgia State University; coauthor, "Biobehavioral Roots of Language: A Comparative Perspective of Chimpanzee, Child, and Culture."

Primate behavior, evolution of behavior, learning theory, and developmental psychology are her current teaching interests. In the course of her work, she applied her psychology background to study language-acquisition abilities in chimpanzees.

Language Research Center
3401 Panthersville Road
Decatur, Georgia 30034 USA

Rose A. Sevcik, Ph.D. and Assistant Research Professor, Georgia State University; coauthor, "Biobehavioral Roots of Language: A Comparative Perspective of Chimpanzee, Child, and Culture."

The development of language and communication skills as well as the relationship of those skills to both cognition in great apes and cognition in children, especially children with disabilities, are her research interests.

Language Research Center
3401 Panthersville Road
Decatur, Georgia 30034 USA

Barbara B. Smuts, Ph.D., Neuro and Biobehavioral Sciences, Stanford University; Professor, Department of Psychology and Anthropology, University of Michigan; coauthor, "Social Relationships of Female Chimpanzees: Diversity Between Captive Social Groups."

Evolution and development of long-term social relationships are her specific research interests. In the course of her work, she conducted field studies on the social behavior of chimpanzees, baboons, and bottlenose dolphins.

University of Michigan
Department of Psychology
580 Union Drive
Ann Arbor, Michigan 48109 USA

Yukimaru Sugiyama, Professor, Primate Research Institute, Kyoto University, Japan; Editor-in-Chief (1985–91) *Primate Research;* coauthor, "Grooming Relationships in Two Species of Chimpanzee."

Population ecology, social behavior, and the culture of primates are his research interests. In the course of his work, he conducted field study of macaques, langurs, and chimpanzees in such diverse locations as Japan, India, Uganda, Guinea, and Cameroon.

Kyoto University
Primate Research Institute
Inuyama, Aichi, 484 Japan

Geza Teleki, Ph.D. Anthropology and Primatology, Pennsylvania State University; Research Scientist; author of "Frontpiece."

African wildlife, biology, ecology, and conservation are his interests. He specializes in launching new field projects in Africa.

3819 48th Street, N.W.
Washington, D.C. 20016 USA

Michael Tomasello, Professor of Psychology; Adjunct Professor of Anthropology, Emory University; Affiliate Scientist, Yerkes Regional Primate Research Center; Past-Director, Emory Program in Cognition and Development; Editorial

Board, *Child Development, Journal of Child Language,* and *First Language;* author, "The Question of Chimpanzee Culture."

The skills of communication and cultural learning in both children and chimpanzees with a special focus on the cognitive and social-cognitive processes that underlie these skills are the focus of his research interests.

Emory University
Department of Psychology and
Yerkes Primate Center
Atlanta, Georgia 30332 USA

Caroline E.G. Tutin, Co-director of Station d'Etudes des Gorilles et Chimpanzés, Centre International de Rescherches Médicales de Franceville, Gabon; Research Fellow, Department of Molecular and Biological Sciences, University of Sterling, Scotland; Scottish Primate Research Group; author, "Reproductive Success Story: Variability Among Chimpanzees and Comparisons with Gorillas."

The interrelationships between the primates and the rain forest especially the behavior and ecology of sympatric chimpanzees and gorillas in the Lopé Reserve, Gabon, are her major research interests. She has studied sexual behavior of chimpanzees at Gombe and socio-ecology of savanna-living chimpanzees at Assirik. With Michaël Fernandez, she conducted a nationwide census of chimpanzees and gorillas in Gabon, which was the first such project on apes to be done in Africa.

SEGC
BP 7847
Libreville, Gabon

Frans B.M. de Waal, Ph.D. Biology, University of Utrecht, Netherlands; Professor in Psychology and Research Professor at the Yerkes Regional Primate Center, Emory University, Atlanta, Georgia; author, "Chimpanzee Adaptive Potential: A Comparison of Social Life Under Captive and Wild Conditions." and "Overview—Culture and Cognition."; coauthor, "Introduction—The Challenge of Behavioral Diversity."

His main research interests are conflict resolution, coalition formation, reciprocal altruism, and social cognition in a variety of primate species in captivity, including chimpanzees and bonobos.

Emory University
Psychology Department
Atlanta, Georgia 30322 USA

Frances J. White, Behavioral Ecologist, Assistant Professor, Department of Biological Anthropology and Anatomy, Duke University; coauthor, "Party Size in Chimpanzees and Bonobos: A Reevaluation of Theory Based on Two Similarly Forested Sites."

How social and ecological factors influence social organization in primates are her main interests. In the course of her work, she conducted research on lemurs, dolphins, and bonobos.

Duke University
Department of Biological Anthropology
and Anatomy
Box 90383
Durham, North Carolina 27708 USA

Richard W. Wrangham, Ph.D., Zoology (Animal Behavior), Cambridge University, England; Professor, Department of Anthropology, Harvard University; author, "Overview—Ecology, Diversity and Culture;" coauthor, "Introduction—The Challenge of Behavioral Diversity," "Party Size in Chimpanzees and Bonobos: A Reevaluation of Theory Based on Two Similarly Forested Sites," "The Significance of Terrestrial Herbaceous Foods for Bonobos, Chimpanzees, and Gorillas," and "Diversity of Medicinal Plant Use by Wild Chimpanzees."

Research focus is the influence of ecology on the evolution of primate social behavior. Species studied have been chimpanzees in Gombe and Kibale, vervet monkeys, and gelada baboons.

Harvard University
Peabody Museum Department
of Anthropology
11 Divinity Avenue
Cambridge, Massachusetts 02138 USA

A four-year-old chimpanzee displays signs of self-recognition.
Photo by D. Bierschwale.

Index